T0245111

DOUBLE-DIFFUSIVE CONVECTION

Double-diffusive convection is a mixing process driven by the interaction of two fluid components that diffuse at different rates. This phenomenon has important ramifications in oceanography and in numerous other fields, from crystal growth to magma chambers and stellar interiors. Nevertheless, several aspects of double-diffusive convection still remain unclear and controversial.

Leading expert Timour Radko presents the first systematic overview of the classical theory of double-diffusive convection, in a coherent narrative which brings together the disparate literature in this developing field. The book begins by exploring idealized dynamical models and illustrating key principles through examples of oceanic phenomena. Building on the theory, it then explains the dynamics of structures resulting from double-diffusive instabilities, such as the little-understood phenomenon of thermohaline staircases. The book also surveys non-oceanographic applications, such as industrial, astrophysical and geological manifestations, and discusses the climatic and biological consequences of double-diffusive convection.

Providing a balanced blend of fundamental theory and real-world examples, this is an indispensable resource for academic researchers, professionals and graduate students in physical oceanography, fluid dynamics, applied mathematics, astrophysics, geophysics and climatology.

TIMOUR RADKO teaches courses in ocean dynamics, circulation analysis and wave motion at the Oceanography Department of the Naval Postgraduate School. Previously, he worked as a research scientist at the Department of Earth, Atmospheric and Planetary Sciences (EAPS) at the Massachusetts Institute of Technology. He has been active in the area of double-diffusive convection for over fifteen years and was closely involved in developing the theory surrounding this topic. Dr. Radko has authored numerous papers on physical oceanography and fluid mechanics, and has received the prestigious NSF CAREER award in 2006, the NPS Merit Award for Research in 2008, and the Schieffelin (2010) and Griffin (2011) Awards for Excellence in Teaching.

DOUBLE-DIFFUSIVE CONVECTION

TIMOUR RADKO

CAMBRIDGE
UNIVERSITY PRESS

University Printing House, Cambridge CB2 8BS, United Kingdom

One Liberty Plaza, 20th Floor, New York, NY 10006, USA

477 Williamstown Road, Port Melbourne, VIC 3207, Australia

4843/24, 2nd Floor, Ansari Road, Daryaganj, Delhi - 110002, India

79 Anson Road, #06-04/06, Singapore 079906

Cambridge University Press is part of the University of Cambridge.

It furthers the University's mission by disseminating knowledge in the pursuit of education, learning and research at the highest international levels of excellence.

www.cambridge.org
Information on this title: www.cambridge.org/9781108445832

First published 2013
First paperback edition 2017

A catalogue record for this publication is available from the British Library

Library of Congress Cataloging in Publication data
Radko, Timour.
Double-diffusive convection / Timour Radko.
pages cm
Includes bibliographical references and index.
ISBN 978-0-521-88074-9 (hardback)
1. Oceanic mixing. 2. Turbulence. 3. Salinity. I. Title.
GC299.R34 2013
551.46′2 – dc23 2013017280

ISBN 978-0-521-88074-9 Hardback
ISBN 978-1-108-44583-2 Paperback

In memory of Melvin Stern. Half a century ago he had this idea . . .

Contents

Color plates section is between pages 176 and 177.

Preface

Ever since I attended lectures on double-diffusion delivered by the founder of the field, a humble genius by the name of Melvin Stern, there has been no doubt in my mind that double-diffusive convection is the most intriguing subject of fluid dynamics. Even the manner in which double-diffusion was discovered is unusual. It represents a rare instance where a major physical phenomenon was predicted without the motivation of preceding field observations. In 1959, Melvin Stern, using little more than his imagination and physical intuition, formulated a theory for salt fingers – thin, alternating fluid filaments that appear when warm and salty water overlies fresh and cold. The subsequent publication of *The Salt Fountain and Thermohaline Convection* (Stern, 1960) officially marked the birth of a new field of fluid mechanics – double-diffusive convection.

Double-diffusion operates in a counterintuitive way; it is a mixing process that makes dense fluid denser and light fluid lighter. It is driven by the difference in the molecular diffusivities of heat and salt – something that is completely ignored in the vast majority of oceanographic theories and numerical models. The dynamics of double-diffusion are still surrounded by controversy. For instance, the most dramatic signature of oceanic double-diffusion comes from its ability to form stepped structures in vertical temperature and salinity profiles. These so-called thermohaline staircases have been observed in the ocean for forty years and are routinely generated in the laboratory. Yet, in recent literature on the subject, one still finds at least half a dozen different hypotheses for their origin. Extreme opposing views have been expressed on the global consequences of double-diffusion. Overall, it is hard to name any other field in fluid mechanics that can excite such passion amongst the experts and, at the same time, would be found so regrettably confusing by non-specialists.

Nevertheless, the growth of interest in double-diffusion is apparent. In ocean science, the last few years have seen marked improvements in the reliability of microstructure measurements. New observations, most notably the Salt Finger

Tracer Release Experiment, have made it possible to better quantify the role of double-diffusive mixing in water-mass transformation – and hence its impact on the climate system. It has been shown that vertical salt-finger diffusivities in the central thermocline often exceed, by as much as an order of magnitude, the diffusivity of overturning gravity waves, the other primary candidate for internal mixing. The significance of lateral mixing induced by double-diffusion has also become increasingly clear. Thermohaline intrusions, so spectacularly manifested in recent seismic observations, can be essential in explaining the elusive link between stirring of the ocean by mesoscale eddies and the ultimate dissipation of temperature and salinity variance by molecular diffusion. Concurrent developments in non-oceanographic realms underscore the breadth of the subject. A surge of interest in astrophysical applications has produced credible theories for double-diffusive control of the composition of main-sequence stars and giant planets. A seemingly countless variety of geological double-diffusive phenomena continue to stimulate explorations of the relevant dynamics. Chemical double-diffusion is another area in which research activities have markedly intensified in the past decade, triggered by a series of imaginative experimental studies of multicomponent reacting systems.

At the same time, it is generally realized that our insight into the physics and consequences of double-diffusion remains inadequate and that more effort should be invested in all aspects of the problem. Given all the motivations and intellectual challenges of double-diffusive research, one cannot help but wonder why we have not witnessed an explosion of knowledge and new ideas similar to what has happened in many other, less critical branches of geophysical fluid dynamics. My answer to this could be biased and personal. I believe that a significant obstacle to the development of our field is that not one single text has been devoted entirely to double-diffusion. A newcomer, trying to build up intuition and a broad understanding of the subject, is left searching for information in dozens of specialized articles. Of course, brief discussions of double-diffusive convection can be found in some outstanding general fluid dynamical books, such as *Buoyancy Effects in Fluids* (Turner, 1979) and *Ocean Circulation Physics* (Stern, 1975). However, they provide only basic information and make no attempt to systematically explore the rich dynamics of double-diffusive structures. Almost fifteen years have passed since I completed my graduate studies. However, I still remember a feeling of frustration that there wasn't one self-contained treatise that would efficiently guide a student through all the knowledge acquired from numerous laboratory experiments, fields programs and theoretical studies. This oversight becomes particularly striking when we recall that the sister subject of double-diffusion – thermal convection – has been discussed in at least thirty dedicated texts.

When a commissioning editor for Cambridge University Press suggested that I write a book on double-diffusion, I was delighted and flattered. This proposition has strengthened my conviction that progress in our field can be greatly accelerated by the double-diffusive book, an organized and self-contained entity accessible to a wide range of scientists with different levels of expertise and backgrounds. The timing appeared to be most appropriate for such an undertaking. All branches of fluid dynamics continuously evolve, but recent shifts in our understanding of double-diffusion have been particularly profound. Theoretical and modeling advances, motivated by recent oceanographic field programs, have led to new insights into the origin and dynamics of thermohaline staircases. The problem of nonlinear equilibration of salt fingers looks increasingly more tractable, with fresh ideas defining the way towards physically based parameterizations of double-diffusive mixing. At the same time, many traditional views, such as the significance of large-scale shear and finger/wave interaction, have to be critically reevaluated. Clearly, the extant descriptions of double-diffusion need considerable updating, as does our perception of its role in ocean mixing.

This book is designed as a comprehensive review of the field that combines the basic theory with major up-to-date findings, touches upon ongoing research, and offers some suggestions for future developments. It consists of three distinct parts. The first part (Chapters 1–5) presents the fundamental theory of double-diffusion at a level suitable for a diligent non-expert with some background in the physical sciences. Chapter 1 opens with a discussion of the basic dynamic principles of double-diffusion, traces the history of events leading to its discovery, and introduces key governing parameters. Chapter 2 summarizes the linear instability theory, and establishes the relevant spatial and temporal scales. While the primary double-diffusive instability operates on scales of several centimeters, it affects the evolution of much larger scales through the vertical fluxes of temperature and salinity. Accordingly, the next three chapters are concerned with the formulation of the flux laws. The properties of vertical transport and the associated nonlinear dynamics depend on the geometry of a double-diffusive system. Specific examples are presented for the unbounded gradient configuration (Chapter 3), the two-layer system (Chapter 4) and the vertically bounded layer model (Chapter 5).

The discussion of flux laws establishes the framework for the second part of this book (Chapters 6–9), where the focus is on structures resulting *from* the primary double-diffusive instability – collective instability waves, intrusions and thermohaline staircases. This component is more specialized and will be of particular interest to oceanographers and applied mathematicians who are already familiar with some convection models. Other readers may still follow the narrative by focusing on relevant observations, numerical results and physical arguments. The presentation weaves together the classical theory with recent developments and can be used as a starting point for graduate students and as a professional reference for active researchers. While the dynamics of collective instabilities (Chapter 6) and thermohaline intrusions (Chapter 7) are reasonably well understood, the origin of thermohaline staircases (Chapter 8) is a more controversial topic. Therefore, Chapter 8 presents several competing views on staircase dynamics; the emphasis being on theoretical and modeling advances made in the past decade. Chapter 9 attempts to unify the analysis of secondary double-diffusive instabilities, discussed separately in Chapters 6–8.

The last, but not least, part of this book (Chapters 10–13) explains the role of double-diffusion within the broader context of environmental and physical sciences. Chapter 10 considers the interaction of double-diffusion with the active oceanic environment, perpetually forced by externally driven shears and turbulence. Here we also examine the most common techniques for identifying double-diffusive signatures in field measurements. Chapter 11 highlights the relatively new and widening interest in the climatic and biological consequences of double-diffusive mixing. Non-oceanographic applications, which include advances in astrophysics, geology, chemistry and engineering, are summarized in Chapter 12. We conclude (Chapter 13) by speculating on the prospects and challenges that face our field in years to come. Chapters 10–13 are more descriptive than the rest of the monograph and, in principle, can be read and understood independently. However, readers who may decide to proceed in this manner should be warned that the assessment of the significance of any phenomenon without a firm grasp of its basic physics is fraught with potential pitfalls.

Any systematic and objective review of double-diffusion inevitably conveys a sense of the singular influence of one person whose ideas permeate all theoretical developments in our field. Not only did Melvin Stern discover double-diffusion but – perhaps even more impressively – he remained the subject's undisputed leader throughout his long and distinguished career. If not fate, then perhaps it was a random stroke of luck that I had Melvin as a graduate adviser and afterward was able to continue our collaboration until his recent passing away. Melvin was a huge inspiration for everyone who worked with him. One realization that came to me upon getting to know Melvin is that science is extremely emotional. It is possible to relate to very abstract concepts on a very personal level. Melvin just loved to do what he did and he was brilliant at it – maybe it is a definition of happiness in a way. I cannot even begin to express how profoundly his ideas influenced my own professional development. I hope that this monograph can reflect, at least to some extent, Stern's vision of our field and his scientific approach.

I am also deeply grateful to everyone who contributed to this project either directly, by offering valuable comments on the earlier version of the text, or through inspiring discussions: Jason Flanagan, Pascale Garaud, William

Merryfield, Barry Ruddick, Ray Schmitt, Bill Smyth, Stephan Stellmach, George Veronis and Anne de Witt. Continuous support of my double-diffusive research by the various branches of the National Science Foundation (Physical Oceanography Program, Fluid Dynamics Program, Division of Astronomical Sciences and Office of Polar Programs) is gratefully acknowledged.

1

General principles

When the density of a fluid is determined by two components, which diffuse at different rates, the fluid at rest can be unstable even if its density increases downward. This simple, although seemingly counterintuitive, idea is the cornerstone of the theory of double-diffusive convection. As with any other instability, double-diffusion requires a finite amount of energy to sustain the growth of perturbations. If the basic state is motionless, instability can be driven by the potential energy of one of the density components. The ensuing convection depends very strongly, in terms of its pattern and dynamics, on whether the destabilizing component is of higher or lower diffusivity. The configuration in which the required energy is supplied by the slower diffuser is called salt fingering; the instability driven by the faster diffuser is known as diffusive convection. Of course, both density components could be concurrently destabilizing. In this case, the total density stratification is unstable and the result is top-heavy convection, a very different and much more violent process, which is beyond the scope of this book.

Because the interest in double-diffusion was originally motivated by oceanic applications, we follow conventional practice and introduce the key concepts in the oceanographic context. For instance, the faster diffuser will be conveniently referred to as temperature (T) and the slower diffuser as salinity (S) – two major components of seawater density. However, it is our intention to present the basic theory of double-diffusion in its most general form. Aside from parameter values and notation, the analysis is generally applicable to a variety of other physical systems, including double-diffusion in geology, astrophysics and metallurgy. Specifics of each field are discussed in Chapter 12.

1.1 Salt fingers

In much of the upper kilometer of tropical and subtropical oceans, warm and salty waters are located above cold and fresh; the mean vertical stratification in

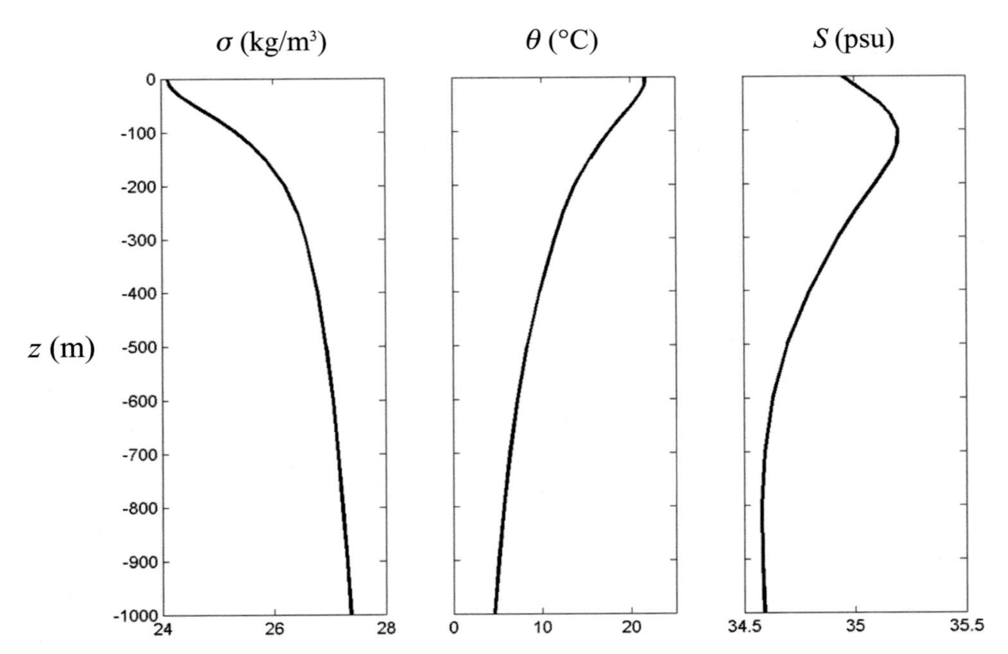

Figure 1.1 The vertical profiles of potential density (left), potential temperature (center) and salinity (right) in the upper kilometer of the ocean, horizontally averaged over the latitude band from 50° S to 50° N. Potential density (σ) and potential temperature (θ) are used to take into account effects of compressibility of seawater. Data are taken from the Levitus world ocean database.

the latitude band from 50° S to 50° N is shown in Figure 1.1. Since the density of seawater decreases with temperature but increases with salinity, the available potential energy of the system is stored in the salinity component. Vertical mixing of salinity tends to lower the center of gravity, thereby releasing potential energy, whilst mixing of temperature does the opposite. If the salinity stratification is losing energy at a higher rate than the temperature gains it, there will be a continuous supply of kinetic energy that could maintain and enhance vertical mixing. In the ocean, the amount of energy contained in the salinity stratification is enormous, and the release of even a small fraction of it can substantially affect the large-scale circulation. But does it really happen? So far, we have only argued that the instability of a two-component bottom-heavy fluid at rest does not contradict the principle of energy conservation – an intriguing, suggestive, but not exactly conclusive statement.

The first paper that hints at the possibility of releasing the potential energy of salt in the gravitationally stable environment appeared in 1956: *An Oceanographical Curiosity: The Perpetual Salt Fountain* by Stommel, Arons and Blanchard. The

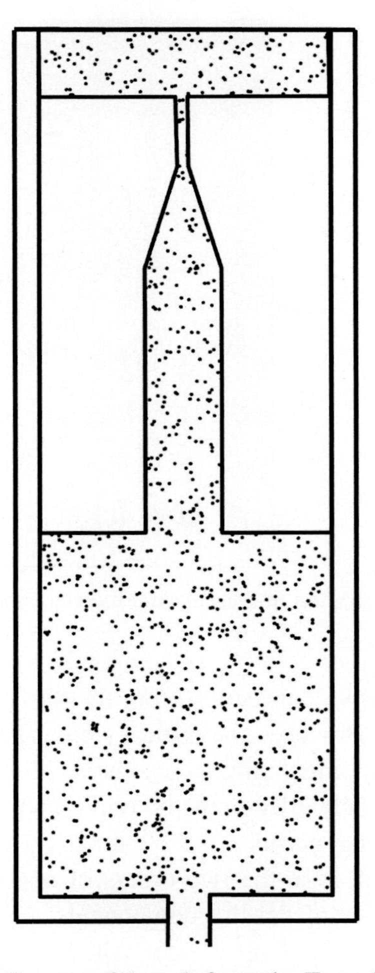

Figure 1.2 Schematic diagram of the salt fountain. From Stommel *et al.* (1956).

authors inserted a narrow heat-conducting pipe into a tank filled with doubly stratified water: warm and salty above cold and fresh. They discovered that if the water in the pipe is pushed upward, the circulation will be maintained for as long as there is a vertical salinity gradient. The reason was attributed to heat conduction through the wall of the pipe. As shown in Figure 1.2, the rising water in the pipe comes into thermal equilibrium with the surrounding fluid while remaining fresher, therefore lighter, and continues to move upward, reinforcing the initial circulation pattern. Stommel *et al.*'s study did not explain the connection between the laboratory experiment and the ocean – the heat-conducting pipe was considered to be essential in driving the circulation, and the defensive term "curiosity" slipped into the title. At this stage, it still seemed like there was a long way to go to take

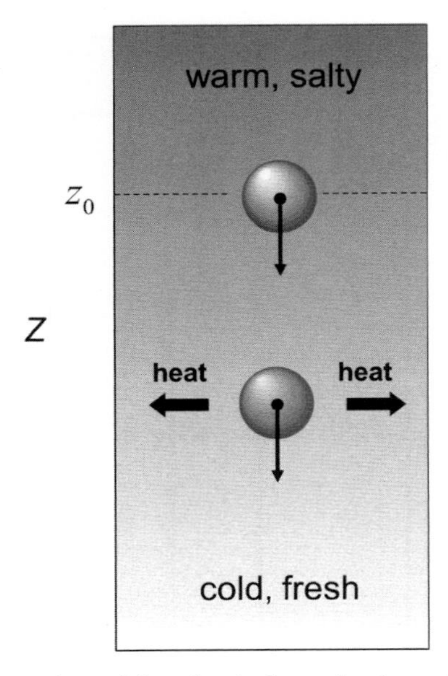

Figure 1.3 Illustration of the physical mechanism of salt fingering.

the idea of double-diffusive convection from the "it is not impossible" to the "it is likely to occur" level. However, the critical step was made only four years later (Stern, 1960). In a surprisingly simple argument, Melvin Stern showed that not only can instability naturally arise in the bottom-heavy stratification, but it should also be very common in the ocean.

The key, Stern argued, is in the two orders of magnitude difference between the molecular diffusivities of density components: $k_T \approx 1.4 \cdot 10^{-7}$ m^2 s^{-1} for temperature and $k_s \approx 1.1 \cdot 10^{-9}$ m^2 s^{-1} for salt. The significance of unequal *T–S* diffusivities can be illustrated as follows. Consider continuously stratified fluid at rest as shown in Figure 1.3; temperature and salinity do not vary horizontally and their vertical gradients are positive. Imagine perturbing this basic state by displacing a small parcel of fluid downward from its equilibrium position (z_0). It rapidly adjusts its temperature (fast diffuser) to that of the surrounding fluid but largely retains its original salinity (slow diffuser). The parcel is saltier than the ambient water at the same level and, since density increases with salinity, heavier. It continues to sink, moving further away from its equilibrium location, which implies that the basic stratification is unstable. The proposed mechanism is not unlike that of the perpetual salt fountain and, in retrospect, it becomes clear that in 1956 Stommel and his collaborators were very close to discovering double-diffusion. They just failed to realize that Stommel's pipe is not necessary to release the potential energy of a doubly stratified system – the low diffusivity of salt can be almost as effective in preserving the salinity of water parcels.

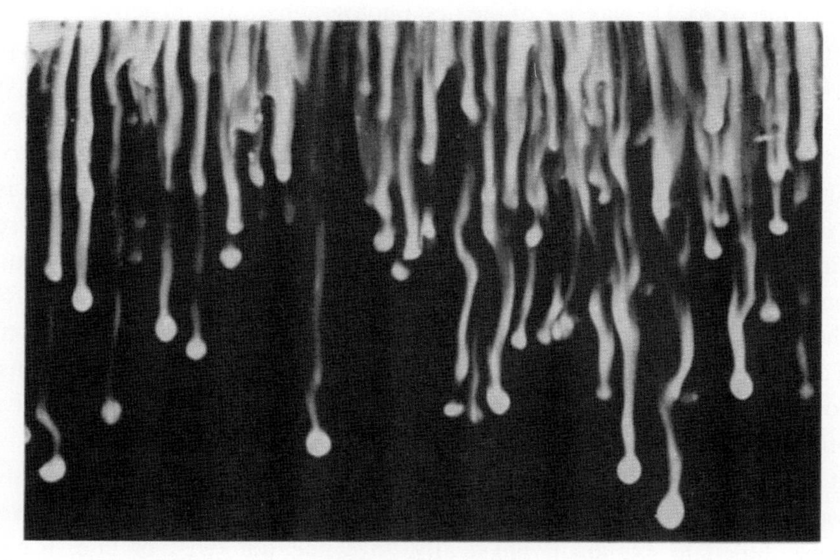

Figure 1.4 Laboratory experiment on fingering convection. An array of salt fingers is created by setting up a stable temperature gradient and pouring salt solution on top. From Huppert and Turner (1981).

Since salt fingering is a fundamentally diffusive process, the spatial scale of convection cells is limited by the range of effective molecular conduction of heat. In the ocean, salt fingers operate on scales of a few centimeters; in the laboratory, even smaller. Often, but not always, salt fingers come in the form of vertically elongated narrow filaments, very much as their colorful name suggests. Figure 1.4 presents an example of a laboratory experiment in which warm salty solution was poured on top of a stable temperature gradient. The emerging pattern consists of long parallel fingers with small round jelly-fish eddies forming at their extremities, a pattern dramatically different from that of ordinary thermal convection.

1.2 The early years: from Jevons to Stommel

While there is no doubt that full credit for the discovery of salt fingers belongs to Melvin Stern, it is interesting to consider some of the earlier missed opportunities. The complete account of the pre-Stern history of double-diffusion, rich with fascinating and lively details of scientific missteps, is given by Schmitt (1995a,b). What is particularly striking in this story is how close science has come to discovering double-diffusion on so many occasions but failed to make one final step.

The first recorded salt-finger experiment was performed by the English–Australian Stanley Jevons. To say that Jevons was a talented fluid dynamicist does not even start to describe his interests – he was a scientist in the broadest sense of the word. He was one of the first photographers and an active researcher in chemistry and meteorology. By the age of twenty-two, Jevons had become a professor of logic, moral and mental philosophy, and political economy. But the coolest thing that he ever did was to accidentally create double-diffusion in the laboratory. After placing warm sugar solute on top of cold and fresh water, he observed an "infiltration of minute, thread-like streams" (Jevons, 1857). In retrospect, it is clear that this phenomenon is double-diffusion: sugar is a relatively slow diffuser and placing it in the upper warmer layer triggers fingering instability. Unfortunately, Jevons misinterpreted his observations. He viewed his experiment as a form of top-heavy convection, missing his opportunity to discover double-diffusion more than a century before Stern. However, we should not feel too sorry for Jevons since he got himself into a rather distinguished group of scientists who almost solved the salt-finger puzzle.

In 1880, Lord Rayleigh, perhaps the most prolific and reputable scientist of his period, reproduced the experiments of Jevons, also observed fingering convection, and also failed to explain it physically. In a peculiar twist of fate, Rayleigh's attempt to understand the origin of the sugar fingers in Jevons' experiments led to the first rigorous stability analysis of a stratified non-diffusive fluid (Rayleigh, 1883). His treatment of the top-heavy configuration describes what is now known as the Rayleigh–Taylor instability. For the bottom-heavy case, Rayleigh predicted the maximum frequency of free oscillations, arriving at the classical expression for the buoyancy frequency $N^2 = -\frac{g}{\rho}\frac{\partial \rho}{\partial z}$. However, the failure to recognize the destabilizing role of thermal diffusion prevented Rayleigh from adding salt fingers to his string of scientific victories.

The next opportunity to discover double-diffusion presented itself in 1906, when Vagn Walfrid Ekman, another giant of fluid dynamics, performed laboratory experiments on the "dead water" phenomenon. To visualize the interface displacements, Ekman used a two-layer system of milk over seawater and observed "a shower of small vortex-rings" – undoubtedly, milk fingers. Ekman came a bit closer than his predecessors to pinning down the elusive phenomenon. He pointed out that when a milk parcel comes into close contact with salty water, it gains, by molecular diffusion, extra salinity. Salt makes the parcel denser than the surrounding fluid and it rapidly sinks. While Ekman was right on target in explaining the physics of milk fingers, he did not realize that the analogous dynamics occur naturally in the ocean, where heat and salt play the same role as salt and milk in his experiments. Apparently Ekman considered the analysis of milk fingering by itself to be overly esoteric and frivolous and did not pursue the subject any further. Another half-a-century had gone by before two crucial steps were made: Stommel's salt fountain idea in 1956 and, finally, Melvin Stern's salt-finger paper in 1960. It is interesting that neither Stommel nor Stern was familiar with the earlier experiments. Stern's

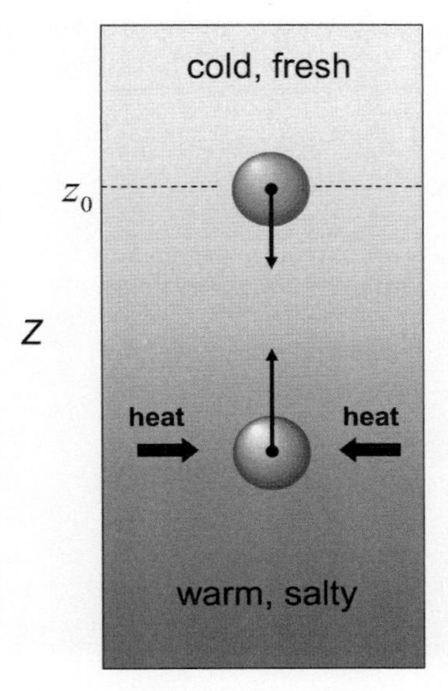

Figure 1.5 Illustration of the physical mechanism of oscillatory diffusive instability.

paper cites only one study (Stommel *et al.*, 1956) and Stommel's paper contains no references at all. The field of double-diffusion was created from scratch.

One can only speculate about possible reasons for a century-long delay between the first experimental realization of fingering and the first physical model. Would Rayleigh have discovered salt fingers if a personal encounter had permitted him to discuss his experiments with Jevons, as suggested by Schmitt (1995a,b)? Was it familiarity with more advanced mathematical methods that helped Stern to make the critical connection, or did the salt fountain idea provide the valuable hint? Could it be that the level of conceptual understanding of hydrodynamic instabilities, developed by the middle of the twentieth century, was a prerequisite for his discovery? We will never know for sure, but I have always suspected that the reason is much simpler: Melvin was just a bit sharper than the rest of the group.

1.3 Diffusive convection

In a footnote to his seminal salt-finger paper – perhaps the most important footnote in the history of fluid dynamics – Stern (1960) suggested the possibility of the oscillatory diffusive instability of cold and fresh water located above warm and salty. The dynamics of diffusive convection can be explained by reversing the arguments used for the fingering case (Section 1.1). The schematic in Figure 1.5 represents a thought experiment in which the diffusively favorable stratification

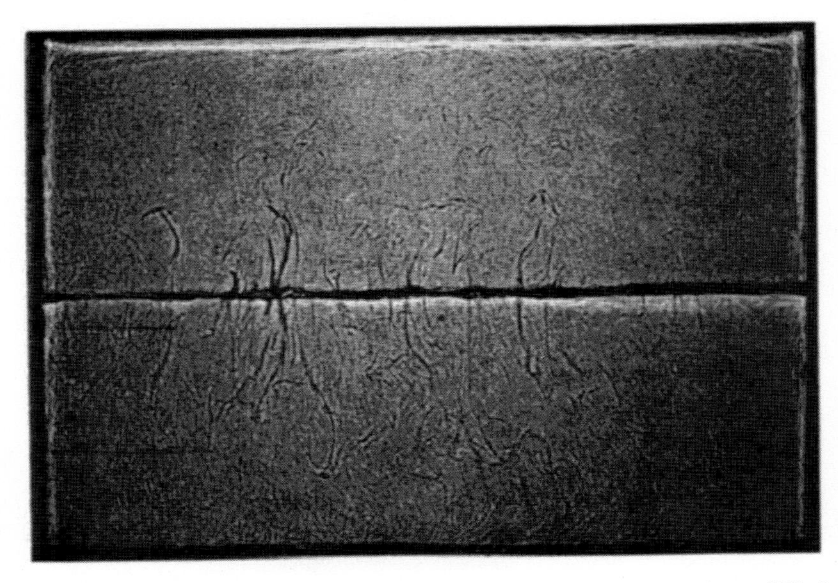

Figure 1.6 Laboratory experiment on diffusive convection. The two-layer diffusive system is created by pouring haline solution on top of a denser sucrose solution. From Turner (1985).

is perturbed by displacing a small parcel downward. As in the salt-finger case (cf. Fig. 1.3), it rapidly adjusts its temperature but retains salinity. However, since the background salinity now increases with depth, the parcel becomes lighter than the ambient fluid at the same level, and the buoyancy force drives it upward. The parcel is not only lighter than its surroundings but, because of heat gain, it is also lighter than it was originally. Thus, on its way back to the point of origin, it experiences a buoyancy force that is greater than on its way down. As a result, the parcel gains some energy and overshoots its original equilibrium position. Above the equilibrium level, the parcel again quickly adjusts its temperature but not its salinity; it is now saltier and therefore heavier than the surrounding fluid. Eventually, gravity forces it downward, back to the original location. The parcel overshoots again, and the process repeats over and over. The energy gain by the particle at each cycle leads to a gradual increase in the amplitude of oscillations, resulting in the so-called "overstable" mode of instability.

Since oscillatory modes could be easily damped by viscous drag, diffusive instability in the uniformly stratified fluid is restricted to a rather narrow range of parameters (quantified in Chapter 2). More common in nature and in laboratory realizations is a stepped configuration consisting of well-mixed layers separated by a thin diffusive interface. Figure 1.6 shows an experiment set up by pouring a layer of salty water on top of a layer of denser sugar solute. In this case, salt is the fast diffuser and sugar the slow one. The more rapid molecular diffusion of salt across the interface produces the downward density flux. The region immediately below (above) the interface becomes denser (lighter), which maintains the top-heavy convection in both mixed layers. Despite the apparent differences, convection in layered systems and overstable oscillations in a continuously stratified fluid are both described by the generic term diffusive convection.

1.4 Scale analysis

The foregoing examples leave us with a sense that double-diffusive convection operates in a most unusual way. All its forms, fingering and diffusive, are driven by the net release of potential energy and therefore fluid necessarily lowers its center of mass: double-diffusive mixing makes the relatively light fluid in upper layers even lighter; the heavy fluid at depth becomes heavier. The primary instability is driven by molecular diffusion, a stabilizing agent in most fluid dynamical problems. The eddy diffusivities of density constituents are different – yet another unexpected consequence of two-component dynamics. Overall, it seems that our physical intuition, built on experience with simple one-component fluids, fails miserably when it comes to double-diffusion. Therefore to fully understand double-diffusive convection, one has to rely, perhaps more strongly than in other branches of fluid dynamics, on formal mathematical results.

The analytical explorations throughout this book are based on the well-known (e.g., Pedlosky, 1979) Boussinesq equations of motion:

$$\begin{cases} \dfrac{\partial \vec{v}}{\partial t} + \vec{v} \cdot \nabla \vec{v} = -\dfrac{\nabla p}{\rho_0} + g\dfrac{\rho - \rho_0}{\rho_0} + \nu \nabla^2 \vec{v}, \\ \dfrac{\partial T}{\partial t} + \vec{v} \cdot \nabla T = k_T \nabla^2 T, \\ \dfrac{\partial S}{\partial t} + \vec{v} \cdot \nabla S = k_S \nabla^2 S, \\ \nabla \cdot \vec{v} = 0, \end{cases} \tag{1.1}$$

where $\vec{v}$ is the (non-divergent) velocity field, p is the dynamic pressure, ν is the kinematic viscosity, ρ is the density and ρ_0 is a reference value. T and S represent two scalar quantities affecting the density of the fluid (e.g., temperature and salinity in the oceanographic context). Molecular diffusivities k_T and k_S are assumed to be uniform but unequal ($k_T > k_S$). Rotational effects are neglected and the fluid is regarded as incompressible. We also assume the linear equation of state:

$$\frac{\rho - \rho_0}{\rho_0} = \beta(S - S_0) - \alpha(T - T_0), \tag{1.2}$$

where (α, β) are the constant expansion/contraction coefficients and (T_0, S_0) are the reference temperature and salinity.

We now proceed to establish, tentatively at first, the scales relevant to double-diffusive convection. Typical temporal and spatial scales are denoted as $\langle t \rangle$ and $\langle L \rangle$; the scales of velocity, temperature, salinity, density and pressure perturbations are $\langle \mathrm{v}' \rangle$, $\langle T' \rangle$, $\langle S' \rangle$, $\langle \rho' \rangle$ and $\langle p' \rangle$ respectively. We are particularly interested in the dependence of these quantities on the background vertical temperature and salinity gradients $\left(\bar{T}_z, \bar{S}_z\right)$ and therefore the temperature scale is expressed as follows:

$$\langle T' \rangle \sim \langle L \rangle \, |\bar{T}_z|. \tag{1.3}$$

Because molecular dissipation plays a central role in double-diffusive convection, the magnitude of the diffusive term in the temperature equation should be comparable to the local rate of change in temperature. Since our focus will be on fully developed instabilities, it is also reasonable to assume that the nonlinear terms are equally important:

$$\frac{1}{\langle t \rangle} \sim \frac{\langle \mathrm{v}' \rangle}{\langle L \rangle} \sim \frac{k_T}{\langle L \rangle^2}. \tag{1.4}$$

For the equation of state (1.2), we expect comparable effects of temperature and salinity on density distribution:

$$\frac{\langle \rho' \rangle}{\rho_0} \sim \alpha \langle T' \rangle \sim \beta \langle S' \rangle. \tag{1.5}$$

Finally, in the momentum equation, we anticipate that the buoyancy force is of the same order as viscous dissipation and the pressure gradient term, which is equivalent to setting

$$g \frac{\langle \rho' \rangle}{\rho_0} \sim \nu \frac{\langle \mathrm{v}' \rangle}{\langle L \rangle^2} \sim \frac{\langle p \rangle}{\rho_0 \langle L \rangle}. \tag{1.6}$$

Drawing together Eqs. (1.3)–(1.6), we arrive at the following magnitudes of key variables:

$$\langle t \rangle \sim \frac{d^2}{k_T}, \ \langle L \rangle \sim d, \ \langle \mathrm{v}' \rangle \sim \frac{k_T}{d}, \ \langle p' \rangle \sim \frac{\rho_0 \nu k_T}{d^2}, \ \langle T' \rangle \sim d |\bar{T}_z|, \ \langle S' \rangle \sim \frac{\alpha}{\beta} \langle T' \rangle, \tag{1.7}$$

where

$$d = \left(\frac{k_T \nu}{g \alpha \left| \bar{T}_z \right|} \right)^{\frac{1}{4}}. \tag{1.8}$$

The combination (1.8) can be interpreted as the nominal length scale expected for primary double-diffusive instabilities. A typical value of $\bar{T}_z$

in the mid-latitude thermocline is about 0.01 °C m^{-1} and therefore (1.7) and (1.8) yield $\langle t \rangle \sim 10^3$ s, $\langle L \rangle \sim 0.01$m, $\langle v' \rangle \sim 10^{-5}$ m s^{-1}, and $\langle T' \rangle \sim 10^{-4}$ °C. Such scales place oceanic double-diffusive convection among the so-called microstructure processes. It should be emphasized, however, that this estimate pertains only to primary instabilities. These scales do not apply to secondary double-diffusive phenomena – collective instability waves, intrusions and thermohaline staircases – which fall into the "finescale" (~10–100 m) category and will be discussed in Chapters 6–9.

1.5 Non-dimensionalization and governing parameters

In order to systematically explore any hydrodynamic phenomenon, it is essential to determine the key non-dimensional variables that control its pattern and strength. We fully appreciate, for instance, the significance of the Rayleigh number in thermal convection and the Reynolds number in viscous shear flows. For double-diffusion, the optimal choice of governing parameters is not as clear-cut as one would hope. The literature contains two approaches to the analysis of double-diffusive convection, each motivated by a distinct view of the dynamics at play.

The older approach builds on experience with the Rayleigh–Bénard thermal convection problem, in which a fluid confined between two rigid planes is heated from below. If the analogy between double-diffusion and thermal convection is justified, then it is sensible to classify the double-diffusive regimes in terms of Rayleigh numbers based on temperature and salinity variation:

$$R = \frac{g\alpha \Delta T H^3}{k_T \nu}, \quad R_s = \frac{g\beta \Delta S H^3}{k_T \nu}, \tag{1.9}$$

where H is the thickness of a diffusive layer and ΔT, ΔS are the temperature and salinity variations across it. However, for many double-diffusive problems, such a description is misleading. The principal feature of double-diffusion is that, unlike thermal convection, it defines its own internal length scale d in (1.8) – the scale that is independent of H. If H greatly exceeds d, the vertical boundaries become irrelevant for the processes in the interior, and therefore inclusion of H among the principal governing parameters is unwarranted.

In subsequent developments we tend to exploit an alternative approach – the more modern "unbounded gradient layer" model. Instead of focusing on the integral measures of a double-diffusive layer (ΔT, ΔS, H), we ascribe greater physical significance to the local large-scale gradients of temperature and salinity. The temperature and salinity fields are separated into the linear background stratification ($\bar{T}_z = const$, $\bar{S}_z = const$) and perturbations (T', S'):

$$\begin{cases} T = \bar{T}(z) + T', \\ S = \bar{S}(z) + S'. \end{cases} \tag{1.10}$$

In most models considered here, the background state is at rest and therefore $\vec{v}' = \vec{v}$. On many occasions, analytical development can be simplified by non-dimensionalizing the governing equations using the scales established in (1.7). The dimensional variables are replaced by their non-dimensional counterparts as follows:

$$\begin{cases} (x, y, z) \to d \cdot (x, y, z), \\ \vec{v} \to \dfrac{k_T}{d} \cdot \vec{v}, t \to \dfrac{d^2}{k_T} \cdot t, p' \to \dfrac{\rho_0 \nu k_T}{d^2} p', \\ \alpha T' \to \alpha \left|\bar{T}_z\right| d \cdot T', \beta S' \to \alpha \left|\bar{T}_z\right| d \cdot S', \end{cases} \tag{1.11}$$

where the nominal double-diffusive scale d is given in (1.8). The expansion/contraction coefficients (α, β) are incorporated in (T', S'), and $\alpha \left|\bar{T}_z\right| d$ is used as the scale for both temperature and salinity. The background gradients $\left(\bar{T}_z, \bar{S}_z\right)$ are consistently treated as dimensional variables. We shall refer to the transformation (1.11) as the standard system of non-dimensionalization. Considerations of clarity require dimensional or non-dimensional treatment for different topics and the type of variables will be explicitly specified.

For the finger case ($\bar{T}_z > 0$, $\bar{S}_z > 0$), expressing the Boussinesq equations (1.1) in terms of (T', S') and non-dimensionalizing the result yields:

$$\begin{cases} \dfrac{1}{Pr}\left(\dfrac{\partial \vec{v}}{\partial t} + \vec{v} \cdot \nabla \vec{v}\right) = -\nabla p + (T' - S')\vec{k} + \nabla^2 \vec{v}, \\ \dfrac{\partial T'}{\partial t} + \vec{v} \cdot \nabla T' + w = \nabla^2 T', \\ \dfrac{\partial S'}{\partial t} + \vec{v} \cdot \nabla S' + \dfrac{w}{R_\rho} = \tau \nabla^2 S', \\ \nabla \cdot \vec{v} = 0, \end{cases} \tag{1.12}$$

where $Pr = \frac{\nu}{k_T}$ is the Prandtl number, $\tau = \frac{k_S}{k_T}$ is the diffusivity ratio, and $R_\rho = \frac{\alpha \bar{T}_z}{\beta \bar{S}_z}$ is the density ratio.

For the diffusive case ($\bar{T}_z < 0$, $\bar{S}_z < 0$) the equivalent set of non-dimensional equations is

$$\begin{cases} \dfrac{1}{Pr}\left(\dfrac{\partial \vec{v}}{\partial t} + \vec{v} \cdot \nabla \vec{v}\right) = -\nabla p + (T' - S')\vec{k} + \nabla^2 \vec{v}, \\ \dfrac{\partial T'}{\partial t} + \vec{v} \cdot \nabla T' - w = \nabla^2 T', \\ \dfrac{\partial S'}{\partial t} + \vec{v} \cdot \nabla S' - \dfrac{w}{R_\rho} = \tau \nabla^2 S', \\ \nabla \cdot \vec{v} = 0. \end{cases} \tag{1.13}$$

The only difference between (1.12) and (1.13) appears in the signs of w terms – terms representing the influence of the background T–S gradients. However, in studies of diffusive convection it is more common to use, instead of R_ρ, the diffusive density ratio $R_\rho^* = \frac{1}{R_\rho} = \frac{\beta \bar{S}_z}{\alpha \bar{T}_z}$. Adopting different conventions for diffusive and fingering cases is not as unreasonable as it first seems. This way, the density ratio always exceeds unity for the bottom-heavy stratification and increases (decreases) with the gradient of the stabilizing (destabilizing) component.

The non-dimensional systems (1.12) and (1.13) suggest that the dynamics of unbounded double-diffusive systems, fingering or diffusive, are controlled by three key parameters: Pr, τ, and R_ρ. The Prandtl number and the diffusivity ratio reflect the properties of a particular fluid; for seawater in the mid-latitude upper ocean, $Pr \sim 7$ and $\tau \sim 10^{-2}$. The density ratio R_ρ, on the other hand, is determined by local environmental conditions and can be highly inhomogeneous. Therefore, a central problem in double-diffusive convection involves prediction of its control by the background density ratio. Physical interpretation of the density ratio is straightforward – it measures the degree of compensation between temperature and salinity gradients in terms of their effects on density stratification. Thus, for instance, $R_\rho = 1$ corresponds to full T–S compensation, resulting in the uniform background density ($\bar{\rho}_z = 0$). Large values of R_ρ imply that the density distribution is controlled by its thermal component.

1.6 Turner angle

The density ratio is closely related to another common measure of the stratification pattern – the Turner angle (Ruddick, 1983). While the Turner angle (Tu) generally serves the same purpose as R_ρ, it can offer a more convenient and transparent interpretation of the environment in terms of susceptibility to double-diffusion. Tu can be defined as a polar angle in the $(\alpha \bar{T}_z, \beta \bar{S}_z)$ plane, measured relative to the $\alpha \bar{T}_z = -\beta \bar{S}_z > 0$ ray, as indicated in Figure 1.7. This definition can be cast in a more explicit form:

$$Tu = 135^\circ - \arg(\beta \bar{S}_z + i\alpha \bar{T}_z), \tag{1.14}$$

which, in turn, can be used to express the T–S gradients in terms of the Turner angle:

$$\begin{cases} \alpha \bar{T}_z = A \sin(Tu + 45^\circ), \\ \beta \bar{S}_z = -A \cos(Tu + 45^\circ), \end{cases} \tag{1.15}$$

where $A = \sqrt{(\alpha \bar{T}_z)^2 + (\beta \bar{S}_z)^2}$. Since the types of instability are determined by the signs of background T–S gradients, rather than by their absolute values, various instabilities occupy distinct sectors in the parameter space of Figure 1.7. Thus,

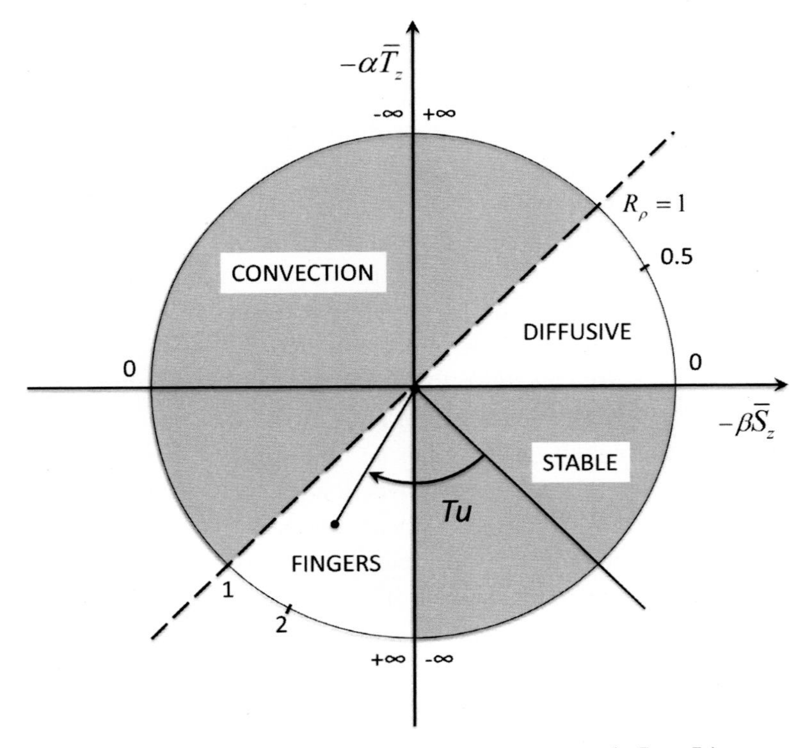

Figure 1.7 Schematic diagram of the Turner angle in the $\left(\alpha\bar{T}_z, \beta\bar{S}_z\right)$ parameter space. After Ruddick (1983).

for $Tu = 0$, temperature and salinity gradients are both stabilizing and contribute equally to the density gradient. As Tu is increased, the absolute value of the salinity gradient decreases, vanishing at $Tu = 45°$. For $45° < Tu < 90°$, the salinity gradient is destabilizing but its magnitude is less than that of the stabilizing temperature gradient. In this region, stratification is susceptible to salt fingering. As we further increase Tu, the system enters into the gravitationally unstable regime ($90° < Tu < 270°$). The region predisposed to diffusive convection is located at $-90° < Tu < -45°$ and the stratification is doubly stable for $-45° < Tu < 45°$.

As apparent from (1.15), there is a direct connection between the Turner angle and the density ratio:

$$R_\rho = -\tan(Tu + 45°). \tag{1.16}$$

However, Ruddick (1983) notes several advantages of focusing on Tu rather than on R_ρ in the analysis of water-mass composition. For instance, complications associated with the singularity of the density ratio in the nearly homogeneous salinity stratification ($\bar{S}_z \to 0$) do not arise if Tu is used instead. Another undesirable

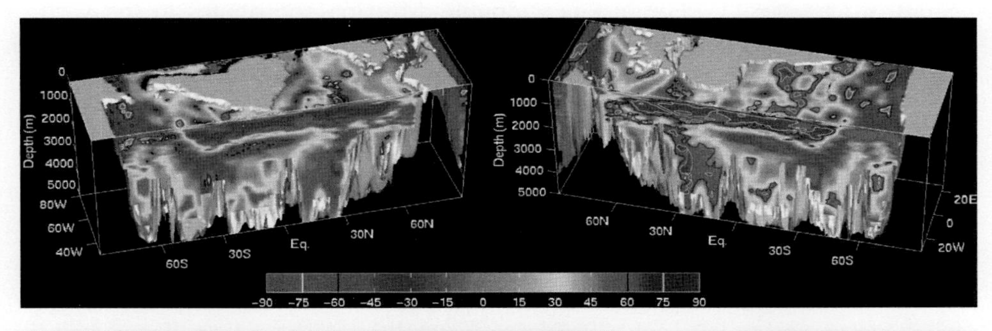

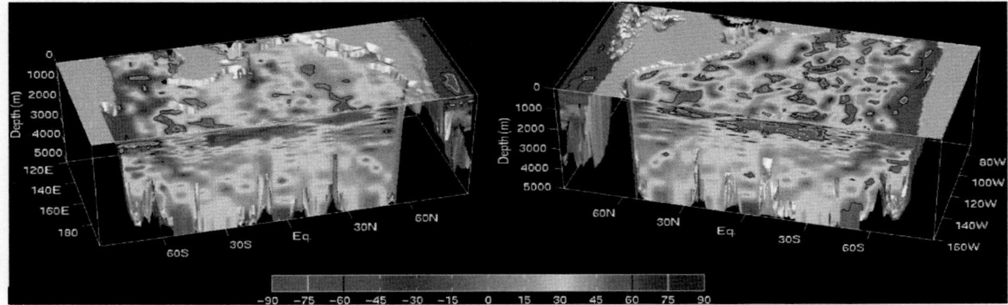

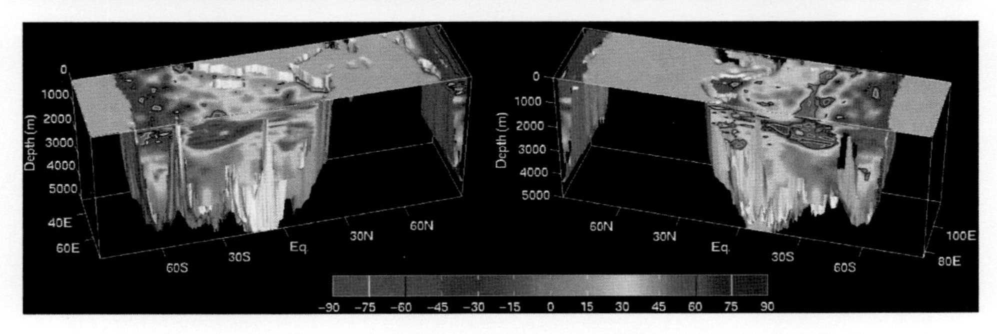

Figure 1.8 Distribution of the Turner angle in the Atlantic (top), Pacific (center) and Indian (bottom) Oceans. From You (2002). See color plates section.

feature of the density ratio is its ambiguity. Concurrently reversing the signs of T–S gradients does not affect the density ratio, although the background stratification and the character of resulting instabilities change dramatically. The Turner angle, on the other hand, uniquely defines the signs of $\bar{T}_z$ and $\bar{S}_z$. Finally, Tu replaces the infinite scale of R_ρ by a finite range of 2π.

Figure 1.8 shows the distribution of the Turner angle in the Atlantic, Pacific and Indian Oceans. In all basins, conditions favoring double-diffusion are quite common. In the world ocean, about 30% of the volume is finger favorable and 15% is diffusively stratified. Double-diffusion is particularly widespread in the main thermocline, which has significant large-scale implications (discussed in greater detail in Chapter 11). For instance, double-diffusion contributes to the vertical transport of heat and carbon dioxide in the thermocline, affecting air–sea fluxes and thereby the global climate. The prevalence of double-diffusion in the upper ocean also leads to enhanced mixing of nutrients, which directly controls biological productivity of the ocean.

It should be emphasized, however, that the classification of vertical stratification as finger favorable for $\bar{T}_z > 0$, $\bar{S}_z > 0$ and diffusive for $\bar{T}_z < 0$, $\bar{S}_z < 0$ offers only a crude and preliminary assessment of double-diffusive characteristics of the environment. These conditions are necessary but not always sufficient. A more precise description of double-diffusive instabilities is based on linear stability analysis, which is discussed next.

2

The linear instability problem

This chapter summarizes the well-known (e.g., Baines and Gill, 1969; Walin, 1964; Nield, 1967) results of linear stability analysis for a basic state with uniform temperature and salinity gradients. There is a consensus that the choice of boundary conditions has little effect on linear stability characteristics as long as the size of a double-diffusive layer greatly exceeds the nominal double-diffusive scale d in (1.8). Therefore, our review is restricted to the simplest configuration – the unbounded gradient layer model.

2.1 Conditions for instability

Consider a motionless, laterally homogeneous layer, which is vertically stratified in either the fingering ($\bar{T}_z > 0, \bar{S}_z > 0$) or diffusive ($\bar{T}_z < 0, \bar{S}_z < 0$) sense. We assume that the perturbation to the basic state is weak and therefore the non-dimensional governing equations (1.12) and (1.13) are linearized by neglecting the advective terms $\vec{v}' \cdot \nabla T'$, $\vec{v}' \cdot \nabla S'$ and $\vec{v}' \cdot \nabla \vec{v}'$ in the temperature, salinity and momentum equations (perturbations are denoted by primes). Stability of the resulting linear system is examined using the normal modes:

$$(T', S', \vec{v}', p') = \mathrm{Re}\left[(\hat{T}, \hat{S}, \hat{\vec{v}}, \hat{p})\exp(\lambda t + ikx + ily + imz)\right], \qquad (2.1)$$

where λ is the growth rate and $\vec{k} = (k, l, m)$ is the wavenumber; the spatially uniform amplitudes of Fourier components are denoted by hats. When the normal modes (2.1) are substituted in the linearized system and the amplitudes $(\hat{T}, \hat{S}, \hat{\vec{v}}, \hat{p})$ are sequentially eliminated, we arrive at the eigenvalue equations:

salt fingers

$$\lambda^3 + [1 + \tau + Pr]\kappa^2\lambda^2 + \left[(\tau + Pr + \tau Pr)\kappa^4 + Prk_H^2\kappa^{-2}\left(1 - R_\rho^{-1}\right)\right]\lambda + \tau Pr\kappa^6 - Prk_H^2\left(R_\rho^{-1} - \tau\right) = 0,$$

diffusive convection

$$\lambda^3 + [1 + \tau + Pr]\,\kappa^2\lambda^2 + \left[(\tau + Pr + \tau Pr)\kappa^4 - Prk_H^2\kappa^{-2}\left(1 - R_\rho^{-1}\right)\right]\lambda$$
$$+ \tau Pr\kappa^6 + Prk_H^2\left(R_\rho^{-1} - \tau\right) = 0, \qquad (2.2)$$

where $\kappa = |\vec{k}|$ and $k_H = \sqrt{k^2 + l^2}$. For instability to occur, $\mathrm{Re}(\lambda) > 0$ for some wavenumbers $(\vec{k})$.

Our stability analysis becomes analytically tractable when we take advantage of a simple algebraic property of (2.2) – the growth rate equations are invariant with respect to the transformation

$$\lambda \to a\lambda, \quad \kappa^2 \to a\kappa^2, \quad k_H^2 \to a^3 k_H^2, \qquad (2.3)$$

where a is an arbitrary positive constant. Choosing $a = \frac{\kappa}{k_H} > 1$ in (2.3), we can transform any wavenumber $\vec{k}$ to the one with $\kappa = k_H$ or, equivalently, with $m = 0$. The new wavenumber represents a vertically oriented (z-independent) harmonic, also known as the elevator mode. The original growth rate can be recovered by dividing λ of the transformed mode by $a > 1$. A corollary of the invariance relation (2.3) is that the stability of our system needs only to be examined, without loss of generality, for the elevator modes – a small subset of all Fourier harmonics in (2.1). The growth rates of tilted modes with finite m are smaller than those of their vertical counterparts but have the same sign.

After restricting analysis to the rapidly growing elevator modes ($m = 0$), the instability conditions are obtained by requiring $\mathrm{Re}(\lambda) = 0$ at the marginal instability point, resulting in

$$\begin{aligned} &1 < R_\rho < \frac{1}{\tau} && \text{for salt fingering,} \\ &1 < R_\rho^* = R_\rho^{-1} < \frac{Pr + 1}{Pr + \tau} && \text{for diffusive convection.} \end{aligned} \qquad (2.4)$$

As previously, we consider only gravitationally stable (bottom-heavy) configurations, which accounts for the lower limit of unity for the density ratios in (2.4).

In the oceanographic (heat–salt) case, the diffusivity ratio is $\tau \approx 0.01$, and therefore (2.4) suggests a wide range of finger-favorable density ratios ($1 < R_\rho < 100$). These conditions are met, for instance, in more than 90% of the Atlantic thermocline. On the other hand, the range of diffusive convection ($1 < R_\rho^* < 1.14$) is extremely narrow. Both predictions, however, should be interpreted cautiously. As we shall see later on (Chapters 3–5), finite amplitude effects, not considered in linear theory, have a profound effect on the distribution and intensity of double-diffusive convection. For salt fingering, nonlinearity constrains the amplitude of perturbations and makes finger-driven mixing rather ineffective for $R_\rho > 2$. The diffusive range predicted by (2.4) can also be misleading. In the ocean, diffusive convection appears more frequently in the form of diffusive layering and the conditions for its maintenance ($R_\rho^* < 10$) are less restrictive than for linear oscillatory instability.

It is instructive to express the instability conditions (2.4) in terms of the Turner angle (Section 1.6). For salt fingers (heat–salt case) the linear instability range is $45.57° < Tu < 90°$, which constitutes 99% of the $\bar{T}_z > 0$, $\bar{S}_z > 0$ sector (Fig. 1.7). This implies that stratification patterns in which temperature and salinity increase upward are most likely to support fingering instability. The situation is markedly different for diffusive convection. The linearly unstable region is $-90° < Tu < -86.26°$, which means that only a small fraction (8.3%) of the diffusive sector ($\bar{T}_z < 0$, $\bar{S}_z < 0$) is susceptible to small-amplitude oscillatory instability.

2.2 Growth rates and spatial scales

Having established conditions for the onset of double-diffusive instabilities, we now proceed to analyze growth rates and preferred spatial patterns along the lines of Schmitt (1979a). Our focus is on fingering (rather than diffusive) instability, which is motivated by the utility and relevance of linear stability analysis. Stability theory successfully explains several key characteristics of salt fingers inferred from field measurements and laboratory experiments, while the connection between primary diffusive instabilities and observed structures is less clear. Following the conventional approach in linear stability models, we concentrate on the unstable mode with the largest growth rate. This methodology is based on a sensible assumption: If a weak initial perturbation contains a full spectrum of normal modes, then the fastest growing mode is more likely to reach the level of nonlinear equilibration first and thereby determine the pattern of fully developed instability.

Since the largest growth rates are always attained by the vertically homogeneous elevator modes, we assume $m = 0$, which reduces the fingering growth rate equation in (2.2) to

$$\lambda^3 + [1 + \tau + Pr]\, k_H^2 \lambda^2 + \left[(\tau + Pr + \tau Pr)\kappa^4 + Pr\left(1 - R_\rho^{-1}\right)\right]\lambda + \tau Pr k_H^6 - Pr k_H^2 \left(R_\rho^{-1} - \tau\right) = 0. \tag{2.5}$$

The most unstable mode is determined by maximizing $\mathrm{Re}(\lambda)$ over the three roots of the cubic in (2.5) for given (R_ρ, τ, Pr). The typical dependence of λ on k_H is shown in Figure 2.1. The growth rate is real, which signifies the direct mode of instability, characterized by monotonic growth of perturbations. The basic state is unstable for a finite range of not-too-large wavenumbers and the growth rate pattern is characterized by a well-defined maximum λ_{max} at the wavenumber k_{max}.

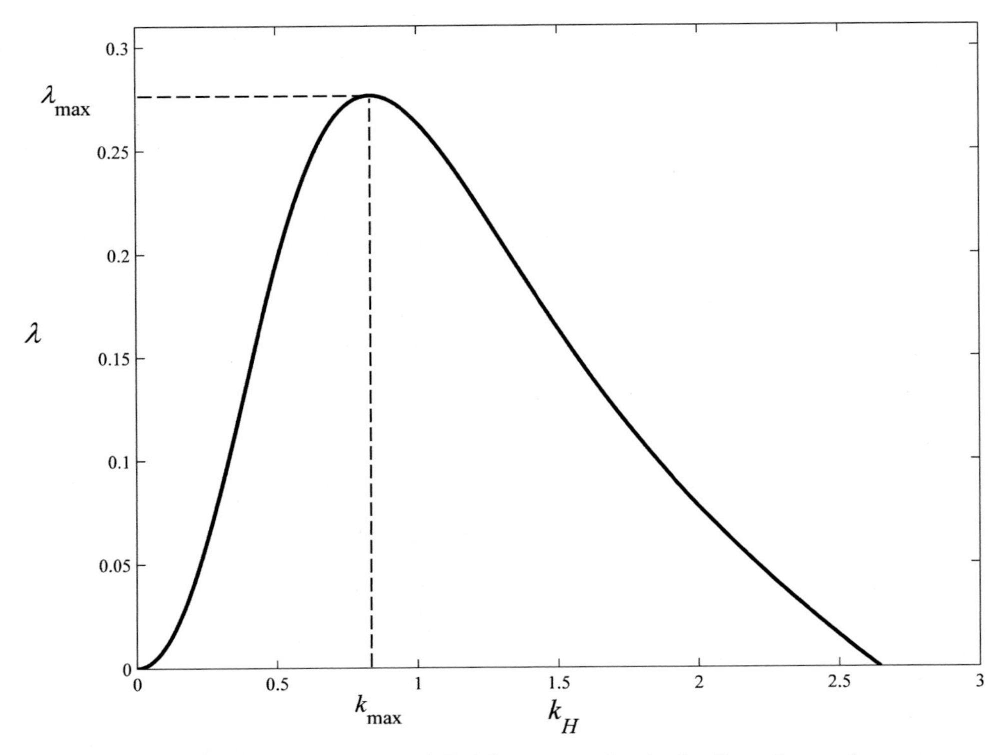

Figure 2.1 The growth rate (λ) of the elevator modes in the fingering regime as a function of the horizontal wavenumber (k_H) for $(R_\rho, Pr, \tau) = (2, 7, 0.01)$. Note the well-defined maximum ($\lambda_{\max}$) at $k_H = k_{\max}$.

This maximal growth rate is plotted as a function of density ratio in Figure 2.2. As expected, $\lambda_{\max}$ monotonically decreases with increasing R_ρ – fingering is most intense when the destabilizing salinity gradient is large and is adversely affected by the stabilizing gradient of temperature. The growth rates in Figure 2.2 are presented in standard non-dimensional units (1.11) and also in the dimensional form for $\bar{T}_z = 0.01\ ^\circ\mathrm{C\ m^{-1}}$, which is the typical stratification in the main mid-latitude thermocline. Salt fingers grow on the time scale of 10–20 minutes and therefore they are able to rapidly adjust to changing background conditions.

Figure 2.3 presents the wavenumber $k_{\max}$, along with the corresponding dimensional wavelength $L_{\dim}$, as a function ofR_ρ. The fastest growing finger wavelength is on the order of a few centimeters and is only weakly dependent on the density ratio. A number of studies (Magnell, 1976; Gargett and Schmitt, 1982) find close agreement between the finger scales predicted by linear analysis and those inferred from field measurements. Note that the non-dimensional values of $\lambda_{\max}$ and $k_{\max}$ in Figures 2.2 and 2.3 are of order unity, which is consistent with the spatial and

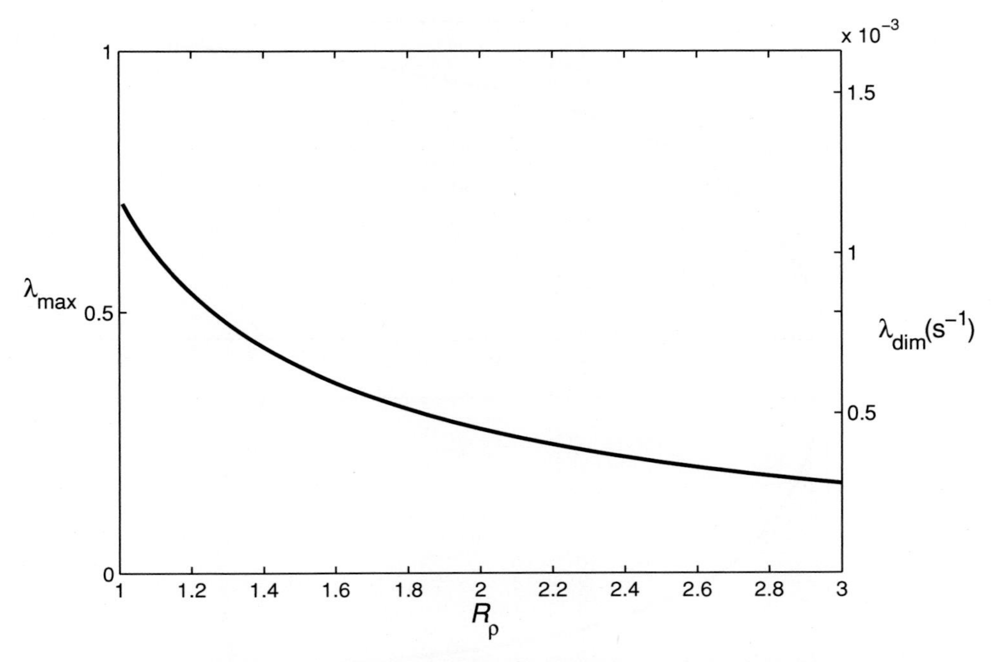

Figure 2.2 The maximal growth rate of salt fingers as a function of the density ratio. Non-dimensional values are indicated on the left axis. The right axis shows dimensional growth rates evaluated for the background temperature gradient of $\bar{T}_z = 0.01\ ^\circ\mathrm{C\,m^{-1}}$.

temporal scaling established in (1.7). Figure 2.4 shows the growth rate as a function of k_H and R_ρ, indicating the sensitivity of primary instability to both the finger scale and the background stratification. Overly thin fingers are damped by viscosity and ultimately by salt diffusion. Wide fingers do not lose enough heat laterally to fully engage double-diffusive mechanics and therefore they cannot grow as fast as intermediate width fingers.

It should be emphasized that, for a given background state, λ is uniquely determined by k_H. Hence, any combination of horizontal wavenumbers (k, l) satisfying

$$k^2 + l^2 = k_{\max}^2 \tag{2.6}$$

results in the identical maximal growth rate $\lambda_{\max}$, as does any linear superposition of normal modes satisfying (2.6). Thus, a variety of horizontal patterns have the same selective advantages and a-priori equal chances of controlling the fully developed state. This degeneracy of linear double-diffusive systems leads to very interesting and challenging theoretical problems regarding the horizontal planform selection of salt fingers (Proctor and Holyer, 1986; Schmitt, 1994a; Radko and Stern, 2000).

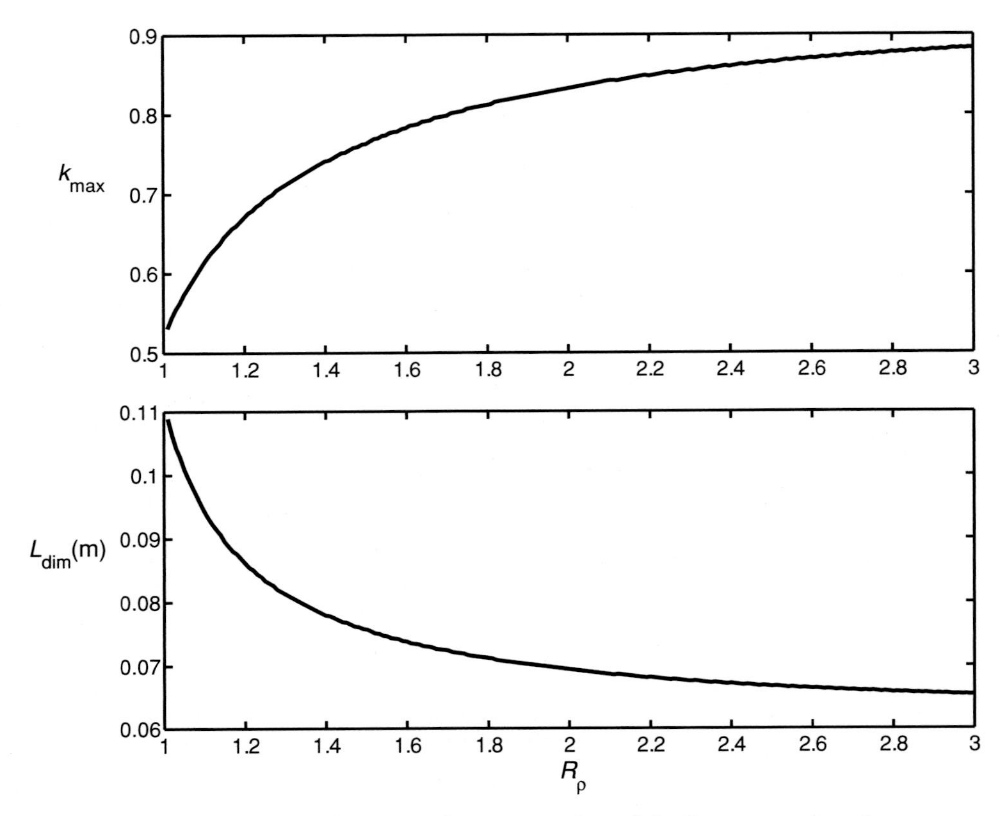

Figure 2.3 The top panel presents the wavenumber of the fastest growing elevator mode as a function of the density ratio. The bottom panel shows the corresponding dimensional wavelength evaluated for $\bar{T}_z = 0.01\ ^\circ\mathrm{C\,m^{-1}}$.

The preferred planform can only be predicted by invoking fundamentally nonlinear considerations (Chapter 5).

The stability properties of salt fingers should be contrasted with those of diffusive convection. An example of a diffusive solution is presented in Figure 2.5, where we plot real and imaginary components of λ_{max} as functions of R_ρ^*. Unlike in the salt-finger case, where λ is real, $\mathrm{Im}(\lambda)$ now significantly exceeds $\mathrm{Re}(\lambda)$. Since $\mathrm{Im}(\lambda)$ is associated with the oscillatory pattern of motion, diffusive modes are best described as periodic waves of gradually increasing amplitude.

2.3 The flux ratio

One of the most widely used and best understood characteristics of double-diffusive convection is represented by the flux ratio, which measures the fraction of the

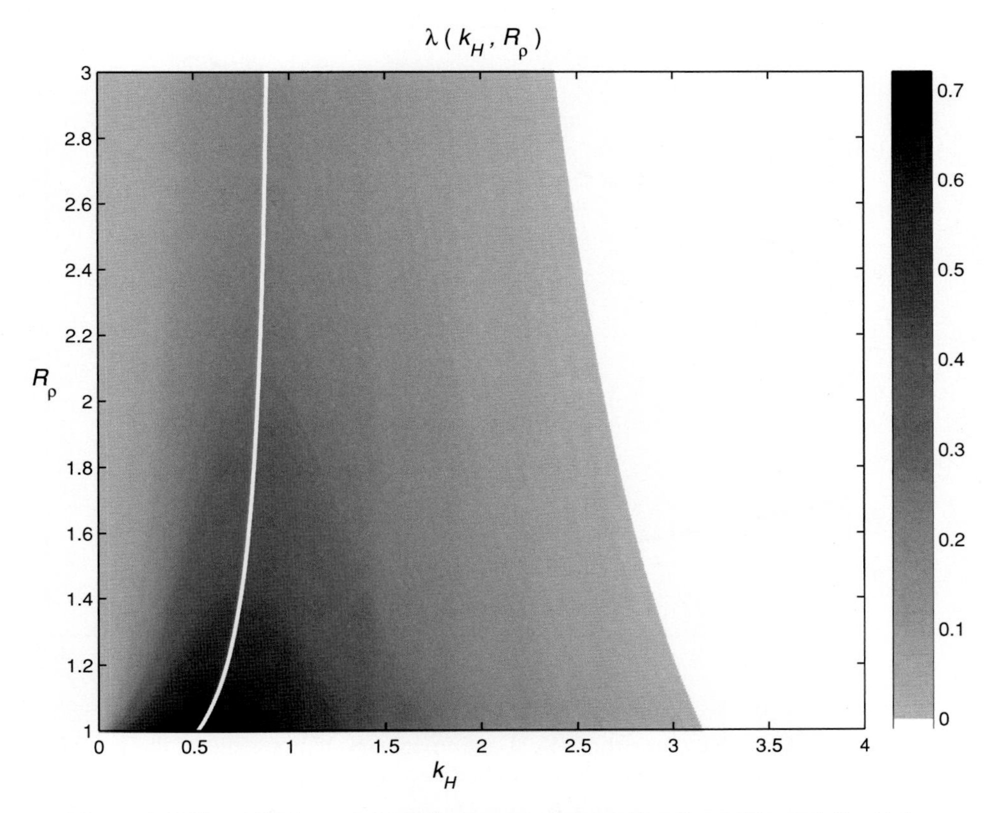

Figure 2.4 The growth rate of the elevator modes as a function of k_H and R_ρ. Only amplifying modes are shown. The white curve indicates the wavenumber of the fastest growing mode ($k_{\max}$).

destabilizing component's potential energy loss gained by the stabilizing substance. In the fingering regime ($\bar{T}_z > 0$, $\bar{S}_z > 0$) the flux ratio takes the form

$$\gamma = \frac{\alpha F_{T\,\mathrm{dim}}}{\beta F_{S\,\mathrm{dim}}}, \tag{2.7}$$

where $(F_{T\,\mathrm{dim}}, F_{S\,\mathrm{dim}})$ are the dimensional vertical fluxes of temperature and salinity. The corresponding expression for the diffusive case ($\bar{T}_z < 0$, $\bar{S}_z < 0$) is given by

$$\gamma^* = \frac{\beta F_{S\,\mathrm{dim}}}{\alpha F_{T\,\mathrm{dim}}}. \tag{2.8}$$

The T–S fluxes in (2.7) and (2.8) include the contribution from transient eddies as well as molecular transport:

$$\begin{aligned} F_{T\,\mathrm{dim}} &= \overline{w'_{\mathrm{dim}} T'_{\mathrm{dim}}} + k_T \bar{T}_z, \\ F_{S\,\mathrm{dim}} &= \overline{w'_{\mathrm{dim}} S'_{\mathrm{dim}}} + k_S \bar{S}_z, \end{aligned} \tag{2.9}$$

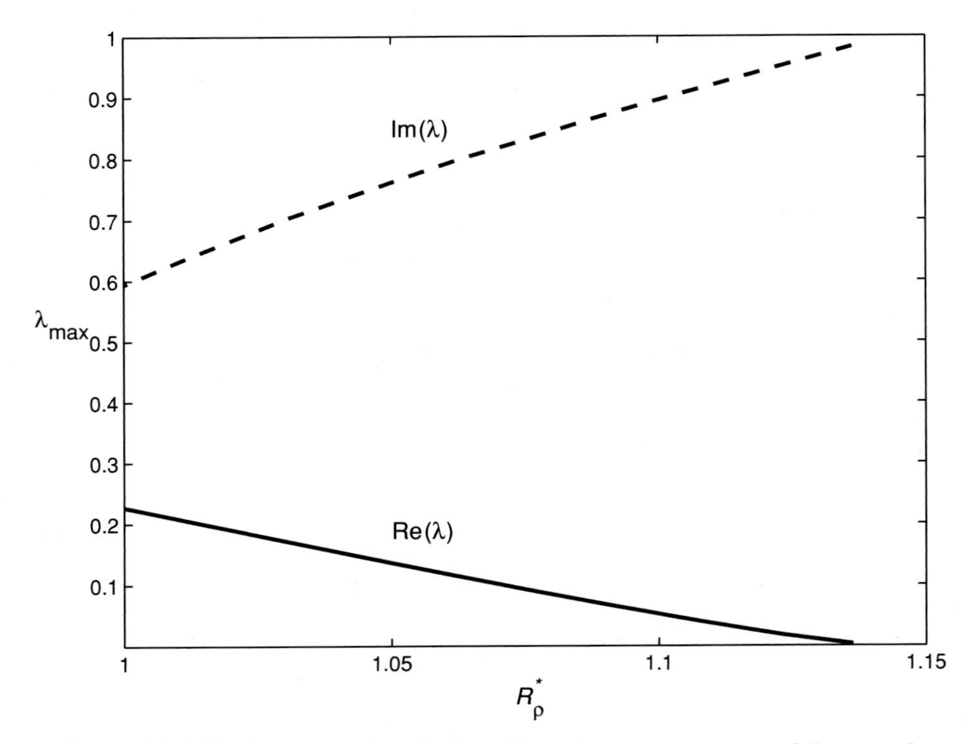

Figure 2.5 Diffusive convection. Real and imaginary components of the growth rate of the fastest growing mode are plotted as a function of the density ratio.

where the overbars denote spatial averages. However, very often, particularly in oceanographic applications, fluxes are dominated by their eddy-induced components ($\overline{w'_{\text{dim}} T'_{\text{dim}}}, \overline{w'_{\text{dim}} S'_{\text{dim}}}$). Neglecting molecular transport in (2.9) and applying the standard double-diffusive non-dimensionalization (1.11), we reduce the expression for the flux ratio to

$$\gamma \approx \frac{\overline{w'T'}}{\overline{w'S'}}. \tag{2.10}$$

From a theoretical standpoint, an attractive feature of the flux ratio is that typical values of γ can be estimated – and its properties rationalized – by means of linear stability analysis. For the z-independent elevator modes, the eddy-induced temperature and salinity fluxes are given by

$$\begin{aligned} \overline{w'T'} &= \frac{1}{2}\hat{w}\hat{T}, \\ \overline{w'S'} &= \frac{1}{2}\hat{w}\hat{S}. \end{aligned} \tag{2.11}$$

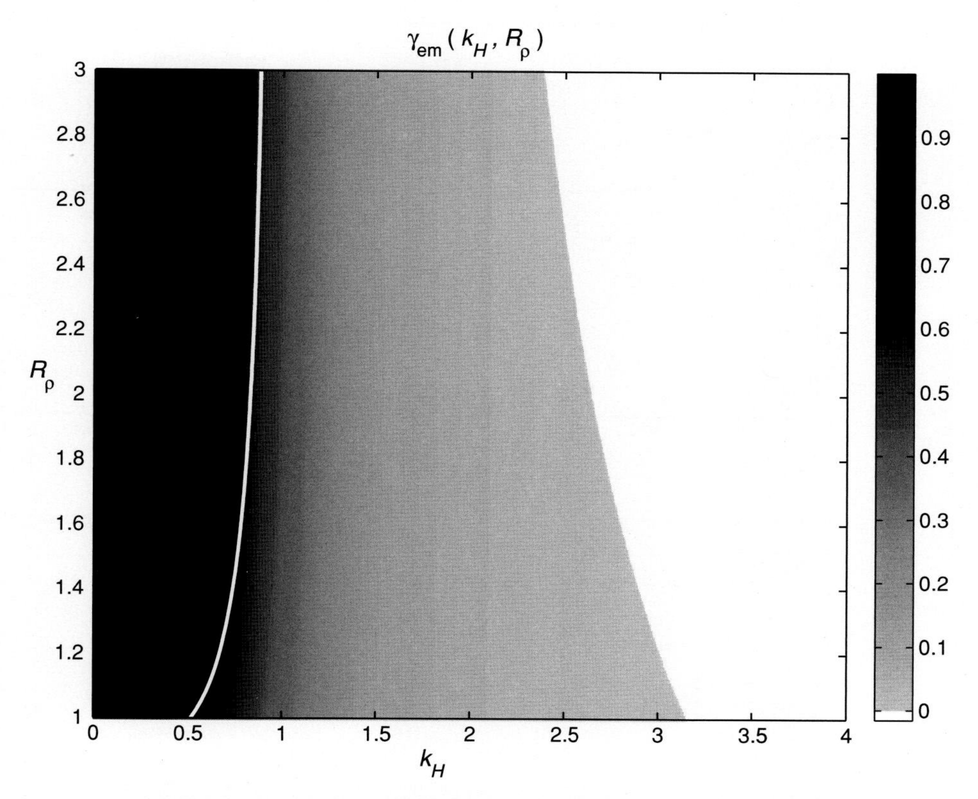

Figure 2.6 Flux ratio of the elevator mode as a function of the wavenumber and the density ratio. Only amplifying modes are shown. The white curve indicates the wavenumber of the fastest growing mode (k_{max}).

Therefore, the flux ratio of elevator modes reduces to $\gamma_{em} = \frac{\hat{T}}{\hat{S}}$. While the linear theory cannot determine the individual amplitudes of temperature ($\hat{T}$) and salinity ($\hat{S}$), each mode is characterized by a unique value of their ratio.

The pattern of flux ratio in the fingering regime is presented in Figure 2.6, where γ_{em} is plotted as a function of k_H and R_ρ for the amplifying ($\lambda > 0$) modes. Note that the flux ratio is always less than unity, which is consistent with the energetics of the system. In order to support the growth of unstable modes, the potential energy lost by the haline stratification should exceed the energy gained by the thermal stratification. Note the monotonic decrease of the flux ratio with increasing wavenumber (k_H). This dependence can be readily rationalized by focusing on the difference in thermal dissipation of thin (high k_H) and thick (low k_H) fingers (Schmitt, 1979a). Thin fingers rapidly lose their temperature anomaly due to lateral diffusion and therefore can support only limited eddy heat flux. As a result, the flux

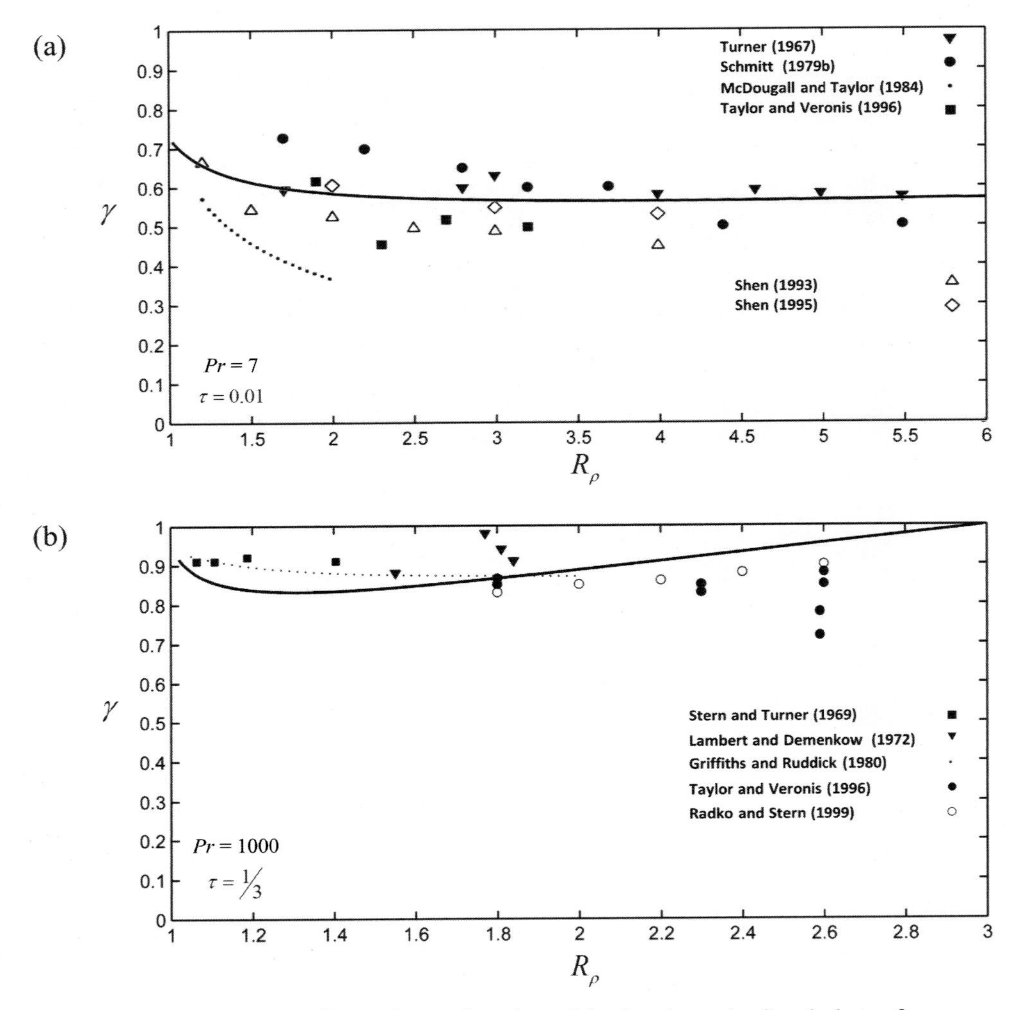

Figure 2.7 Salt-finger flux ratio as a function of the density ratio. Symbols are from laboratory and numerical experiments and the solid curves represent the theoretical estimates based on the fastest growing finger model. Calculations are made for (a) the oceanographic (heat–salt) parameters $(\tau, Pr) = (0.01, 7)$ and (b) salt–sugar parameters $(\tau, Pr) = (1/3, 10^3)$. After Kunze (2003).

ratio (2.7) is low. Wide fingers, on the other hand, can carry a larger temperature anomaly and will thus have a higher flux ratio. The preferred (fastest growing) fingers operate on intermediate scales and are characterized by a mid-range flux ratio.

Figure 2.7 shows the flux ratio of the fastest growing elevator modes ($k_H = k_{max}$) along with the flux ratios evaluated from a series of laboratory and numerical experiments. Presented are two cases. The oceanographic heat–salt system $(\tau, Pr) = (0.01, 7)$ is in Figure 2.7a. Both experimental and theoretical values of the flux ratio in Figure 2.7a are in the $\gamma = 0.5-0.8$ range. The flux ratio decreases with R_ρ for $1< R_\rho < 3$ and then remains nearly flat with a shallow minimum at $R_\rho \approx 4$. For larger density ratios (not shown) the flux ratio of the elevator mode gradually increases with R_ρ towards $\gamma_{\text{em}} = 1$ at $R_\rho \rightarrow \tau^{-1}$. The second calculation (Fig. 2.7b) was made for $(\tau, Pr) = (1/3, 10^3)$. This parameter set corresponds to the laboratory experiments in which heat and salt are replaced by aqueous solutions of sugar and salt as diffusing substances. Such experiments were introduced (Stern and Turner, 1969) to avoid problems associated with the heat loss across the walls of an experimental container, and sugar–salt systems still remain popular in laboratory studies (e.g., Krishnamurti, 2003, 2009). The salt–sugar flux ratio ($\gamma \sim 0.9$) is considerably higher than that for heat and salt. This tendency can be seen clearly in experiments and it is nicely reflected by the fastest growing mode theory. The pattern of the salt–sugar $\gamma(R_\rho)$ relation is also non-monotonic. As the density ratio increases from unity, the flux ratio first decreases, reaches minimum in the interior of the salt-finger interval, and then starts to increase towards unity. The location of the minimum, however, is shifted for the salt–sugar case to a much lower density ratio ($R_{\text{min}} \approx 1.3$) than for heat and salt. The quantitative differences in the flux ratio for heat–salt and sugar–salt systems underscore the sensitivity of salt fingering to the molecular characteristics of working fluids. The dependence of double-diffusive fluxes and structures on (τ, Pr) will be examined in greater detail in our review of non-oceanographic applications (Chapter 12).

Overall, for both heat–salt and salt–sugar systems, the predictions of the elevator mode theory are consistent with the corresponding laboratory and numerical experiments. This result is by no means trivial. Laboratory experiments are characterized by a broad spectrum of fully developed nonlinearly interacting modes that are not represented in the single-mode theory. The experimental validation of the theoretical model is highly valuable as it demonstrates the central role played by the fastest growing elevator modes in fingering dynamics.

2.4 Effects of horizontal gradients

A natural extension of the stability analysis is afforded by inclusion of horizontal background temperature and salinity gradients. For simplicity, we illustrate key effects using the two-dimensional (x, z) model, which does not lead to loss of generality – the three-dimensional linear stability problem can be reduced to its two-dimensional equivalent using a simple algebraic transformation (Holyer, 1983).

In order to construct a well-defined steady state, horizontal gradients are assumed to be density compensating:

$$\alpha \bar{T}_x = \beta \bar{S}_x. \tag{2.12}$$

For $\bar{T}_z > 0$, the non-dimensional governing equations for perturbation quantities take the form

$$\begin{cases} \frac{1}{Pr}\left(\frac{\partial \vec{v}}{\partial t} + \vec{v}\cdot\nabla\vec{v}\right) = -\nabla p + (T' - S')\vec{k} + \nabla^2\vec{v}, \\ \frac{\partial T'}{\partial t} + \vec{v}\cdot\nabla T' + Gu + w = \nabla^2 T', \\ \frac{\partial S'}{\partial t} + \vec{v}\cdot\nabla S' + Gu + \frac{w}{R_\rho} = \tau\nabla^2 S', \\ \nabla\cdot\vec{v} = 0, \end{cases} \tag{2.13}$$

where $G = \frac{\bar{T}_x}{\bar{T}_z}$ represents the slope of isotherms in the basic state. The influence of horizontal gradients is reflected by the Gu terms in the temperature and salinity equations, terms that do not appear in the corresponding system for the horizontally uniform basic state (1.12).

As previously, system (2.13) is linearized and its stability is examined using the normal modes (2.1), which leads to the growth rate equation:

$$\begin{aligned} &\lambda^3 + [1 + \tau + Pr]\kappa^2\lambda^2 + \left[(\tau + Pr + \tau Pr)\kappa^4 + Prk^2\kappa^{-2}\left(1 - R_\rho^{-1}\right)\right]\lambda \\ &\quad + \tau Pr\kappa^6 - Prk^2\left(R_\rho^{-1} - \tau\right) + PrGkm(1 - \tau) = 0. \end{aligned} \tag{2.14}$$

Typical values of the slopes of isotherms in the ocean are quite low ($G \sim 10^{-3}$). Hence, if the vertical stratification supports fingering instability, the addition of lateral gradients has very limited effect on linear stability characteristics. The maximal growth rates and preferred wavelengths of fingers change by a fraction of a percent. However, fundamentally new effects come into play when vertical stratification is double-diffusively stable.

Figure 2.8a presents the growth rate pattern for $R_\rho = -1$ ($Tu = 0$). An extended region in the (k, m) space is occupied by the unstable ($\lambda > 0$) modes. Since vertical temperature and salinity gradients are both stabilizing, the growing modes in Figure 2.8a can only be attributed to the horizontal stratification. In contrast to the horizontally uniform case (Section 2.2), the fastest growing modes in Figure 2.8a are of finite vertical extent $H_{\text{dim}} \sim \frac{2\pi}{m_{\max}} d \sim 2\text{m}$. They grow much slower than salt fingers (cf. Fig. 2.2) and operate on significantly larger spatial scales. The orientation of wave fronts in the fastest growing mode is nearly horizontal, with slopes of $s = -\frac{k_{\max}}{m_{\max}} \sim 2\cdot 10^{-4}$, and therefore these instabilities are best described as molecularly driven intrusions. Growing modes are found only for $s > 0$, which means that intrusions in the doubly stable ($\bar{T}_z > 0$, $\bar{S}_z < 0$) vertical gradient are oriented in the same sense as salinity contours and opposite to that of isotherms. Figure 2.8b presents an analogous example for a diffusively stratified ($\bar{T}_z < 0$, $\bar{S}_z < 0$) fluid. The diffusive density ratio in this case is $R_\rho^* = 3$ and therefore horizontally

(a)

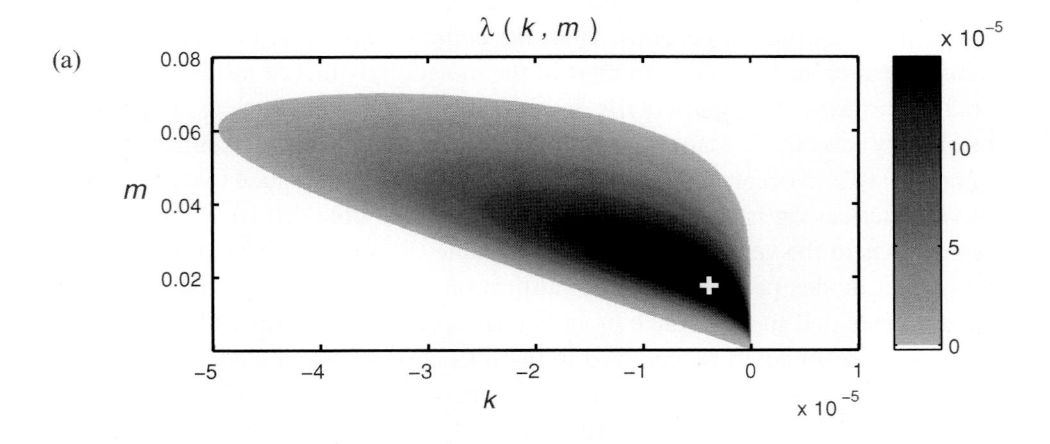

(b)

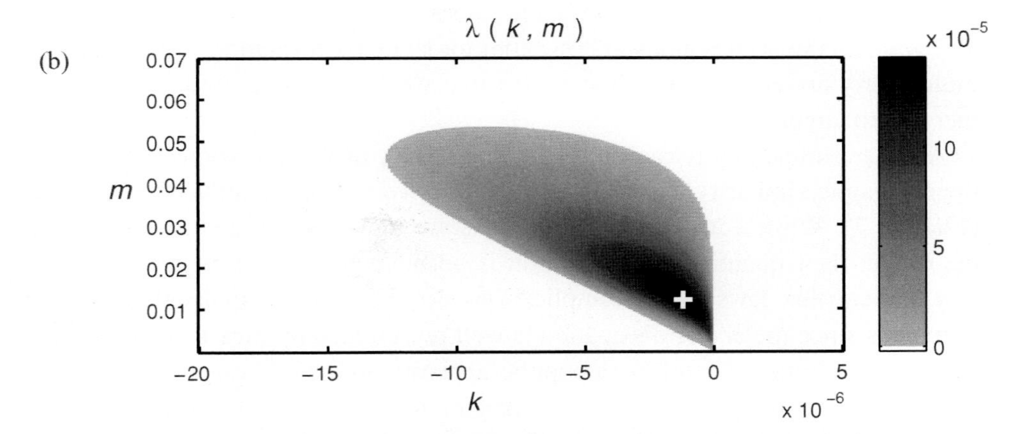

Figure 2.8 Destabilizing effects of horizontal gradients for the heat–salt system $(\tau, Pr) = (0.01, 7)$. Growth rate is plotted as a function of wavenumbers (k, m) for (a) doubly stable ($\bar{T}_z > 0$, $\bar{S}_z < 0$, $R_\rho = -1$) and (b) diffusive ($\bar{T}_z < 0$, $\bar{S}_z < 0$, $R_\rho^* = 3$) vertical stratification. In both cases, $G = \bar{T}_x/\bar{T}_z = 10^{-3}$. Only amplifying modes are shown and the maximal growth rates are indicated by the plus signs.

homogeneous stratification would be linearly stable according to (2.4). However, the inclusion of lateral gradients destabilizes the system. As in the doubly stable case (Fig. 2.8a), the instability comes in the form of interleaving currents. The wave fronts are slightly inclined relative to the horizontal in the same sense as the background isotherms.

The examples in Figure 2.8 illustrate a generic property of horizontally non-uniform double-diffusive systems. Lateral stratification, no matter how weak, always generates unstable interleaving modes. It should be noted in this regard that horizontal T–S gradients are universally present in the ocean, although their magnitude is highly inhomogeneous. Thus, the entire ocean supports double-diffusion in one form or another. Growth rates of the molecularly driven intrusions are generally low (Fig. 2.8), particularly in comparison with the vertical (fingering or oscillatory) double-diffusive instabilities. Nevertheless, these instabilities serve an important role in ocean dynamics. For instance, it has been argued that molecularly driven interleaving is essential for the initiation of double-diffusive convection in regions where the vertical stratification is stable (Stern, 2003). Amplifying nearly horizontal modes perturb vertical stratification, eventually producing local vertical gradients that support much more intense vertical double-diffusive convection. This mechanism could be involved in the formation of Arctic and Antarctic thermohaline staircases – step-like patterns in vertical temperature and salinity profiles (Chapter 8). High-latitude diffusive staircases are observed in regions with generally moderate large-scale density ratios of $2 < R_\rho^* < 10$, which are linearly stable with respect to vertical double-diffusion. Stern (2003) proposed that weak lateral T–S gradients in such regions are essential for staircase formation – they generate molecularly driven intrusions that evolve into small-scale steps and subsequently merge into larger layers.

Another compelling reason to study molecularly driven intrusions is related to their dynamic similarity to intrusions driven by vertical eddy-induced T–S fluxes (Chapter 7). While eddy-induced intrusions are more intense and widespread in the ocean, their quantitative description is problematic in view of highly uncertain vertical flux laws. Such complications do not arise for molecularly driven intrusions since molecular dissipation is well represented by Fick's flux laws with known diffusivities. Therefore, a popular approach to the analysis of interleaving involves theoretical and laboratory investigation of molecularly driven intrusions (e.g., Thorpe *et al.*, 1969; Kerr, 1992). Such studies are often accompanied by imaginative, albeit fundamentally qualitative, extrapolations of the results to the problem of eddy-induced interleaving.

This discussion concludes our summary of linear models of double-diffusive convection. In this review we restrained from venturing into numerous generalizations of the classical theory. For instance, we have not considered extensions to multicomponent ($n > 2$) systems (Griffiths, 1979; Terrones and Pearlstein, 1989; Liang, 1995) and cross-diffusion effects, in which molecular diffusion of one diffusing substance is affected by the gradient of others (Hurle and Jakeman, 1971; McDougall, 1983; Liang *et al.*, 1994). Not considered here are temperature-dependent viscosity and diffusivity, which can affect the stability properties of both fingering and diffusive systems (Tanny *et al.*, 1995). While such generalizations are undoubtedly of theoretical interest, they do not fundamentally change our understanding of double-diffusion, particularly for the oceanic (heat–salt) case.

Therefore, readers are referred to the original papers for details of the modified linear stability analysis. The content of this chapter has been selected to provide the essential background for the analysis of observations and for the nonlinear theory that is considered next. Nonlinear effects in double-diffusive convection are illustrated using three elementary geometric configurations: (i) the unbounded gradient model, (ii) the two-layer system and (iii) the model of a vertically bounded double-diffusive layer. Each model gets a chapter.

3

The unbounded gradient model

A fundamental problem in double-diffusive convection concerns the equilibrium transport of temperature, salinity, chemical tracers and momentum. The quantification of double-diffusive fluxes and their dependencies is an essential step in linking the microstructure dynamics with larger scales of motion – a problem that has motivated numerous laboratory and field experiments, as well as theoretical and numerical models. Linear stability theory for double-diffusion (Chapter 2) is well developed and fully understood. However, it does not explain the equilibration mechanisms, or predict the saturation amplitude, of the unstable perturbations. The problem of equilibration is complicated and principally nonlinear. Only recently have we started to see the first signs of a consensus between theorists, modelers and observationalists with regard to the typical magnitudes of double-diffusive transport of heat and salt in the ocean and its variation with environmental parameters. Our discussion of nonlinear effects starts with the unbounded gradient system, which is used to explain the dynamics of double-diffusive mixing in smooth background temperature and salinity gradients. We focus on theoretical models, which attempt to predict the equilibrium transport from first principles. However, their success is judged in terms of consistency with observations and simulations.

3.1 Flux-gradient laws

The unbounded gradient models are based on an assumption that, in relatively smooth background gradients of temperature and salinity, the intensity of double-diffusive convection is controlled by the local *T–S* gradients and, to a lesser extent, by the background shear. The separation from physical boundaries and the scales of inhomogeneity of the background flow are assumed to be too large to affect local properties of double-diffusive convection. This assumption may not apply for diffusive layering, which is better represented by interfacial models (Chapter 4). Therefore, the vast majority of unbounded models are focused on the salt-finger case and we shall follow suit. As with any idealization, the unbounded fingering model should be applied to observations and laboratory experiments with caution. Examples can be constructed in which the intensity of salt fingers is not uniquely determined by the background gradients but affected by such parameters as the vertical extent of the double-diffusive region (e.g., Radko and Stern, 2000). However, such counter-examples are rather rare and somewhat artificial. What makes the unbounded model of salt fingers particularly robust and relevant in the oceanographic context is a clear scale separation that generally exists between salt fingers and larger features of circulation.

The ultimate goal of the unbounded model is to formulate explicit relationships between the vertical eddy transport of temperature and salinity and their local gradients:

$$\begin{cases} \overline{w'_{\text{dim}} T'_{\text{dim}}} = F_{T\,\text{dim}}\left(\frac{\partial \bar{T}}{\partial z}, \frac{\partial \bar{S}}{\partial z}, k_T, k_S, \nu\right), \\ \overline{w'_{\text{dim}} S'_{\text{dim}}} = F_{S\,\text{dim}}\left(\frac{\partial \bar{T}}{\partial z}, \frac{\partial \bar{S}}{\partial z}, k_T, k_S, \nu\right), \end{cases} \tag{3.1}$$

were $(\overline{w'_{\text{dim}} T'_{\text{dim}}}, \overline{w'_{\text{dim}} S'_{\text{dim}}})$ are the dimensional eddy-induced fluxes. The problem can be simplified based on dimensional reasoning: any dimensionless number can only be determined by the other dimensionless parameters. As discussed in Chapter 1, the key dimensionless governing parameters in the unbounded model are limited to the density ratio (R_ρ), diffusivity ratio (τ) and Prandtl number (Pr). Following the analogy with thermal convection, the intensity of double-diffusive mixing is commonly quantified by the Nusselt number, defined as the ratio of the eddy-induced and conductive heat fluxes:

$$Nu = \frac{-\overline{w'_{\text{dim}} T'_{\text{dim}}}}{k_T \frac{\partial \bar{T}}{\partial z}} = Nu(R_\rho, \tau, Pr). \tag{3.2}$$

The same non-dimensional parameters control another key measure of vertical transport, the flux ratio (Chapter 2):

$$\gamma = \frac{\alpha \overline{w'_{\text{dim}} T'_{\text{dim}}}}{\beta \overline{w'_{\text{dim}} S'_{\text{dim}}}} = \gamma\left(R_\rho, \tau, Pr\right). \tag{3.3}$$

While the estimate of the flux ratio (3.3) can be made on the basis of linear theory, prediction of the $Nu(R_\rho)$ dependence – the core problem of double-diffusive convection – is impossible without taking nonlinearity into account. An alternative form for the flux-gradient laws (3.2) and (3.3) can be given in terms of the eddy diffusivities of heat and salt (K_T, K_S):

$$\begin{cases} K_T = -\dfrac{\overline{w'_{\mathrm{dim}} T'_{\mathrm{dim}}}}{\bar{T}_z} = k_T Nu, \\ K_S = -\dfrac{\overline{w'_{\mathrm{dim}} S'_{\mathrm{dim}}}}{\bar{S}_z} = K_T \dfrac{R_\rho}{\gamma} = \dfrac{k_T R_\rho}{\gamma} Nu. \end{cases} \tag{3.4}$$

Thus, for any given fluid, the eddy diffusivities in the unbounded gradient model are determined entirely by the density ratio.

Double-diffusive convection theory places much lesser emphasis on large-scale momentum transfer than on T–S transfer. Numerical simulations indicate that, unlike diffusivity, the eddy viscosity of salt fingers is actually less than its molecular counterpart (Stern *et al.*, 2001). Thus, even neglecting eddy viscosity completely may have relatively minor ramifications for the large-scale dynamics. This suggestion, however, has been confirmed only in the oceanographic heat–salt context. Limited evidence suggests that the eddy viscosity due to double-diffusion may be essential (Radko, 2010) for the dynamics of low Prandtl number fluids, such as occur in astrophysical systems (Chapter 12).

3.2 Secondary instabilities: Stern–Kunze constraint and Holyer modes

Numerous attempts have been made to deduce flux-gradient laws from first principles. The first and, perhaps, the most influential idea was presented by Stern (1969), who suggested that the linear growth of salt fingers is arrested when the Stern number

$$A = \frac{F_{\rho\,\mathrm{dim}}}{\nu \bar{\rho}_z}, \tag{3.5}$$

where $F_{\rho\,\mathrm{dim}}$ is the vertical density flux, reaches O(1). Stern's physical argument is compelling. When A exceeds unity, the unbounded salt-finger system becomes unstable with respect to the so-called collective instability, the term referring to the spontaneous excitation of gravity waves by salt fingers. Stern suggested that collective instability – to be considered in greater detail in Chapter 6 – could disrupt salt fingers, thereby arresting amplification of primary growing modes.

At first glance, various pieces of indirect evidence seem to validate Stern's hypothesis. Oceanographic measurements (Hebert, 1988; Inoue *et al.*, 2008) indeed suggest the order one values of A. Kunze (1987) gave an alternative argument in support of the Stern number constraint. He speculated that the well-known Richardson number criterion for dynamic instability of horizontal inviscid parallel flows (Richardson, 1920; Howard, 1961; Miles, 1961) can be extended to salt fingers. The Richardson number is defined as the ratio of the squared buoyancy frequency and the squared velocity shear. For instability to occur, it is required to be less than 1/4. Kunze showed that the Richardson number of 1/4 based on the salt-finger scales is equivalent to the Stern number of unity. His model further assumes that the growth of salt fingers is arrested when they become susceptible to their secondary dynamic instabilities and therefore $A \sim 1$ sets the equilibrium condition.

However, despite the very promising start, the theory of finger equilibration evolved into one of the most controversial topics in double-diffusive convection. The first seeds of doubt as to the relevance of the A-based theory were planted when laboratory experimentalists started to use sugar and salt as diffusive substances. Such experiments became particularly popular for practical reasons. In heat–salt experiments it is difficult to avoid – or accurately take into account – the heat loss across the walls of the tank. The parameters in sugar–salt experiments ($\tau \approx 1/3$, $Pr \approx 10^3$) differ from the oceanographically relevant heat–salt case ($\tau \approx 0.01$, $Pr \approx 7$) but the dynamics appear to be similar. The first experiment of this nature was performed by Stern and Turner (1969) and was followed by numerous studies offering more detailed and precise measurements. Stern numbers realized in sugar–salt experiments are extremely low. Lambert and Demenkow (1972) reported values as small as $A = 2 \cdot 10^{-3}$, Griffiths and Ruddick (1980) and Krishnamurti (2003) found that, typically, $A \sim 0.1$–0.01, bringing into question the generality of Stern's constraint.

On the theoretical side, a critical advance was made by Holyer (1984), who performed a formal linear stability analysis for the vertical steady salt fingers. She discovered the direct, relatively small-scale – comparable to the salt-finger width – secondary instabilities, which typically grow much faster than the collective instability modes. Furthermore, unlike collective instabilities, Holyer modes grow regardless of the (finite) amplitude of salt fingers. A corollary of these findings is that salt fingers are more likely to be disrupted by the Holyer mechanism, rather than by collective instabilities. The schematic in Figure 3.1 illustrates the action of secondary instabilities, distorting the vertical fingers and ultimately arresting their growth.

Numerical simulations have shed extra light on the mechanics of equilibration. Two-dimensional experiments in Whitfield *et al.* (1989) performed with various molecular parameters (Pr, τ) resulted in Stern numbers that varied by a factor of 3400. A series of simulations by Shen (1993) in which the density ratio was systematically varied whereas (Pr, τ) were kept constant revealed that the Stern number (A) is rapidly decreasing with R_ρ. This finding was confirmed by recent three-dimensional simulations in Traxler *et al.* (2011a), who showed that the Stern number decreases by two orders of magnitude from $R_\rho = 1.2$ to $R_\rho = 10$. Diagnostics of the numerical simulations in Radko and Stern (1999) revealed that the

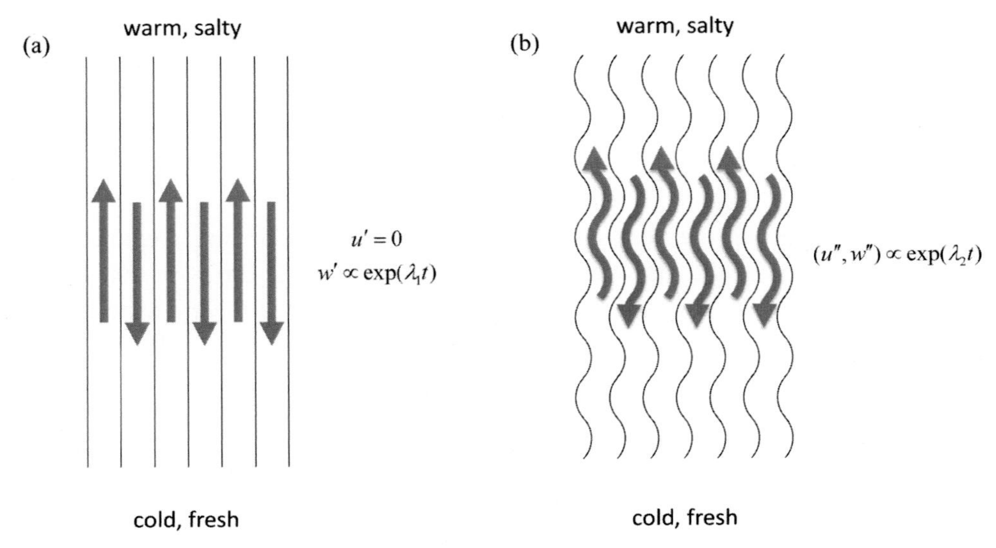

Figure 3.1 Schematic diagram illustrating development of primary salt-finger instabilities in the form of vertical elevator modes (a) and their ultimate equilibration by secondary instabilities (b) acting to distort fingers, thereby reducing the vertical transport of heat and salt. From Schmitt (2012).

equilibration occurs by means of the nonlinear interaction of primary instabilities with Holyer modes. The equilibrium heat and salt fluxes recorded in simulations (e.g., Stern *et al.*, 2001; Traxler *et al.*, 2011a) and even diagnosed from oceanic field measurements (St. Laurent and Schmitt, 1999) monotonically decrease with the density ratio, whereas the opposite trend is expected based on the Stern–Kunze constraint. In view of all the numerical, laboratory and observational evidence, it is tempting to conclude that the order-one values of A frequently observed in the ocean are a peculiar coincidence.

It should be emphasized at this point that, while numerical studies contest the Stern–Kunze constraint for the equilibrium amplitude of salt fingers, the collective instability itself is revealed very clearly in simulations (Stern *et al.*, 2001; Stern and Simeonov, 2002; Stellmach *et al.*, 2011). Likewise, there is no reason to doubt that the intensity of collective instability is controlled by the Stern number. What *is* highly questionable is the link between the Stern number and the amplitude saturation of salt fingers, which can grow even in the presence of very active collective instabilities. Concerns with regard to the Stern–Kunze constraint motivate development of alternative models of equilibration. While a comprehensive nonlinear theory of salt fingers is still lacking, double-diffusers have started to fill some gaps in the understanding of equilibration and have already successfully treated various limits. Some promising results in this direction are summarized next.

3.3 Weakly nonlinear models

Traditionally, saturation mechanisms for various instabilities of fluids are identified and explained by developing weakly nonlinear asymptotic models. Specific examples are discussed and summarized in numerous textbooks (e.g., Drazin and Reid, 1981; Godreche and Manneville, 2005) and the procedure is briefly outlined below. Inspired and guided by the treatment of Rayleigh–Bénard convection (Malkus and Veronis, 1958), weakly nonlinear models focus on the marginally unstable regime. The governing equations are expanded in a small parameter measuring the strength of the instability. This quantity, denoted by ε hereafter, is usually defined as the difference in one of the key parameters between the current configuration and the one with zero growth rate. After collecting terms of the same order in ε, one is left with a hierarchy of simpler, analytically solvable problems. It is often possible, by sequentially solving these balanced equations, to relate the amplitude of instability to ε and thus to the governing parameters. Another precious product of weakly nonlinear models is related to the insight they bring into the dynamics of equilibration. Complexities of the original governing equations usually preclude the direct association of individual terms with distinct mechanisms. In contrast, each balance in the weakly nonlinear models usually offers straightforward physical interpretation, making it possible to trace the chain of events leading to equilibration. The price paid for tractability is the limited accuracy outside of the weakly unstable range and, for most intents and purposes, it is a fair price.

The weakly nonlinear models for salt fingers have been explored in several studies (Proctor and Holyer, 1986; Stern and Radko, 1998; Radko and Stern, 1999, 2000; Stern and Simeonov, 2004; Radko, 2010). In the unbounded model, the key parameter is the density ratio and the marginally unstable point (Chapter 2) corresponds to $R_\rho = \tau^{-1}$. Thus, a sensible choice for the expansion parameter in the weakly nonlinear model could be

$$\varepsilon = R_\rho^{-1} - \tau. \tag{3.6}$$

The limit $\varepsilon \to 0$ has been explored most recently by Radko (2010). The resulting weakly nonlinear theory predicted that (i) the equilibrium fluxes of heat and salt are proportional to ε^2 and (ii) the equilibration occurs through the interaction of salt fingers with Holyer (1984) instabilities. More specifically, theory attributes the equilibration of salt fingers to a combination of two processes: the triad interaction and spontaneous development of the mean vertical shear. The non-resonant triad interactions control the equilibration of linear growth for moderate and large values of Prandtl number (Pr) and for slightly unstable parameters. For small Pr and/or rigorous instabilities, the mean shear effects become essential. These theoretical predictions are consistent with the two-dimensional direct numerical simulations

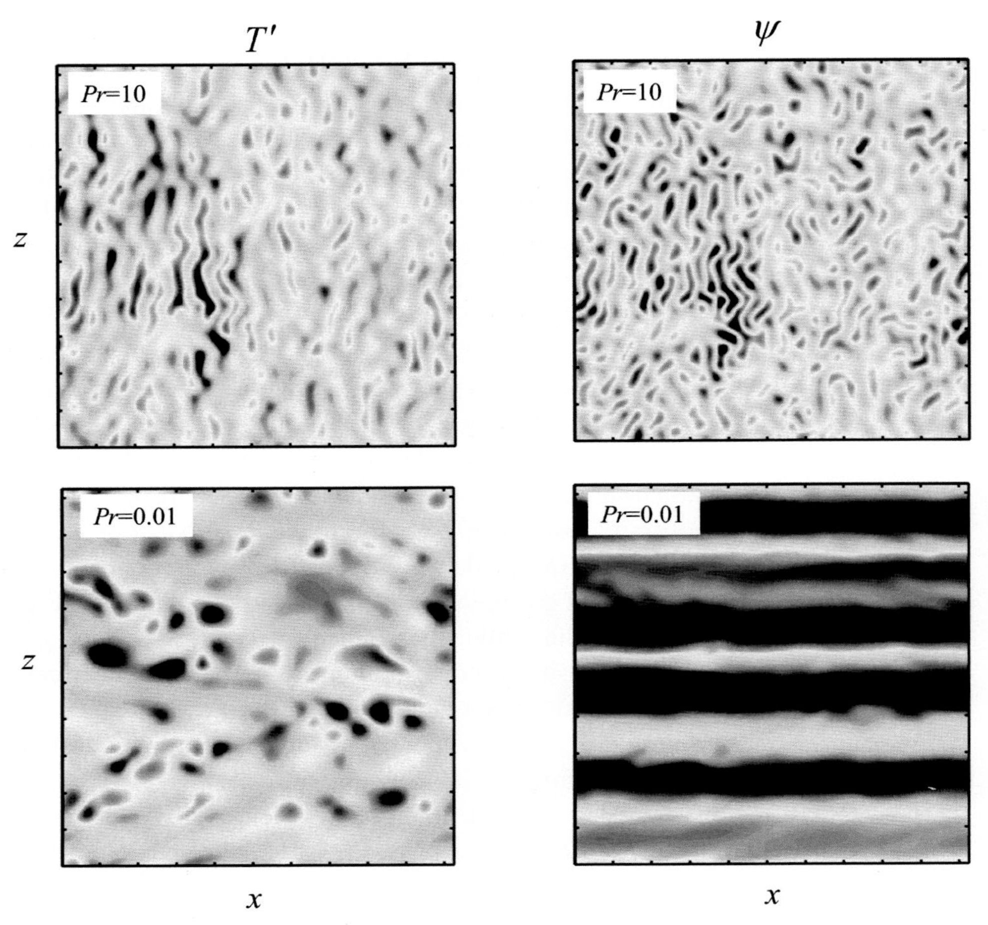

Figure 3.2 Variability of the salt-finger patterns. Instantaneous temperature (left) and streamfunction (right) fields from direct numerical simulations (DNS) are presented for $Pr = 10$ (top) and for $Pr = 0.01$ (bottom). In all cases $R_\rho = 2.8$ and $t = 1/3$. Red color corresponds to high values and low values are shown in blue. From Radko (2010). See color plates section.

(DNS) shown in Figure 3.2. For $Pr = 10$ (Fig. 3.2, top) salt fingers appear in the form of narrow vertically oriented filaments. The pattern changes dramatically for smaller $Pr = 0.01$ (Fig. 3.2, bottom). The temperature field is now dominated by disorganized eddies with comparable vertical and horizontal dimensions. The streamfunction ψ, defined by $(u, w) = (-\psi_z, \psi_x)$, assembles into well-defined horizontal bands, representing a vigorous parallel shear flow.

Insightful as they might be, weakly nonlinear models offer a mostly qualitative description of heat–salt fingering. Salt fingers in the ocean are strongly nonlinear. Much of the mid-latitude thermocline is characterized by $1 < R_\rho < 3$, a small fraction of the net salt-finger favorable range $1 < R_\rho < \tau^{-1} \approx 100$. Typical oceanic parameters are greatly separated from the marginal instability point ($R_\rho = \tau^{-1}$), being in fact closer to the opposite end of the salt-finger range, the boundary between salt fingers and static instability ($R_\rho = 1$). Unfortunately, the general analytical theory for strongly nonlinear instabilities is virtually non-existent. Users have to develop tractable approximations on a case-by-case basis. The next section summarizes several representative attempts to conceptualize fully nonlinear processes in double-diffusive convection. Understandably, the following theories may not be as rigorous, conventional, physical, transparent, or even elegant as their weakly nonlinear counterparts. The development of strongly nonlinear models is motivated by only one consideration – their relevance.

3.4 Phenomenological and empirical models

Ironically, it was perhaps the profound influence of the collective instability theory (Stern, 1969) on double-diffusive research in the 1970s and 80s that led to the relatively slow development of the fully nonlinear equilibrium theory for salt fingers. New models, and even the realization that new models are needed, did not come about until gradually accumulating evidence gave a reason to question the early views. A few equilibrium models – the viable alternatives to the collective instability theory – are summarized next.

Similarity solutions

An important step in terms of developing a physically based mixing model for salt fingers was made by Shen (1995), who performed two-dimensional simulations in a configuration compatible with the unbounded gradient model. Shen noticed that primary finger instabilities rapidly give way to more irregular patterns consisting of isolated "blobs," which move vertically and interact with each other (Fig. 3.3). These blobs are products of Holyer (1984) secondary instabilities of the vertical elevator modes. Shen suggested that the equilibration level is controlled by the dynamics of individual blobs, rather than by larger-scale collective instability waves, and developed a simple empirical theory for the equilibrium transport. His model assumed that (i) the size of blobs is comparable to the typical finger width, (ii) the vertical velocity of a blob can be estimated from the advective–diffusive balance, and (iii) the vertical transport is affected by the interactions between upward and downward propagating blobs. Shen's (1995) theory was found to be qualitatively consistent with simulations and observations. In particular, it predicted a rapid decrease of the vertical transport with the density ratio. This model

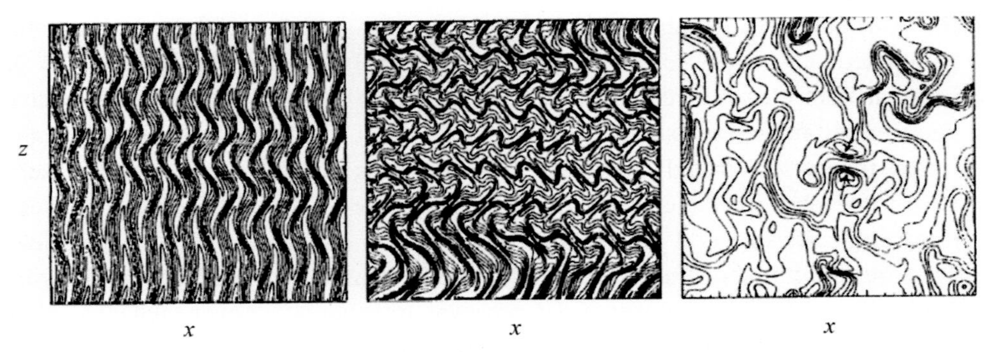

Figure 3.3 The simulated temperature field in fingering convection illustrating the flow evolution from incipient instability to disorganized convection. From Shen (1995).

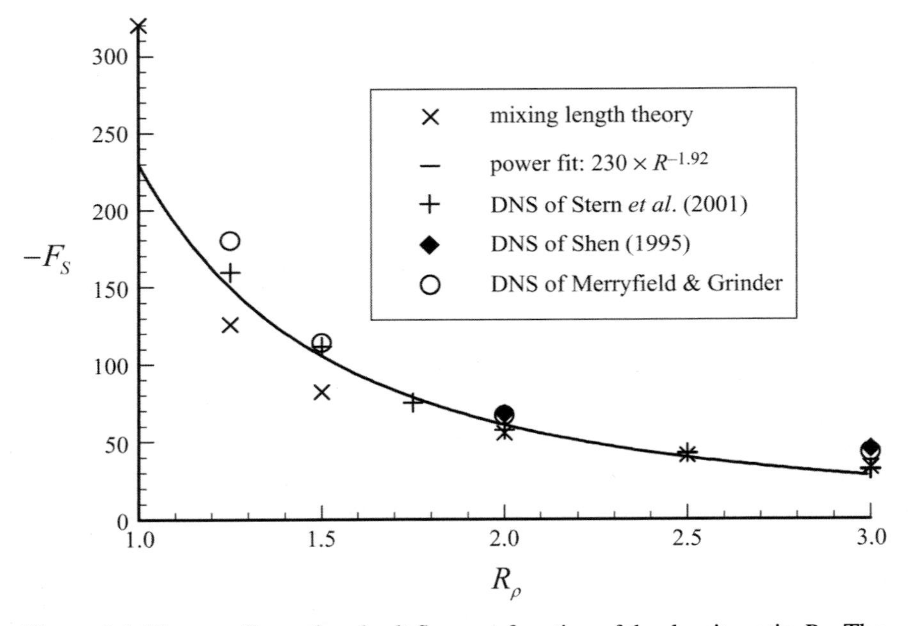

Figure 3.4 The non-dimensional salt flux as a function of the density ratio R_ρ. The prediction based on the mixing length theory is compared with the DNS results. From Stern and Simeonov (2005).

was refined and extended by Stern and Simeonov (2005), who suggested that the relevant vertical scale of temperature and salinity perturbations – the mixing length – is set by the size of the fastest growing Holyer instabilities. Figure 3.4 compares the flux prediction based on the mixing length theory with various numerical simulations in the unbounded configuration, revealing a reasonable agreement between all estimates.

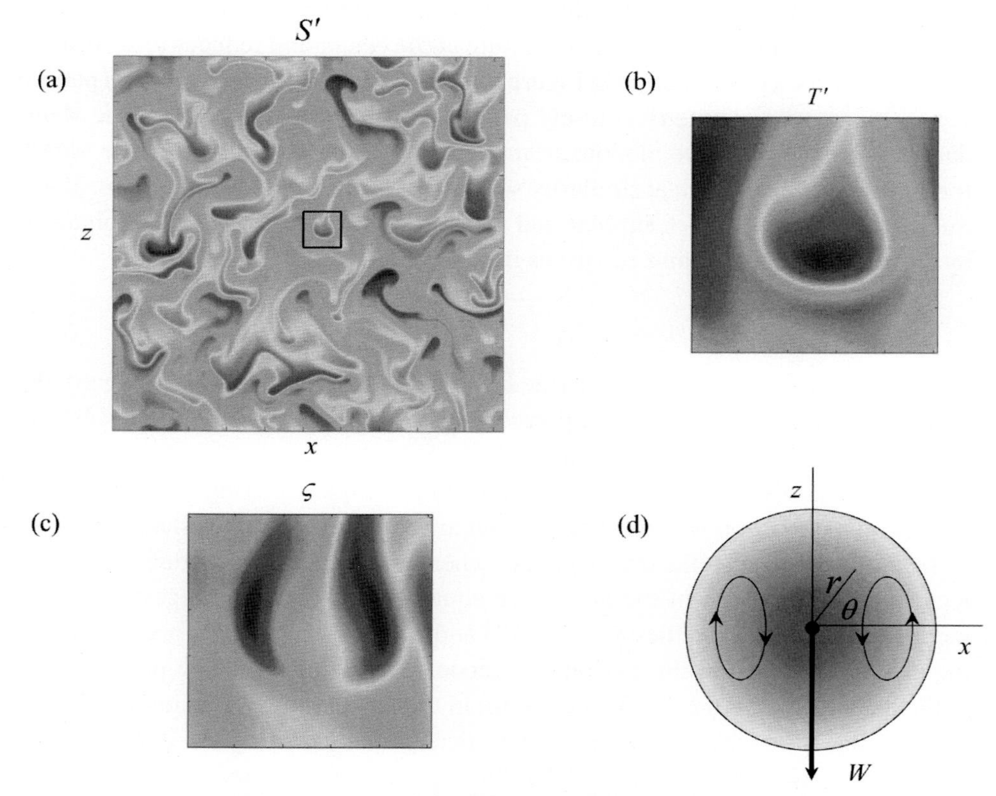

Figure 3.5 (a) Instantaneous salinity field in the numerical experiment with $R_\rho = 1.2$, $Pr = 7$ and $\tau = 1/3$. Red color corresponds to high values of S' and low values are shown in blue. Distribution of temperature (b) and vorticity (c) in the square area marked in (a), which contains a well-defined double-diffusive modon. (d) Schematic diagram of the analytical similarity solution for the downward propagating modon. From Radko (2008). See color plates section.

Radko (2008) attempted to explain the dynamics of coherent structures arising in double-diffusive convection by developing an explicit similarity solution for vertically propagating dipolar vortices – the double-diffusive modons. This nomenclature was adopted to emphasize the dipolar structure of the vorticity field and certain similarity to the "modon" solution proposed by Stern (1975) as a prototype of coherent oceanic vortices. Such structures are prevalent in the regime in which density stratification is close to neutral and fingering instability is extremely vigorous, as shown in the two-dimensional numerical experiment in Figure 3.5. Figure 3.5a presents the perturbation salinity pattern in the entire computational domain. The magnified images of one of the double-diffusive modons, looking particularly sweet and juicy, are shown in Figure 3.5b,c. Its temperature pattern in Figure 3.5b indicates that the modon is roughly circularly symmetric; the temperature anomaly reaches maximum at the center and reduces to zero at the edge. The vorticity distribution in Figure 3.5c reveals the interior circulation pattern consisting of two symmetric, closely packed counter-rotating patches. The abundance of double-diffusive modons in numerical simulations motivated the search for corresponding analytical similarity solutions (see the schematic in Figure 3.5d). Such solutions – compact, dipolar and rectilinearly propagating – were obtained by expanding the governing equations in powers of the small parameter

$$\delta = \sqrt{1 - R_\rho^{-1}} \to 0, \tag{3.7}$$

and thus formally pertained only to the low density ratio regime. In this regard, the strongly nonlinear limit (3.7) is opposite to the large R_ρ regime explored by the weakly nonlinear theory in Section 3.3.

The modon solution is the basis of a phenomenological mixing model in which salt-finger convection is represented by an array of vertically translating double-diffusive modons, as illustrated in the schematic diagram in Figure 3.6a. The leading-order balances of the governing equations for small δ suggest that each modon carries T–S anomalies T', $S' \propto \delta^{-\frac{1}{4}}$ and moves with speed $W \propto \delta^{-\frac{3}{4}}$. Hence, the vertical transport in the modon-flux model (Fig. 3.6a) should be proportional, at the leading order, to δ^{-1}. The next term in the asymptotic expansion is of order one and the following terms are asymptotically small. Hence, for density ratios somewhat above unity, the vertical salt flux can be approximated by

$$- F_S \approx a \frac{1}{\sqrt{1 - R_\rho^{-1}}} + b, \tag{3.8}$$

where a and b are adjustable constants that can be determined by numerical experiments or field data. The prescription (3.8) was successfully tested by two-dimensional numerical simulations (Figs. 3.6b,c). Here, the variables are non-dimensionalized using the standard system (1.11) and the fluxes are plotted as a function of R_ρ. Note that the transport of heat and salt in fingering convection is downward and therefore the magnitude of fluxes is given by $(-F_T, -F_S) = (Nu, Nu/\gamma)$. The corresponding dimensional eddy diffusivities take the following form:

$$K_T \approx k_T \left(a \frac{1}{\sqrt{1 - R_\rho^{-1}}} + b \right), \quad K_s \approx \frac{R_\rho k_T}{\gamma} \left(a \frac{1}{\sqrt{1 - R_\rho^{-1}}} + b \right).$$

The applicability of this parameterization is yet to be examined in a broader context, including field measurements and simulations that are not necessarily limited to low density ratios (3.7).

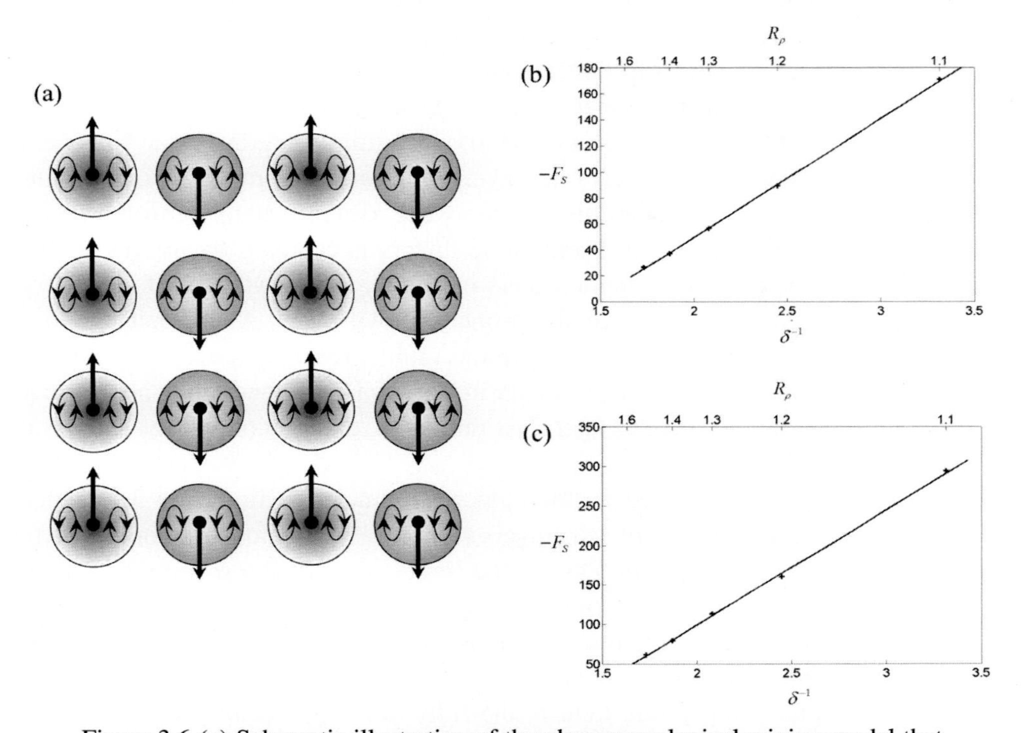

Figure 3.6 (a) Schematic illustration of the phenomenological mixing model that represents the fully developed fingering convection by a closely packed array of double-diffusive modons. The non-dimensional flux of salt (F_S) inferred from DNS is plotted as a function of $\delta^{-1} = \frac{1}{\sqrt{1-R_\rho^{-1}}}$ for $\tau = 1/3$ (b) and $\tau = 1/12$ (c). Numerical data conform to the linear $F_S(\delta^{-1})$ relation suggested by the theory. From Radko (2008).

The growth rate balance

In terms of the ability to represent the widest parameter range (R_ρ, Pr, τ), perhaps the most general equilibrium model to date has been proposed by Radko and Smith (2012). These authors based their theory on the "growth rate balance." As weak, initially linear, salt fingers grow, they develop secondary instabilities (Fig. 3.1). Radko and Smith hypothesized that the equilibration of salt fingers occurs when the growth rates of primary (λ_1) and secondary (λ_2) instabilities become comparable:

$$\lambda_2 = C\lambda_1. \tag{3.9}$$

The coefficient C in (3.9) is a dimensionless order-one constant that can be calibrated on the basis of simulations or experiments. The primary growth rate is determined by the background gradients of temperature and salinity, as well as by molecular characteristics (diffusivities and viscosity). The secondary instability is also affected by these quantities, but, additionally, it depends very strongly on the amplitude of primary salt fingers. Thus, for any given background state and value of C, the growth rate balance (3.9) implicitly determines the equilibrium amplitude of salt fingers. The physical reasoning behind the growth rate balance is straightforward. When salt fingers start to emerge from random small-scale perturbations, their secondary instabilities are too weak to significantly perturb primary modes – the evolution of small-amplitude fingers is captured by the linear theory. However, as the amplitude of fingers increases, the growth rate of secondary instabilities significantly exceeds the primary growth rates. As a result, the secondary instabilities start to gain in magnitude, rapidly reaching the level of primary modes. When the amplitude of secondary instability is large enough, it nonlinearly suppresses the growth of salt fingers. At this stage, the system reaches statistical equilibrium.

This scenario is supported by the numerical simulation in Figure 3.7, which shows the typical evolution of salt fingers from small-amplitude random noise to statistically steady convection. During the initial stage of linear growth ($t = 20$ in Fig. 3.7a) the flow field is dominated by vertically uniform salt fingers. This pattern is consistent with the linear stability theory, which predicts that the most rapidly growing perturbations take the form of vertical elevator modes. Figure 3.7b presents the temperature field in the fully developed equilibrium regime ($t = 50$). While the previously dominant elevator mode is still visible, it is now comparable in magnitude to the irregular transient structures produced by secondary instabilities, which act to distort the vertical fingers, adversely affecting their growth. The rapid transition in the flow pattern is reflected in the heat and salt flux records (Fig. 3.7c). As previously, the standard system of nondimensionalization (1.11) is used. The initial stage of the simulation is characterized by the exponential growth of fluxes, which is followed by their equilibration at $t \sim 30$. After equilibration, the intensity of salt fingering remains statistically steady.

Figure 3.8 presents the heat flux (a), salt flux (b) and the flux ratio (c) as functions of the density ratio, evaluated from the growth rate balance theory (3.9) for $C = 2.7$. The theoretical predictions (solid curves) are plotted along with the corresponding DNS results (plus signs) and with the explicit empirical fit to the numerical data (dashed curves):

$$\begin{cases} -F_S = \frac{a_S}{\sqrt{R_\rho - 1}} + b_S, & (a_S, b_S) \quad = (135.7, -62.75), \\ \gamma = a_\gamma \exp(-b_\gamma R_\rho) + c_\gamma, & (a_\gamma, b_\gamma, c_\gamma) = (2.709, 2.513, 0.5128), \\ F_T = \gamma F_S. \end{cases} \quad (3.10)$$

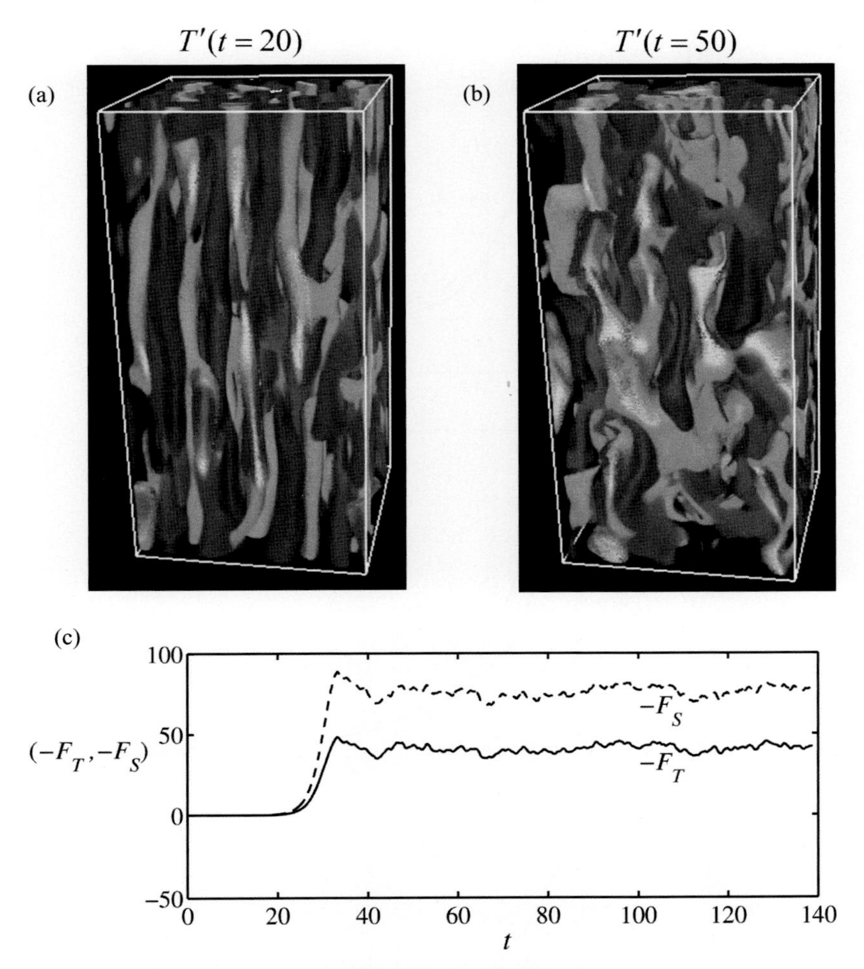

Figure 3.7 Equilibration of salt fingers in the numerical experiment with $Pr = 7$, $\tau = 0.01$ and $R_\rho = 1.9$. The temperature fields are shown for (a) the early stage of linear growth at $t = 20$ and (b) the fully equilibrated state at $t = 50$. Red/green corresponds to high values of T and low values are shown in blue. (c) Time record of the temperature (solid curve) and salinity (dashed curve) fluxes. From Radko and Smith (2012). See color plates section.

In addition to the heat–salt case, Radko and Smith (2012) examined other combinations of the governing parameters, varying Pr by four orders of magnitude and τ by a factor of thirty. In each regime, the growth rate balance theory adequately describes the (rapidly decreasing) dependence of fluxes on density ratio. The calibrated coefficients C are fairly stable. In all cases, the values of C are limited to a relatively narrow range $1.6 < C < 2.7$. The theoretical prediction with $C = 2$ in (3.9) explains all the numerical fluxes to within a factor of two.

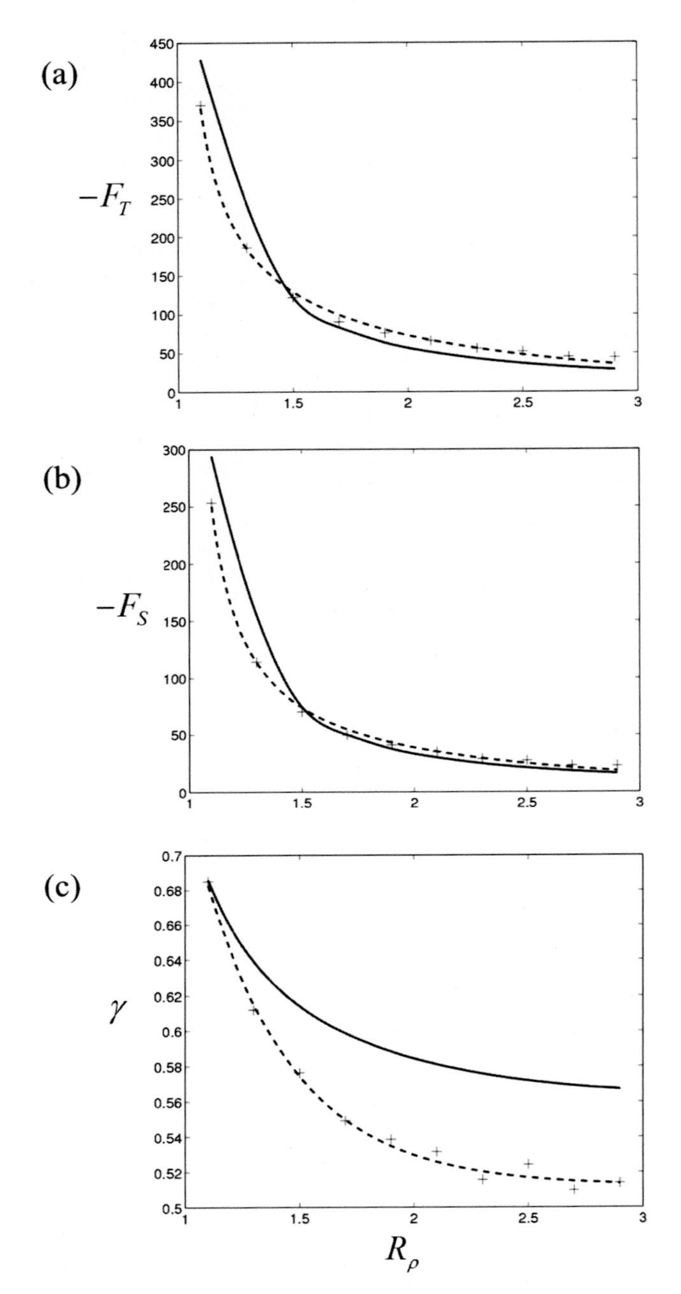

Figure 3.8 Comparison of the theoretical growth rate balance model for $C = 2.7$ (solid curve) with DNS (plus signs). The heat flux, salt flux and flux ratio are shown in (a), (b) and (c) respectively. The dashed curves represent the empirical expressions (3.10). From Radko and Smith (2012).

Empirical parameterizations

An important role in the double-diffusive literature is played by various ad hoc parameterizations, which are based entirely on external observational (Schmitt, 1981; Zhang *et al.*, 1998) or numerical data (Stern *et al.*, 2001; Stern and Simeonov, 2004; Kimura *et al.*, 2011). While such parameterizations do not address the physical principles of equilibration, they serve an important practical purpose – representing salt fingers in large-scale numerical and theoretical models. The simplicity of the proposed expressions for eddy diffusivities becomes an important consideration in the formulation of such parameterizations. Table 3.1 presents a list of various parameterizations that have been used in double-diffusive studies. It is encouraging to see qualitative agreement between various models in terms of predicting the rapid decrease of the heat and salt diffusivities with increasing density ratio. The mismatch in the diffusivity values is generally limited to a factor of two to three. It is also worth noting that even the earliest model (Schmitt, 1981) – essentially an educated guess of the flux law – adequately captures the pattern of heat–salt diffusivities inferred from recent simulations and observations.

Promising attempts have also been made (Canuto *et al.*, 2002, 2008) to extend the second-order closure models, commonly used in turbulence research, to incorporate double-diffusive mixing. Such models use governing equations to relate the second-order moments of dynamic variables. Since these relations are generally insufficient to close the problem, additional assumptions are made to bring the resulting solutions into agreement with the available data. Therefore, a certain degree of empiricism involved in the development of such models seems to be inevitable.

3.5 Numerical simulations

The classic masterpiece *Anna Karenina* by Leo Tolstoy starts with an observation, "All happy families resemble one another, each unhappy family is unhappy in its own way." If this principle were extended to modeling communities, we would have to conclude that groups of double-diffusers are truly happy as their efforts and products are mutually consistent and often resemble each other.

The most natural and common numerical configuration used to represent the unbounded model assumes periodic, in all spatial dimensions, boundary conditions for temperature, salinity and velocity perturbations of the uniform background gradients. Two-dimensional DNS of this type were performed by Whitfield *et al.* (1989), Shen (1995), Stern and Radko (1998), Merryfield (2000), Merryfield and Grinder (2002), Stern and Simeonov (2002,

Table 3.1 *Various double-diffusive parameterizations of salt diffusivity and flux ratio as a function of density ratio*

Model, type	Salt diffusivity (K_S)	Flux ratio (γ)	Coefficients	Basis
Schmitt (1981), SF	$\dfrac{K_0}{1+\left(\frac{R_\rho}{R_c}\right)^n}+K_\infty$	0.7	$K_0 = 10^{-3}\,\mathrm{m^2\,s^{-1}}$ $K_\infty = 5\cdot 10^{-6}\,\mathrm{m^2\,s^{-1}}$ $R_c = 1.7$ $n = 32$	Obs.
Merryfield *et al.* (1999), SF Zhang *et al.* (1998), SF	$\dfrac{K_0}{1+\left(\frac{R_\rho}{R_c}\right)^n}+K_\infty$	0.7	$K_0 = 10^{-2}\,\mathrm{m^2\,s^{-1}}$ $K_\infty = 3\cdot 10^{-5}\,\mathrm{m^2\,s^{-1}}$ $R_c = 1.6$ $n = 6$	Obs.
Large *et al.* (1994), SF	$K_f\left[1-\left(\dfrac{R_\rho-1}{R_\rho^0-1}\right)^2\right]^p,$ $1 < R_\rho < R_\rho^0$ $0,\ R_\rho < R_\rho^0$	0.7	$K_f = 10^{-3}\,\mathrm{m^2\,s^{-1}}$ $R_\rho^0 = 1.9$ $p = 3$	Obs.
Merryfield and Grinder (2002), SF	$0.15\dfrac{1-\tau R_\rho}{R_\rho-\gamma}$	0.6		DNS
Simeonov and Stern (2004), SF	$\dfrac{158}{\gamma R^2}k_T$	0.62		DNS
Kimura *et al.* (2011), SF	$4.38\cdot 10^{-5}R_\rho^{-2.7}Ri^{0.17}$	$0.7R_\rho^{-0.3}$		DNS
Radko and Smith (2012), SF	$\left(\dfrac{a_S}{\sqrt{R_\rho-1}}+b_S\right)k_T R_\rho$	$a_\gamma\exp(-b_\gamma R_\rho)+c_\gamma$	$(a_S, b_S) = (135, -62.75)$ $(a_\gamma, b_\gamma, c_\gamma) = (2.709, 2.513, 0.5128)$	DNS
Fedorov (1988), DC	$0.909\nu\dfrac{\gamma^*}{R_\rho^*}$ $\times\exp\left(4.6\exp[-0.54(R_\rho^*-1)]\right)$	$\gamma^* = 1.85-0.85R_\rho^*,\ 1 < R_\rho^* < 2$ $\gamma^* = 0.15,\ R_\rho^* > 2$		Obs., Expts.
Kelley (1984, 1990), DC	$\gamma^* C R_a^{1/3}k_T$	$\gamma^* = \dfrac{R_\rho^*+1.4(R_\rho^*-1)^{3/2}}{1+14(R_\rho^*-1)^{3/2}}$	$C = 0.0032\exp\left[4.8(R_\rho^*)^{-0.72}\right]$ $R_a = 0.25\cdot 10^9 R_\rho^{1.1}$	Obs., Expts.

Salt-finger models are denoted by SF and diffusive models as DC.
The parameterizations are based on observations (Obs.), direct numerical simulations (DNS) and laboratory experiments (Expts.).

2005), Radko (2008, 2010) and Denissenkov (2010). Three-dimensional DNS are more challenging computationally but nevertheless are becoming increasingly common (Radko and Stern, 1999; Stern *et al.*, 2001; Denissenkov and Merryfield, 2011; Stellmach *et al.*, 2011; Traxler *et al.*, 2011a; Mirouh *et al.*, 2012; Radko and Smith, 2012). The unbounded configuration is ideally suited for the use of spectral methods based on Fourier basis functions (Canuto *et al.*, 1987), and such models have dominated the DNS of double-diffusion. With minor modifications, spectral models can be applied to studies of double-diffusion in external shear (Smyth and Kimura, 2007; Kimura and Smyth, 2011). The latter studies will be discussed in greater detail in Chapter 10.

A particularly encouraging aspect of fingering simulations is related to the pronounced lack of sensitivity to details of the numerical setup. The size and aspect ratio of the computational domain have very limited effects on the intensity and patterns of double-diffusion as long as the box size significantly exceeds the typical salt-finger scale. Radko and Stern (1999), for instance, observed that an increase in the height of the salt-finger zone by a factor of ten changes the vertical *T–S* transport by only a small percentage. Boundary conditions also do not affect the major characteristics of salt fingering. Calculations in which the salt-finger zone is sandwiched between relatively homogeneous reservoirs (i.e., Kimura *et al.*, 2011) produce vertical diffusivities consistent with models assuming uniform background stratification (Stern *et al.*, 2001; Traxler *et al.*, 2011a). We note in passing that the robust, modeler-friendly nature of double-diffusion is in great contrast with the properties of thermal convection in a triply periodic system – the so-called homogeneous Rayleigh–Bénard problem (Borue and Orszag, 1996; Calzavarini *et al.*, 2006). In homogeneous thermal convection, the fastest growing modes span the entire domain and depend sensitively on the aspect ratio of the computational box, as well as on other numerical details. The typical length scale of eddies in the fingering regime is set by the diffusive length scales and is independent of the box size, which accounts for the dramatic difference between the two problems.

The generic difficulty in performing numerical simulations of double-diffusive convection is related to the necessity to resolve a wide range of scales. The smallest dynamically significant scale is that of salt dissipation (d_S); the scale of heat dissipation (d_T) is larger by factor of $\tau^{-1/2}$. The heat dissipation scale, in turn, is less than the finger scale (L) by a factor of $Nu^{-1/4}$. Problems in addressing the interplay between salt fingers and waves, shear, convective cells or intrusions necessarily require resolving even larger scales of motion. But even if we temporarily exclude such problems from consideration, the challenge is still formidable. For typical oceanographic conditions ($\tau \sim 0.01$, $Nu \sim 100$) the scale of salt dissipation is about thirty times smaller than the salt-finger width. Thus, a simulation representing several fingers requires computational grids on the order of 1000^3. Simulations of this order have only recently become possible (Traxler *et al.*, 2011a; Stellmach *et al.*, 2011). Earlier three-dimensional simulations either used a diffusivity ratio that is substantially higher than the heat–salt value or failed to resolve the salt-dissipation scale.

However, to be fair to the early studies, it should be mentioned that none of these shortcomings resulted in major quantitative errors. As discussed in Stern *et al.* (2001) and Radko (2008), use of a moderate diffusivity ratio is not expected to alter the fundamental physics and characteristics of salt fingering, as long as τ remains significantly less than unity. For instance, the four-fold decrease in the diffusivity ratio, from $\tau = 1/12$ to $\tau = 1/48$, results in about a 15% increase in vertical *T*–*S* transport. Traxler *et al.* (2011a) note that the impact of resolution is even less. Their computational domain, equivalent in size to $5 \times 5 \times 10$ finger widths, was discretized on a $768 \times 768 \times 1536$ grid, which was barely sufficient to resolve the salt-dissipation scale. Remarkably, the much coarser resolution of 96^3 yields fluxes that are within 20% of their fully resolved counterparts. The lack of sensitivity to the precise values of τ and to the resolution of the salinity scale suggests that salt dissipation plays a rather passive role in fingering dynamics. The mixing intensity is apparently controlled by processes operating on the much larger finger scales (L) and this suggestion is supported by most theoretical models of salt fingers (Sections 3.3 and 3.4). With regard to resolution issues, the modeling of double-diffusion once again proves to be a much more rewarding task than the modeling of homogeneous thermal convection, which is highly sensitive to numerical truncations (Borue and Orszag, 1996; Calzavarini *et al.*, 2006).

To be specific in describing double-diffusive simulations, consider a representative treatment of the unbounded model in the recent numerical study by Traxler *et al.* (2011a). Figure 3.9 presents typical salinity patterns in the fully developed fingering flow. The high density ratio simulations (e.g., Fig. 3.9d) result in relatively laminar vertically elongated fingers. However, as R_ρ decreases, the flow rapidly becomes more disorganized and isotropic (Fig. 3.9b,c). The heat and salt fluxes, non-dimensionalized using (1.11), rapidly decrease with R_ρ (Fig. 3.10a,b) and are consistently higher in three dimensions than in two dimensions. The pattern of the flux ratio (Fig. 3.10c) is generally consistent with the prediction of linear theory (Section 2.3) although some quantitative differences are visible. For instance, linear theory consistently overestimates values of $\gamma(R_\rho)$ and its minimum is shifted towards higher R_ρ in DNS. The two-dimensional DNS are closer to the theoretical prediction than the three-dimensional ones. Finally, the Stern number (Fig. 3.10d) rapidly decreases with density ratio. The values of eddy diffusivities and of other integral characteristics for each R_ρ are listed in Table 3.2.

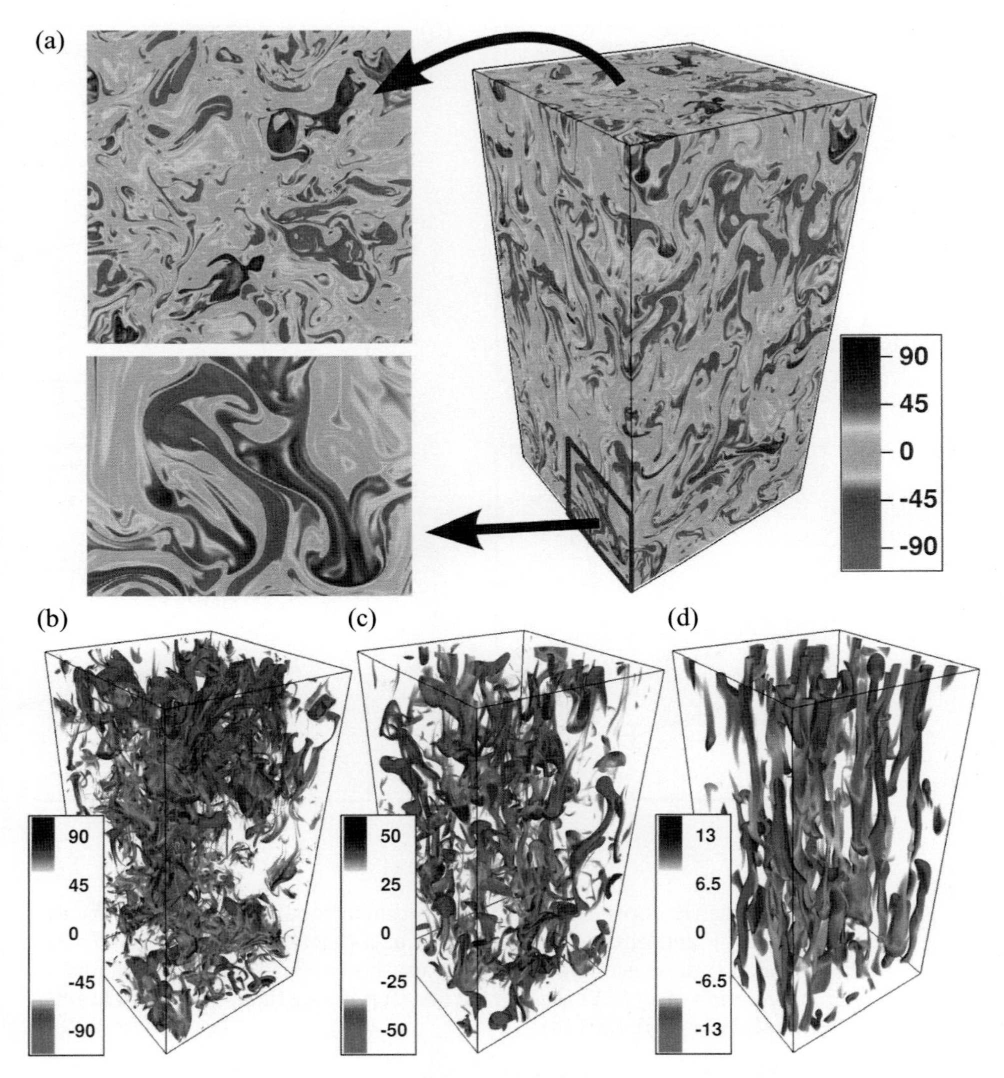

Figure 3.9 Snapshots of the salinity field in simulations of fingering convection for $Pr = 7$, $\tau = 0.01$. (a) Salinity field at $R_\rho = 1.2$, plotted on the three planes of the computational domain. (b)–(d) Volume rendering of the salinity field for $R_\rho = 1.2$, $R_\rho = 2$ and $R_\rho = 10$ (from left to right). From Traxler *et al.* (2011a). See color plates section.

3.6 Laboratory experiments

Modeling the effectively unbounded system in the laboratory requires creating a relatively thick gradient region, where fingers are vertically correlated over a fraction of the entire domain. There are at least two methods for creating such deep gradient regions.

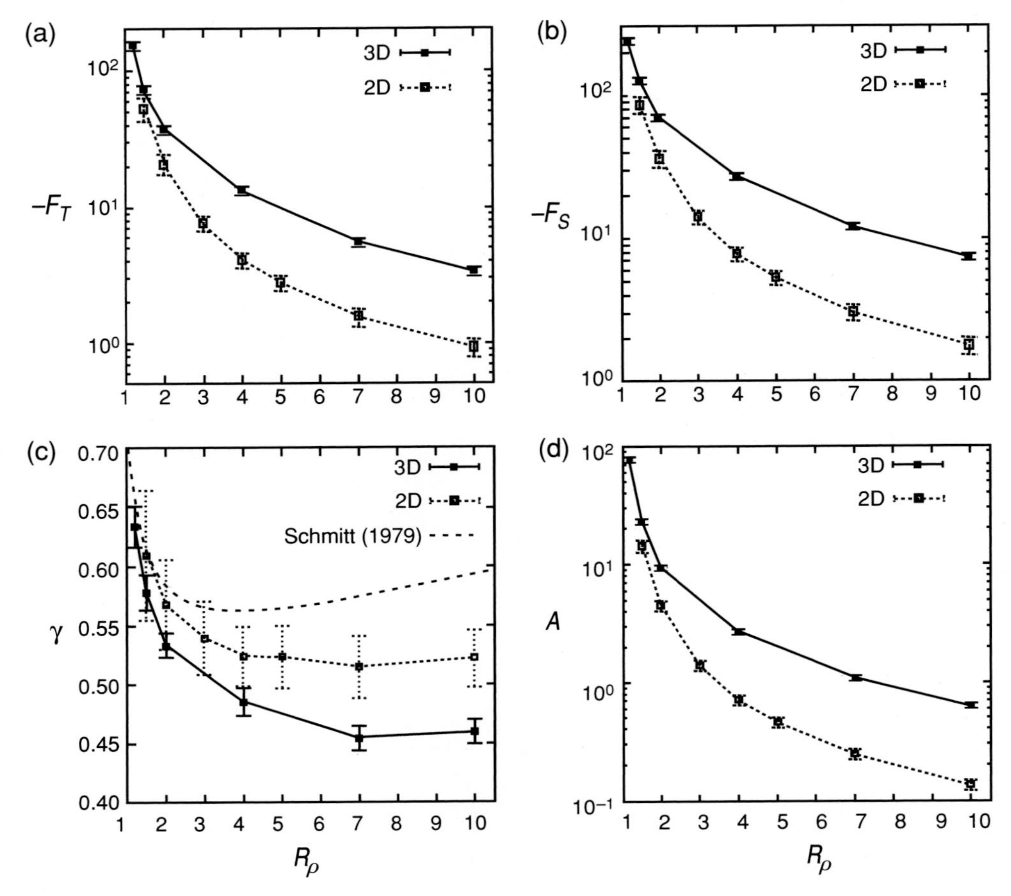

Figure 3.10 Parametric dependence of the non-dimensional fluxes F_T and F_S as well as their ratio γ and of the Stern number A as a function of R_ρ. Results from both three- and two-dimensional simulations are shown. Panel (c) also contains a theoretical prediction of $\gamma(R_\rho)$ based on the fastest growing linear model (Schmitt, 1979a). From Traxler *et al.* (2011a).

For two-solute experiments, such as the sugar–salt system (Stern and Turner, 1969; Linden, 1978; Krishnamurti, 2003), a deep double gradient can be created using the "double-bucket" technique, which is illustrated in Figure 3.11. A pure T-solution (e.g., salt) in one bucket is connected by a siphon to a second bucket at the same elevation containing a pure S-solution (e.g., sugar) of lesser density. The latter is connected by another siphon to the bottom of the experimental tank located at a lower level. When both siphons are opened, S-fluid enters the tank, and T-fluid flows into the S-bucket where it is completely mixed by a mechanical stirrer. The diluted mixture then enters the tank bottom underneath the slightly

Table 3.2 *Summary of fingering heat–salt simulations in a computational domain containing 5 × 5 × 10 fastest growing finger wavelengths*

	$R_\rho = 1.2$	$R_\rho = 1.5$	$R_\rho = 2.0$	$R_\rho = 4$	$R_\rho = 7$	$R_\rho = 10$
Resolution	$768^2 \times 153^6$	$768^2 \times 1536$	$384^2 \times 768$	$384^2 \times 768$	$384^2 \times 768$	$384^2 \times 768$
t_{average}	39.1	57.8	121.2	223.8	422.7	390.1
$\lvert F_T \rvert$	153.5 ± 11.7	73.2 ± 5.7	37.6 ± 2.2	13.3 ± 1.0	5.48 ± 0.38	3.35 ± 0.21
$\lvert F_S \rvert$	241.8 ± 13.1	126.4 ± 7.5	70.3 ± 3.1	27.4 ± 1.5	12.1 ± 0.65	7.29 ± 0.34
γ	0.63 ± 0.02	0.58 ± 0.01	0.53 ± 0.01	0.49 ± 0.01	0.45 ± 0.01	0.46 ± 0.01
$\sqrt{\langle u^2 \rangle}$	14.1 ± 0.4	9.4 ± 0.3	6.5 ± 0.15	3.7 ± 0.1	2.37 ± 0.06	1.82 ± 0.04
K_T [10^{-6} m^2 s^{-1}]	21 ± 2	10 ± 1	5.3 ± 0.3	1.9 ± 0.1	0.77 ± 0.05	0.47 ± 0.03
K_S [10^{-6} m^2 s^{-1}]	41 ± 2	27 ± 2	20 ± 1	15 ± 1	11 ± 1	10 ± 0.5
A	76 ± 3	23 ± 1	9.4 ± 0.3	2.7 ± 0.1	1.1 ± 0.05	0.63 ± 0.03

t_{average} denotes the length of the time interval over which the data have been averaged.
The non-dimensionalization is based on the standard double-diffusive transformation in (1.11).
From Traxler *et al.* (2011a).

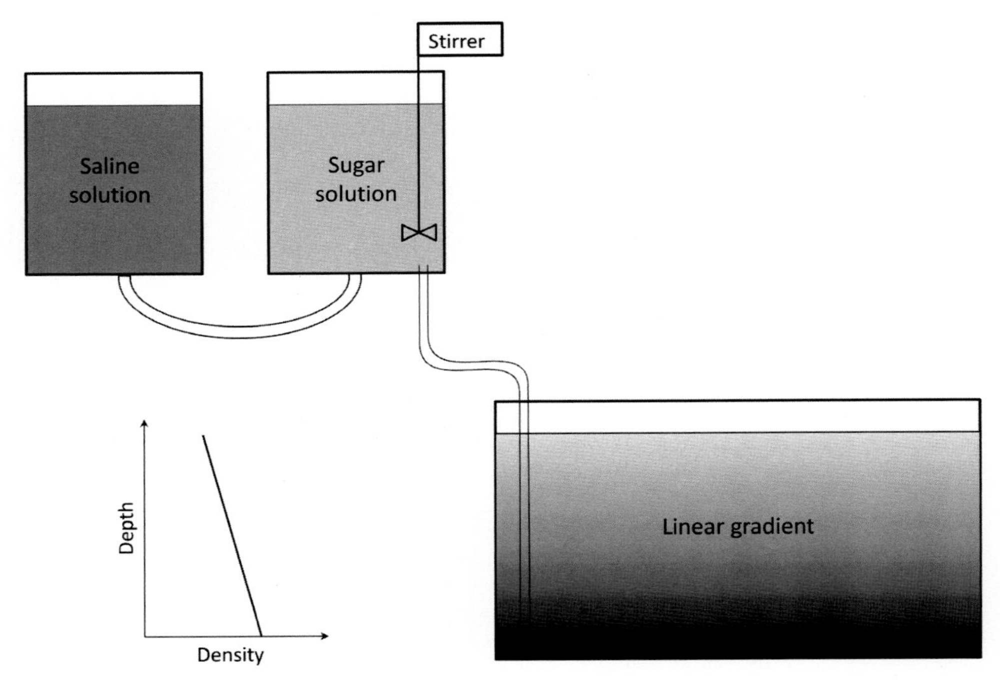

Figure 3.11 Schematic diagram of the double-bucket method for creating extended vertical property gradients in laboratory experiments.

lighter S-fluid, until the tank is filled with pure S-fluid (sugar) at the top, pure T-fluid (salt) at the bottom, and with nearly uniform T–S gradients in-between. For suitable filling conditions and moderately large density ratio, the up-going and down-going salt fingers form and occupy the entire tank (Stern and Turner, 1969). The double-bucket technique has been conveniently used in experimental studies of the interaction between salt fingers and shear (Wells *et al.*, 2001), interaction with intermittent turbulence (Wells and Griffiths, 2003) and effects of localized stirring (Wells and Griffiths, 2002) – topics that will be discussed in greater detail in Chapter 10. If the vertical gradients are not maintained continuously, the system runs down and the background density ratio systematically drifts towards larger and larger values. In many cases, it is possible to take advantage of the inherent time dependence in the run-down experiments. For instance, a single experiment of this type yields a continuous record of salt-finger characteristics as a function of the density ratio.

However, time dependence can occasionally lead to unintended adverse consequences. For instance, the formation and evolution of thermohaline staircases (Krishnamurti, 2003, 2009) and intrusions (Krishnamurti, 2006) occur on temporal scales that greatly exceed the finger scales. Thus, in order to realistically

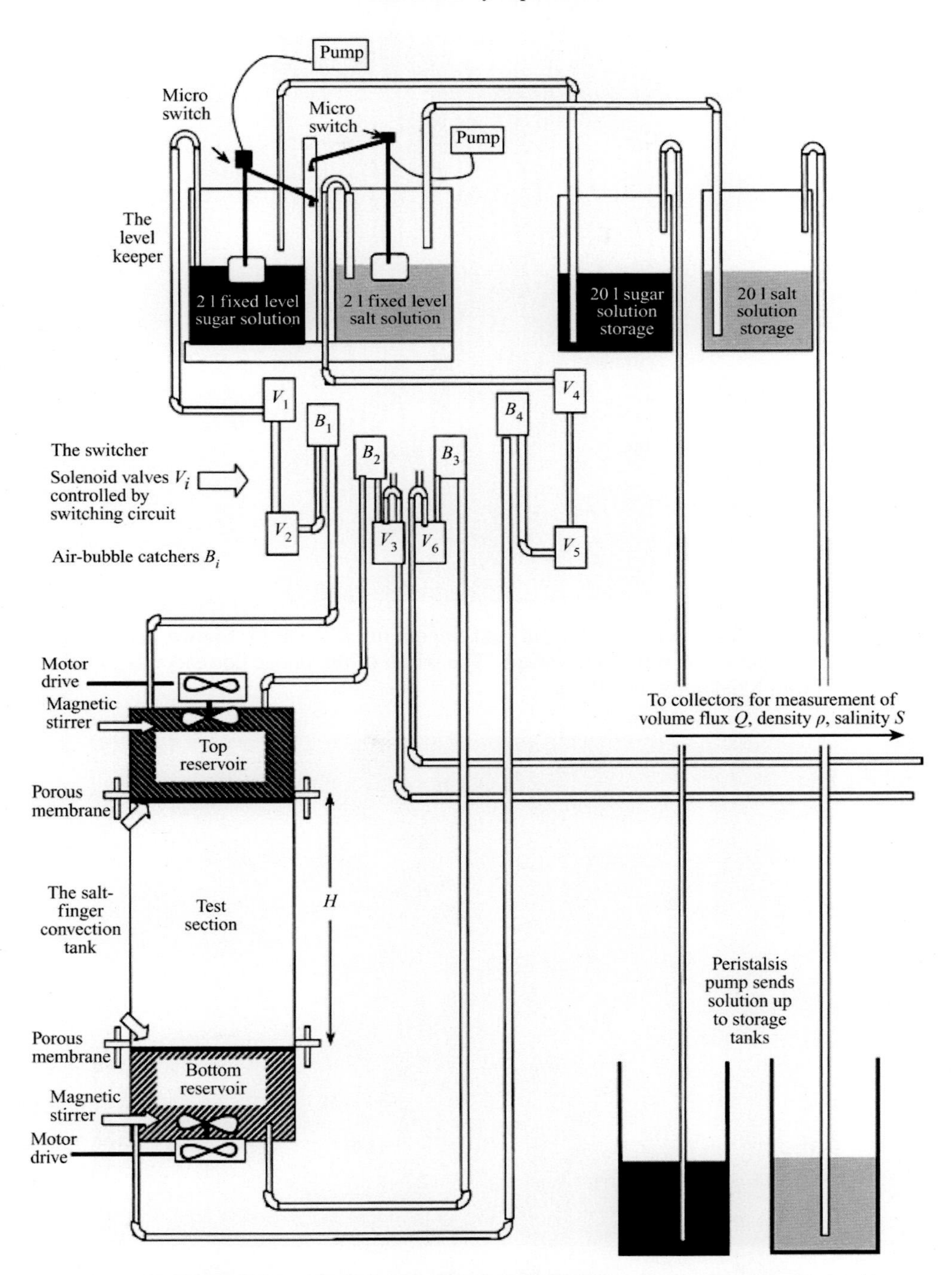

Figure 3.12 Schematic diagram of the apparatus designed to maintain the background stratification during fingering experiments. From Krishnamurti (2003).

Figure 3.13 Shadowgraph of sugar–salt fingers for $R_\rho = 1.25$. Shown here is an excerpt from a 2 m tall shadowgraph. The width of the image corresponds to 8.5 cm. From Krishnamurti (2003).

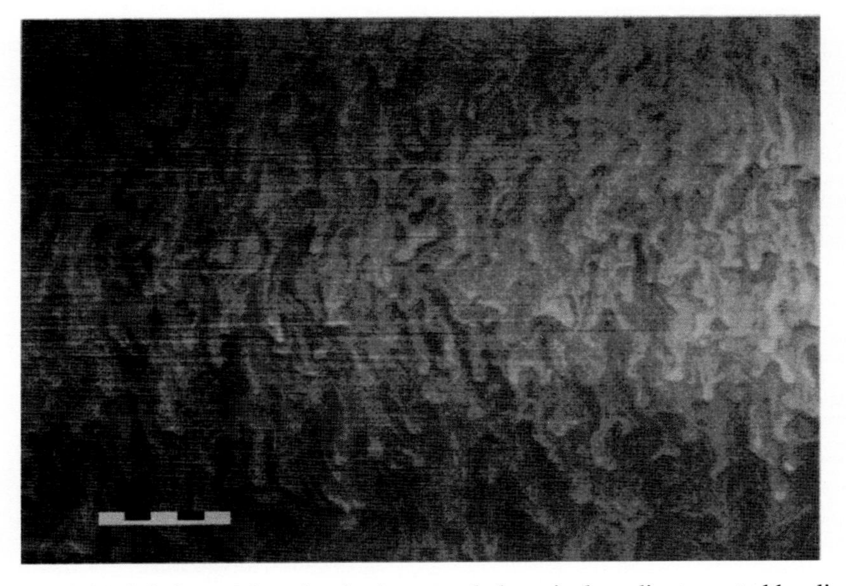

Figure 3.14 Salt fingers forming in the extended vertical gradient created by displacing a metal grid across the interface between two initially homogeneous layers. The scale bar is 5 cm long. From Taylor (1993).

represent these slow secondary double-diffusive structures and to quantify their control by the background T–S gradients, it is imperative to maintain the vertical stratification in the course of an experiment. Such fixed-gradient experiments are more involved technically. The schematic in Figure 3.12 illustrates the experimental setup designed and used by Krishnamurti (2003). Initially, the "test section" is filled with linear salt-finger favorable gradients using the double-bucket method. The working fluid is bounded by two reservoirs of specified salinity and temperature, which are separated from the test section by rigid porous membranes. To maintain constant in time T–S values in the reservoirs, they are periodically flushed. This procedure is equivalent to enforcing the prescribed vertical boundary conditions for the gradient region in the test section. In this manner, the background interior stratification is sustained throughout the duration of an experiment, which can easily last for weeks. A shadowgraph of a thick gradient sugar–salt experiment is shown in Figure 3.13 (Krishnamurti, 2003), revealing vertically elongated fingers, typical of the large Pr/moderate R_ρ regime.

For heat–salt experiments, the double-bucket method may not be as effective as in two-solute systems because of the inevitable heat loss to the environment. An alternative method for creating extended gradient layers was proposed by Taylor (1991, 1993). Initially, a two-layer heat–salt system is created and then the interface between homogeneous layers is mechanically thickened by vertically displacing a metal grid. The result is a deep double-gradient such as shown in Figure 3.14. Taylor (1993) used this technique to study the anisotropy of salt fingers in the gradient layers. One of the conclusions from these experiments is that at relatively low density ratios $1 < R_\rho < 3$, relevant for much of the mid-latitude thermocline, salt fingers take the form of relatively isotropic blobs, whereas the increase in R_ρ makes fingers progressively more elongated. The link between the density ratio and the anisotropy of salt fingers is robust and readily reproducible by DNS (e.g., Fig. 3.9).

4

The two-layer system

The next conceptual model of double-diffusive convection described in this monograph consists of a sharp interface sandwiched between two deep well-mixed layers. Such a configuration, illustrated in Figure 4.1, is particularly common in laboratory studies. Aside from considerations of simplicity and convenience, the two-layer model is motivated by observations of stepped structures in vertical temperature and salinity profiles known as thermohaline staircases – their origin and dynamics will be discussed in Chapter 8. The layered and gradient (Chapter 3) systems should not be regarded as mutually exclusive. For instance, flux-gradient finger models may be applicable locally for the interior region of double-diffusive interfaces. Furthermore, in nature it is often difficult to categorize a system as layered or gradient. Oceanographic observations frequently reveal irregular "steppiness" of T–S profiles (e.g., Schmitt and Evans, 1978), which apparently reflects elements of both layered and gradient dynamics. Nevertheless, the concept of a double-diffusive interface plays a prominent role in the development of double-diffusion theory for both salt-finger and diffusive regimes, warranting a detailed discussion. The central problem in the two-layer model, or for that matter in any double-diffusive system, involves prediction of the vertical transport of diffusing components. We start with the four-thirds interfacial flux law, originally proposed by Turner (1965, 1967) but approached here in a slightly different manner.

4.1 Interfacial flux laws

Consider a system consisting of two nearly homogeneous layers, separated by a high-gradient interface, as shown in Figure 4.1 for the salt-finger (a) and diffusive (b) cases. In both models, homogeneous stratification in the layers is maintained by top-heavy convection, which, in turn, is driven by the unequal transport of heat and salt through the double-diffusive interface. For the salt-finger case (Fig. 4.1a) fluxes of heat and salt are downward (i.e., negative). The contribution of salt flux to

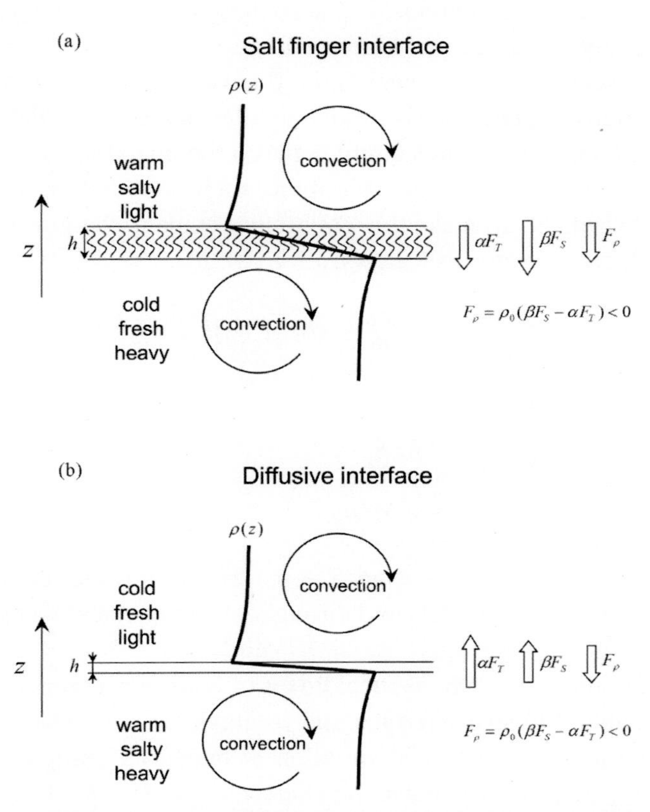

Figure 4.1 Schematic diagrams illustrating the dynamics of two-layer fingering (a) and diffusive (b) systems.

density flux exceeds that of heat, as measured by the flux ratio $\gamma < 1$ (Chapter 2) and therefore the transport of salt controls the direction of the net density flux – it is also downward. Thus, fluid at the top of the lower layer, immediately below the interface, consistently gains density and maintains convection. Fluid immediately above the interface loses density, which drives convection in the upper layer.

The physics of the diffusive system (Fig. 4.1b) is somewhat different. Transport of heat and salt is controlled by molecular diffusion across the interface, particularly for moderate density ratios. Therefore, the direction of density transport is controlled by heat (faster diffuser). As heat diffuses from the warm lower layer into the cold upper layer, fluid immediately above (below) the interface becomes lighter (denser), forcing convection in both layers. Ultimately, the motion in diffusive and salt-finger systems is driven by the release of potential energy stored in the unstably stratified components. As the upper layer becomes lighter and the lower layer denser, the center of mass is displaced downward and the released potential energy is partially converted into kinetic energy.

In both configurations, salt-finger and diffusive, it is natural to assume that the behavior and transport characteristics of a two-layer system are determined by the differences in temperature and salinity between the mixed layers. Furthermore, it is not just the variation in T and S but rather their contribution to buoyancy that controls the dynamics. To make this point more evident, the governing equations (1.1) and (1.2) are rewritten as follows:

$$\begin{cases} \dfrac{d\vec{v}}{dt} = -\frac{\nabla p}{\rho_0} + (b_T + b_S)\vec{k} + \nu\nabla^2\vec{v}, \\ \nabla \cdot \vec{v} = 0, \\ \dfrac{db_T}{dt} = k_T\nabla^2 b_T, \\ \dfrac{db_S}{dt} = k_S\nabla^2 b_S, \end{cases} \tag{4.1}$$

where $b_T = g\alpha(T - T_0)$ and $b_S = g\beta(S - S_0)$ are the buoyancy components due to temperature and salinity. In (4.1) and throughout this chapter, we use dimensional quantities.

Of particular interest are the vertical fluxes of buoyancy components (F_{b_T}, F_{b_S}) across the double-diffusive interface and its thickness (h). These quantities are assumed to be determined by the variation of buoyancy components across the interface (Δb_T, Δb_S) and by molecular characteristics (k_T, k_S, ν):

$$\begin{cases} F_{b_T} = F_{b_T}(\Delta b_T, \Delta b_S, k_T, k_S, \nu), \\ F_{b_S} = F_{b_S}(\Delta b_T, \Delta b_S, k_T, k_S, \nu), \\ h = h(\Delta b_T, \Delta b_S, k_T, k_S, \nu). \end{cases} \tag{4.2}$$

Formulation (4.2) explicitly ignores the possible influence of the mixed layer thickness. Layers are assumed to be sufficiently deep so that the bottom and surface of the working fluid are too separated from the interface to affect processes operating in its immediate vicinity. Also neglected are evolutionary effects, such as the role of the initial thickness of the interface. While these assumptions have to be critically reexamined on a case-by-case basis, it will be shown shortly that the interfacial flux model based on (4.2) is reasonably successful in explaining some observations and laboratory experiments.

The next step involves the dimensional argument, elegantly expressed by the Buckingham (1914) π-theorem: each non-dimensional combination is controlled by other non-dimensional combinations or π-groups. Principles of dimensional analysis and examples of its application have been discussed in numerous fluid dynamical texts (e.g., Barenblatt, 1996; Kundu and Cohen, 2008). In our case, the expression for the quantity of interest – F_{b_T} in (4.2) – involves six independent variables in two fundamental units, time (s) and length (m) : $F_{b_T}[\mathrm{m^2\,s^{-3}}]$; Δb_T, $\Delta b_T[\mathrm{m\,s^{-2}}]$; and $k_T, k_S, \nu[\mathrm{m^2\,s^{-1}}]$. Thus, we can expect a functional relation between four independent π-groups. The non-dimensional combination involvingF_{b_T}, Δb_T and k_T is easily constructed:

$$\pi_1 = \frac{F_{b_T}}{(\Delta b_T)^{\frac{4}{3}}\, k_T^{\frac{1}{3}}}. \tag{4.3}$$

Other non-dimensional combinations of variables in (4.2) are familiar from previous chapters; they are the density ratio $R_\rho = \frac{\Delta b_T}{\Delta b_S}$, the diffusivity ratio $\tau = \frac{k_S}{k_T}$ and the Prandtl number $Pr = \frac{\nu}{k_T}$. Applying Buckingham's theorem to these π-groups yields:

$$\frac{F_{b_T}}{(\Delta b_T)^{\frac{4}{3}} k_T^{\frac{1}{3}}} = C_T(R_\rho, \tau, Pr). \tag{4.4}$$

Function C_T cannot be determined internally by the dimensional theory and requires experimental or numerical calibration. Expressing (4.4) in terms of temperature rather than buoyancy reduces it to

$$\alpha F_T = (k_T g)^{\frac{1}{3}}\, C_T\, (\alpha \Delta T)^{\frac{4}{3}}. \tag{4.5}$$

The same reasoning can be applied to the salt flux:

$$\beta F_S = (k_T g)^{\frac{1}{3}}\, C_S\, (\beta \Delta S)^{\frac{4}{3}}, \tag{4.6}$$

where C_S is a yet unknown function of (R_ρ, Pr, τ). Expressions (4.5) and (4.6) are referred to as the "four-thirds flux laws" for double-diffusive convection (Turner, 1965, 1967, 1979). As in the gradient model (Chapter 3), the vertical transfer properties of the two-layer system are often characterized by the flux ratio:

$$\gamma(R_\rho, \tau, Pr) = \frac{\alpha F_T}{\beta F_S} = \frac{C_T}{C_S} R_\rho^{\frac{4}{3}}. \tag{4.7}$$

When the dimensional argument is applied to the interfacial thickness (h), we arrive at

$$h = \left(\frac{k_T^2}{g}\right)^{\frac{1}{3}} C_h(R_\rho, \tau, Pr)\, (\alpha \Delta T)^{-\frac{1}{3}}. \tag{4.8}$$

To put these results into historical perspective, it should be mentioned that Turner originally arrived at the four-thirds laws as a plausible extension of a classical model for turbulent thermal convection (Priestley, 1954; Howard, 1963). This model assumes a unique power law relation between the Nusselt number $Nu = \frac{F_T}{k_T \Delta T / H}$, a non-dimensional measure of heat transport, and the Rayleigh number $R = \frac{g\alpha \Delta T H^3}{k_T \nu}$, measuring the strength of thermal forcing:

$$Nu \propto R^n. \tag{4.9}$$

The assumption that the heat flux is independent of the depth of the convective zone (H) specifies the exponent $n = \frac{1}{3}$. This prescription for the Nusselt number is equivalent to the four-thirds flux law $F_T \propto \Delta T^{\frac{4}{3}}$, the counterpart of (4.5). While it seems sensible to assume that the turbulent fluxes are not affected by H as long as it is sufficiently large, subsequent studies of single-component convection have revealed systematic deviations from the four-thirds flux law (e.g., Grossmann and Lohse, 2000). In both problems – thermal and double-diffusive convection – the assumed lack of importance of layer depths, initial conditions and evolutionary history can be questioned. Nevertheless, the four-thirds flux law provides a convenient starting point for the discussion of experiments and observations. It has some success in explaining the salient features of layered double-diffusive systems. Caution is advised though for sensitive problems that require precise models of the interfacial fluxes.

4.2 Salt-finger interfaces

A major source of information on the dynamics of interfaces, both fingering and diffusive, has traditionally been provided by laboratory experiments. The first quantitative experiment with salt-finger interfaces was performed by Turner (1967). Figure 4.2 shows a schematic of the experimental setup (right panel) and salt fingers forming on the interfaces (left panel) – this particular experiment contained three mixed layers separated by two interfaces. Turner's work for the first time offered an estimate of the heat–salt flux ratio, which was shown to be nearly uniform with $\gamma \approx 0.56$ over the wide range $2 < R_\rho < 10$. This result was, to large extent, confirmed and reproduced by subsequent laboratory and numerical studies. Technical complications prevented Turner from measuring the flux ratio for lower values of R_ρ, although he speculated that γ should increase with decreasing density ratio, approaching unity at $R_\rho = 1$. More accurate measurements were taken by Schmitt (1979b), whose experiments (i) supported the four-thirds flux law to within the experimental uncertainties, (ii) yielded Stern numbers (3.5) of order one, and (iii) revealed the gentle variation in the flux ratio with R_ρ. Schmitt's experiments were followed by McDougall and Taylor (1984). Changes to the experimental procedure in the latter study made it possible to examine a lower range of density ratios and smaller T–S contrasts between the upper and lower layers – a welcome adjustment towards typical oceanic conditions. The results in McDougall and Taylor (1984) were broadly consistent with Schmitt (1979b), although some quantitative

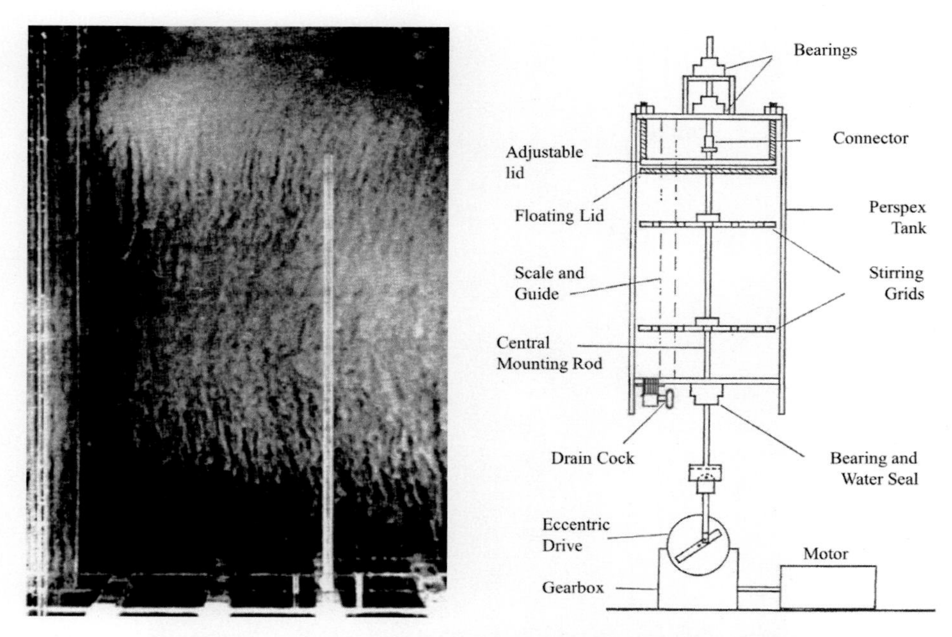

Figure 4.2 Formation of salt fingers across two high-gradient interfaces (left). Schematic of the experimental apparatus is shown in the right panel. From Turner (1967).

differences emerged. The values of Stern number and flux ratios were consistently higher in McDougall and Taylor and the 4/3 exponent in (4.6) was revised downward to 1.23.

The experiments of Taylor and Bucens (1989) revealed in greater detail the spatial and temporal patterns of salt fingers – see the shadowgraphs in Figure 4.3. Salt fingers in a two-layer system are irregular, time dependent and vertically coherent over only a fraction of the interfacial region. Overall, the structure of interfacial fingers is similar to that in extended gradient layers (cf. Fig. 3.14). Interestingly, Taylor and Bucens' study reported systematic quantitative differences in transport characteristics relative to earlier investigations. The disagreement was attributed to the differences in the experimental setup, such as the quiescent initial state in McDougall and Taylor's experiments and the presence of mechanical perturbations in Turner (1967) and Schmitt (1979b). The sensitivity to such details has significant implications for the dynamics of layered double-diffusive systems. The derivation of the four-thirds flux law necessarily assumes a unique solution for fluxes. The variation of *T*–*S* transport in response to very modest changes in configuration suggests that fluxes are not entirely determined by the temperature and salinity contrasts, giving a legitimate reason to question (4.2) and thus the flux laws themselves. This concern becomes particularly worrisome when the

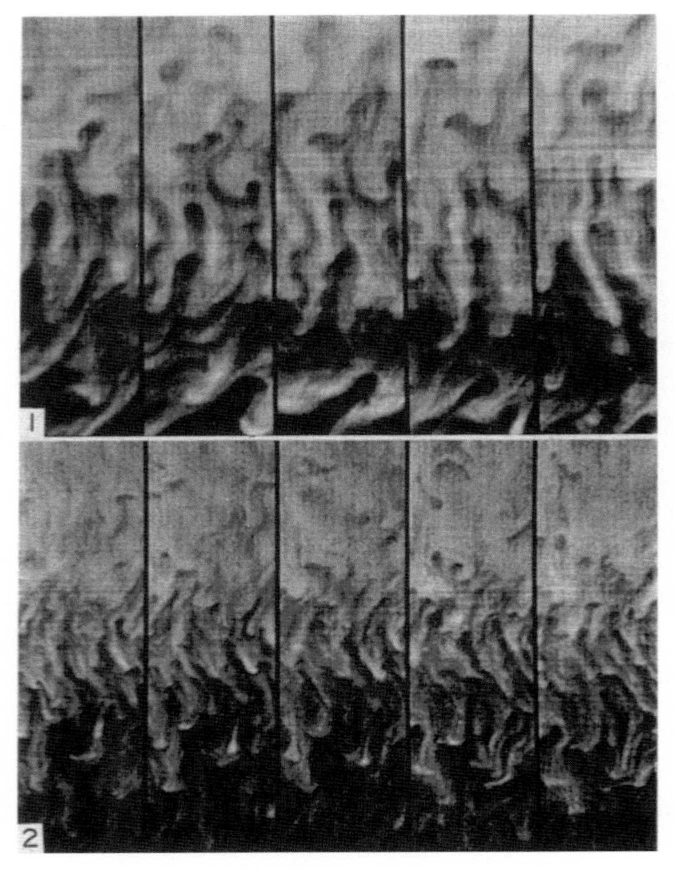

Figure 4.3 Each panel shows successive images of a part of the finger interface in the two-layer laboratory experiment. The upper panel shows the state recorded for $R_\rho = 2.46$ and the lower panel shows the later stage of the same experiment when R_ρ had increased to 2.75. The images in each panel were captured 5 seconds apart. From Taylor and Bucens (1989).

four-thirds law is applied to oceanic staircases. Clearly, the oceanographic situation contains many ingredients that are absent in the laboratory analogue – internal waves, an energetic eddy field and large-scale shear, to name just a few. Therefore, it is not surprising that extrapolation of the laboratory-calibrated four-thirds flux law to the staircase in the Western Atlantic overestimated the vertical transport by an order of magnitude (Kunze, 1987). While this mismatch has never been fully explained, the lack of uniqueness in the formulation of the flux laws is likely to play a role.

An important role in the analysis of the double-diffusive two-layer system is played by laboratory experiments in which heat and salt are replaced by two solutes with different molecular diffusivities. Such experiments were initially proposed to

Figure 4.4 Shadowgraph of a finger interface formed by placing a layer of sucrose solution on top of a denser salt solution. From Turner (1985).

avoid spurious effects due to sidewall cooling and heating. This approach was pioneered by Stern and Turner (1969), who chose sugar and salt as two diffusing components for both two-layer and uniform-gradient (Chapter 3) experiments. This study was followed by progressively more precise measurements by Lambert and Demenkow (1972), Griffiths and Ruddick (1980) and Taylor and Veronis (1996).

The insights brought by the two-layer salt–sugar experiments are difficult to overestimate. The exploration of a completely different, relative to the heat–salt experiments, parameter regime made it possible to identify generic characteristics of double-diffusive convection and fluid-dependent features. The sugar–salt fingers (Fig. 4.4) are typically more regular and vertically elongated than the heat–salt fingers (compare with Figs. 4.2 and 4.3). The more organized and steady flow character of the salt–sugar system permitted visualization of the horizontal cell structure of the interfaces. Shirtcliffe and Turner (1970) used a simple and ingenious optical system (Fig. 4.5a) and discovered that the interfacial fingers have a remarkably regular square cross-section (Fig. 4.5b). In view of the degeneracy of the linear theory with respect to the planform selection (Chapter 2), the preference for the square cells must be due to the nonlinear interactions between various finger modes.

There are numerous quantitative differences in the transport characteristics of heat–salt and salt–sugar systems. For instance, Stern numbers (A) in salt–sugar

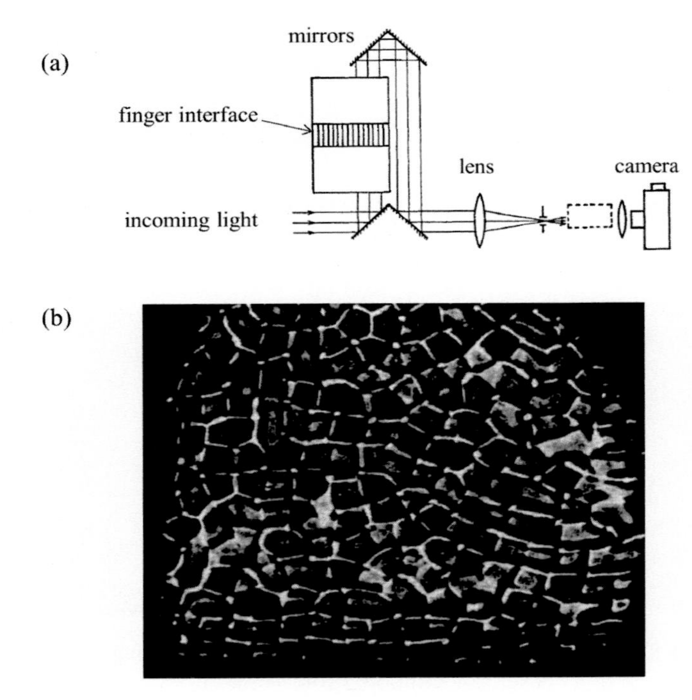

Figure 4.5 (a) The optical system designed to observe the cell structure of fingers in the laboratory experiment. This system was used to capture the image (b) of the horizontal pattern at the mid-plane of the fingering interface. From Shirtcliffe and Turner (1970).

experiments are much lower than in their heat–salt counterparts. Lambert and Demenkow (1972) report values as low as $A = 2 \cdot 10^{-3}$, casting doubt on the generality of Stern's number criterion for the magnitude of salt fingers (a controversial issue discussed in Section 3.2). The coefficient $C'_S = C_S(k_T g)^{\frac{1}{3}} = \frac{\beta F_S}{(\beta \Delta S)^{\frac{4}{3}}}$ of the flux law (4.6) recorded in the salt–sugar experiments (e.g., Griffiths and Ruddick, 1980; Taylor and Veronis, 1996) also differs from the heat–salt values. It rapidly decreases from $C'_S \sim 10^{-4}\,\mathrm{m\,s^{-1}}$ at R_ρ close to unity to $C'_S \sim 10^{-6}\,\mathrm{m\,s^{-1}}$ for $R_\rho \sim 2$. In the heat–salt experiments, the flux law coefficients are typically on the order of $C'_S \sim 10^{-3}\,\mathrm{m\,s^{-1}}$ and exhibit much more limited variation with the density ratio (Schmitt 1979b; McDougall and Taylor, 1984; Taylor and Bucens, 1989).

Of course, sensitivity of C'_S to the molecular characteristics of diffusing substances is perfectly natural and fully consistent with the flux law formulation. However, salt–sugar experiments have also underscored more fundamental uncertainties with regard to the four-thirds law. Taylor and Veronis (1996) compared their estimates of the flux law coefficient with earlier measurements in Griffiths and Ruddick (1980) and documented the systematic offset between the two. This

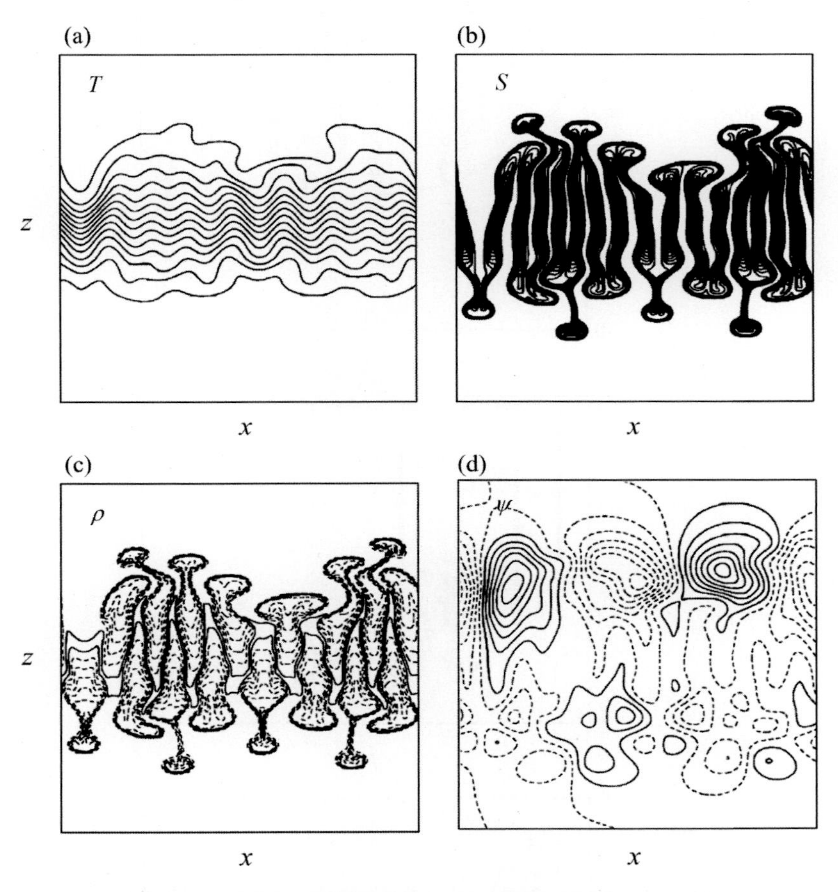

Figure 4.6 Two-dimensional numerical simulation of salt fingering. Contour plots of (a) T, (b) S, (c) ρ and (d) ψ are shown for a system of eight fingers forming across the interface separating two homogeneous layers. From Shen and Veronis (1997).

offset can be interpreted as a manifestation of a non-unique relation between vertical fluxes and the variation of properties between the layers. Echoing a similar concern of Taylor and Bucens (1989) for heat–salt interfaces, Taylor and Veronis (1996) suggested that the four-thirds law should not be applied indiscriminately to fingering convection.

Numerical simulations of the two-layer configuration (Shen, 1993; Shen and Veronis, 1997; Ozgokmen *et al.*, 1998; Paparella, 2000) have been generally successful in reproducing major features of laboratory experiments. The patterns of fingers forming in the interfacial zone – see Shen and Veronis (1997) simulations in Figure 4.6 – resemble their laboratory analogues. Numerical models consistently reproduced typical values of vertical fluxes, the flux ratio, Stern number,

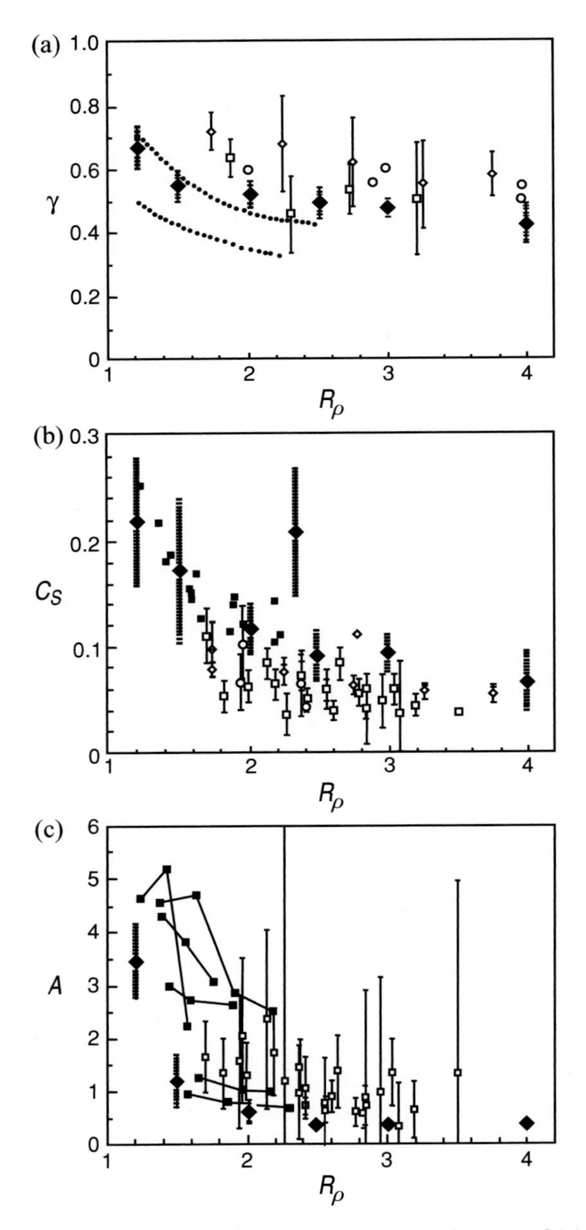

Figure 4.7 Comparison of numerical and laboratory estimates of (a) the flux ratio γ, (b) coefficient of the four-thirds flux law C_S and (c) the Stern number A. Values from the numerical experiments are marked with ♦. The laboratory data are given by ◊ (Schmitt, 1979b), □ (Taylor and Bucens, 1989), ○ (Turner, 1967), ■ (McDougall and Taylor, 1984) and the dotted curves in (a) mark the range of McDougall and Taylor's measurements. From Shen (1993).

Table 4.1 *A summary of two-layer flux laws in the form $F_S \propto (\Delta S)^p$ for salt-finger (SF) and diffusive convection (DC)*

Model	Form of double-diffusion	Flux law exponent (p)	Basis
Schmitt (1979b)	SF	1.24–1.37	Expts.
McDougall and Taylor (1984)	SF	1.23	Expts.
Krishnamurti (2003)	SF	1.18–1.19	Expts.
Ozgokmen *et al.* (1998)	SF	4/3	DNS
Paparella (2000)	SF	1.33–2.0	DNS
Sreenivas *et al.* (2009)	SF	1.32	DNS
Marmorino and Caldwell (1976)	DC	1.32–1.34	Expts.
Kelley *et al.* (2003)	DC	1.27–1.47	Expts.

The exponent p is obtained by fitting the experimental (Expts.) or numerical (DNS) data.

and the pattern of their variation with R_ρ (Fig. 4.7). Unfortunately, simulations have so far not provided a definitive answer regarding the utility of the four-thirds flux law. Sreenivas *et al.* (2009) found their simulations to be well described by (4.6). Ozgokmen *et al.* (1998) reported the overall adequate performance of the four-thirds law, noting certain inconsistencies during the initial phase of system adjustment to the quasi-equilibrium state and the final run-down phase followed by the disintegration of layered structure. Paparella (2000), on the other hand, suggested that the exponent of the flux law consistently exceeds 4/3. Several studies, summarized in Table 4.1, proposed empirical adjustments to the four-thirds law in the form

$$F_S \propto (\Delta S)^p \,. \tag{4.10}$$

The exponent p was estimated by fitting (4.10) to experimental or numerical data. In most cases, the modifications of the four-thirds flux law fall within the statistical error bars. However, at this point we reiterate that the adequate performance of the four-thirds law in a single simulation – or even in a set of similarly designed simulations – does not fully validate it. If the flux law coefficient C_S depends on the setup of an experiment, numerical or laboratory, then the practical use of the four-thirds law may be limited.

4.3 Diffusive interfaces

Despite significant differences in the dynamics of diffusive and salt-finger interfaces, similarities in the historical development of the two problems are apparent.

Figure 4.8 Laboratory experimentation with the two-layer diffusive system. The lower, more saline layer has been colored with fluorescent dye. From Turner (1965).

The first insights into the diffusive case were also brought by laboratory experiments. Figure 4.8 presents Turner's (1965) experiment in which the two-layer system, initially stratified only with respect to salinity, was heated from below. The heating rate was sufficiently weak and the lower layer remained denser than the upper one throughout the duration of the experiment. Turner's study quantified the dependencies of temperature and salinity fluxes on the diffusive density ratio (R_ρ^*). Both heat and salt fluxes monotonically increased with decreasing R_ρ^* – an expected consequence of decreasing density stratification. Figure 4.9a presents the heat flux across the interface normalized by the "solid plane value" – the flux realized in pure thermal convection bounded by rigid horizontal planes with the same temperature variation. The solid plane flux was evaluated using the experimentally calibrated four-thirds flux law (Chandrasekhar, 1961). For relatively high density ratios ($R_\rho^* > 2$), the actual heat flux is less than the corresponding solid plane value, which reflects the tendency of salinity stratification to suppress the vertical motion

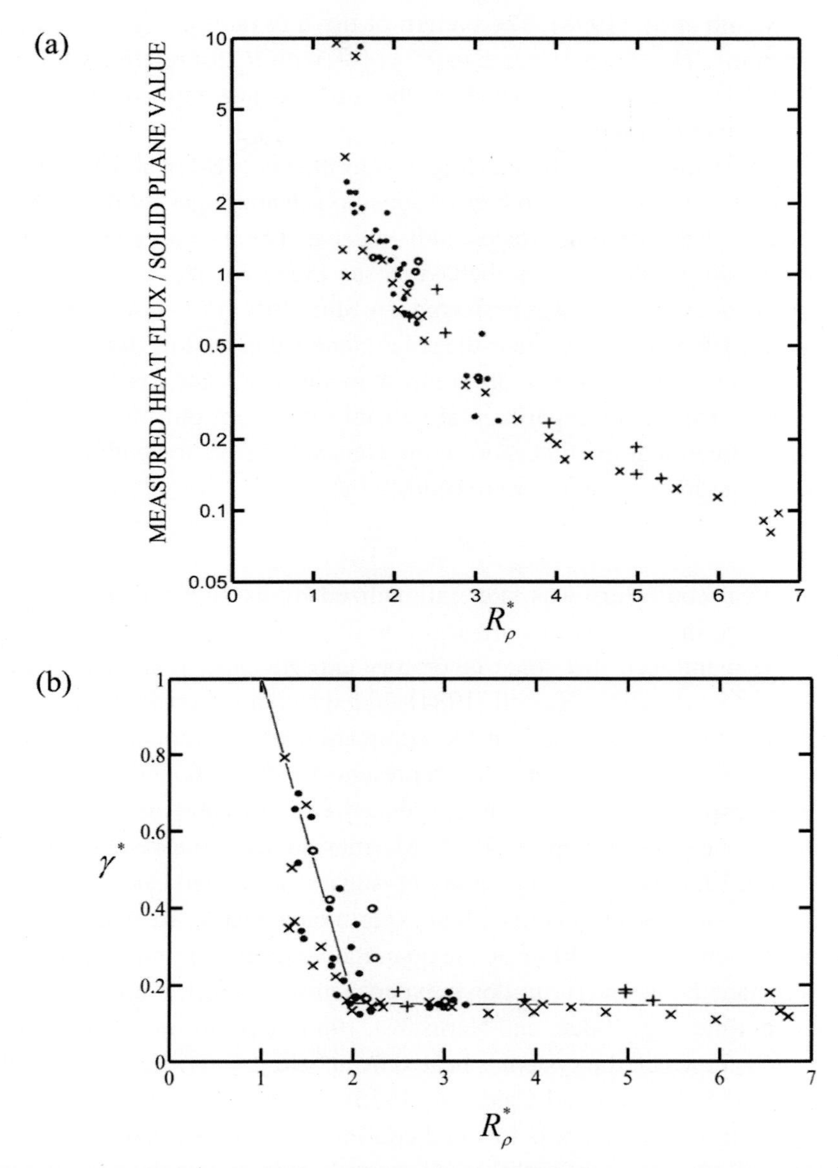

Figure 4.9 (a) The measured heat flux across the diffusive interface in Figure 4.8, normalized by the solid plane value and plotted against the diffusive density ratio. (b) The flux ratio as a function of the density ratio. From Turner (1965).

of fluid and thus heat transfer. However, the opposite is true for $R_\rho^* < 2$: the heat flux exceeds the solid plane value by as much as an order of magnitude. Such a dramatic effect is attributed to the eddy transfer across the interface, which for low R_ρ^* becomes strongly time dependent. The latter effect was examined in greater detail by Stamp *et al.* (1998). The pattern of the flux ratio in Figure 4.9b is even more intriguing. The sharp decrease in $\gamma^* = \frac{\beta F_S}{\alpha F_T}$ with R_ρ^* for relatively low density ratios (variable regime) is followed by the uniform flux ratio of $\gamma^* \approx 0.15$ for $R_\rho^* > 2$ (constant regime).

Undoubtedly inspired by the salt-finger experiments of Stern and Turner (1969), several experimentalists used isothermal aqueous salt and sugar solute to investigate the dynamics of diffusive interfaces. Salt–sugar and heat–salt experiments reveal qualitatively similar patterns and behavior (see Fig. 1.6). The first diffusive two-layer salt–sugar experiment was performed by Shirtcliffe (1973). As in the heat–salt case (Turner, 1965) T-flux exceeds the solid plane value for low density ratios and falls below for high. However, in contrast to the heat–salt case, the flux ratio is nearly uniform for all experimental density ratios, not only the higher range. Shirtcliffe noticed that the flux ratio in the constant regime for both heat–salt and sugar–salt experiments can be approximated by

$$\gamma^* \approx \sqrt{\tau}. \tag{4.11}$$

This empirical conjecture was later rationalized by explicit mechanistic models of the diffusive interface (Linden and Shirtcliffe, 1978; Worster, 2004), although it should be mentioned that some laboratory experiments (Takao and Narusawa, 1980; Turner *et al.*, 1970; Newell, 1984) find systematic departures from (4.11). Stern (1982) concluded, based on the application of variational methods to the diffusive two-layer system, that (4.11) represents the lower bound of the flux ratio.

Diffusive experiments have been reproduced, extended and improved numerous times in both heat–salt (Crapper, 1975; Marmorino and Caldwell, 1976; Newell, 1984; Taylor, 1988; among others) and salt–sugar (Turner and Chen, 1974; Turner, 1985; Stamp *et al.*, 1998) systems. Most experimental results are qualitatively consistent with each other but differ on the quantitative level. An important distinction should be made between (i) run-down experiments, in which T and S are allowed to diffuse in time (e.g., Takao and Narusawa, 1980), (ii) run-up experiments (e.g., Turner, 1965) in which the system is heated from below, and (iii) quasi-equilibrium experiments (Marmorino and Caldwell, 1976) maintained in a steady state by a combination of heating from below and cooling from above. It should be kept in mind that only the last configuration can be truly steady. For the run-up and run-down experiments, an assumption is often made that the system passes through a series of quasi-equilibrium states, and therefore all theoretical steady-state arguments can be applied directly. This assumption is justified for low and intermediate values of density ratio and deep mixed layers, in which case the rate of evolution of the diffusive interface greatly exceeds that of the mixed layers. As density ratio increases, the quasi-steady model becomes progressively more questionable. An interesting time-dependent effect involves the systematic drift of the interface.

The interfacial drift becomes particularly strong at low density ratios and can be attributed either to the imbalance in the turbulent intensities in the adjoining layers (Marmorino and Caldwell, 1976) or to the nonlinearity of the equation of state (McDougall, 1981).

A comprehensive theoretical model of the diffusive interface was developed by Linden and Shirtcliffe (1978) and considerably extended by Worster (2004). The key features of these models are illustrated in Figure 4.10. The diffusive interface is represented by a laminar statically stable core region bounded from above and below by thin unstable boundary layers. Following ideas proposed for thermal convection from a heated plate (Howard, 1964), it is assumed that the boundary layers undergo cyclic growth and convective eruption, which maintains the boundary layers close to the marginally unstable state characterized by the critical Rayleigh number. To close the problem, the temperature and salinity jumps across the boundary layers in the model are also assumed to be compensating in terms of density – that is, density is supposed to vary continuously between the core of the diffusive interface and the mixed layers as shown in Figure 4.10a. Combining these assumptions leads to an explicit expression for the flux ratio (4.11), which elegantly explains the laboratory result. Linden and Shirtcliffe (1978) formulated the steady-state version of the model, deriving an expression for the equilibrium temperature flux:

$$F_T = \frac{1}{\pi^{\frac{1}{3}}} \frac{\left(1 - \tau^{\frac{1}{2}} R_\rho^*\right)^{\frac{4}{3}}}{\left(1 - \tau^{\frac{1}{2}}\right)^{\frac{1}{3}}} F_T^{SP}, \tag{4.12}$$

where F_T^{SP} is the solid plane value. From (4.12) it is apparent that no steady solution exists for $R_\rho^* > \tau^{-\frac{1}{2}}$. The mixed layer convection in this regime is no longer capable of arresting the diffusive spreading of the interface, which monotonically thickens in time (Newell, 1984).

Prediction (4.12) is qualitatively consistent with the laboratory results for the intermediate range of density ratios ($3 < R_\rho^* < 7$ for heat–salt and $1.2 < R_\rho^* < 1.65$ for salt–sugar experiments) but largely fails for lower and higher values of R_ρ^*. Both problems can be traced to the model assumptions. For low R_ρ^* the interface becomes turbulent and therefore the inability of the model to account for the eddy transport of heat and salt across the core leads to an underestimate of fluxes. Failure of the model for large density ratios can also be ascribed to time-dependent effects, albeit of a different kind. At high density ratios the adjustment of fluxes to changing background state is extremely slow and therefore the diffusive system no longer evolves through a series of quasi-steady states. In this regime, each current state depends sensitively on the past history of the system, which renders the steady-state model (4.12) largely inapplicable. The generalization of the Linden and Shirtcliffe

(a)

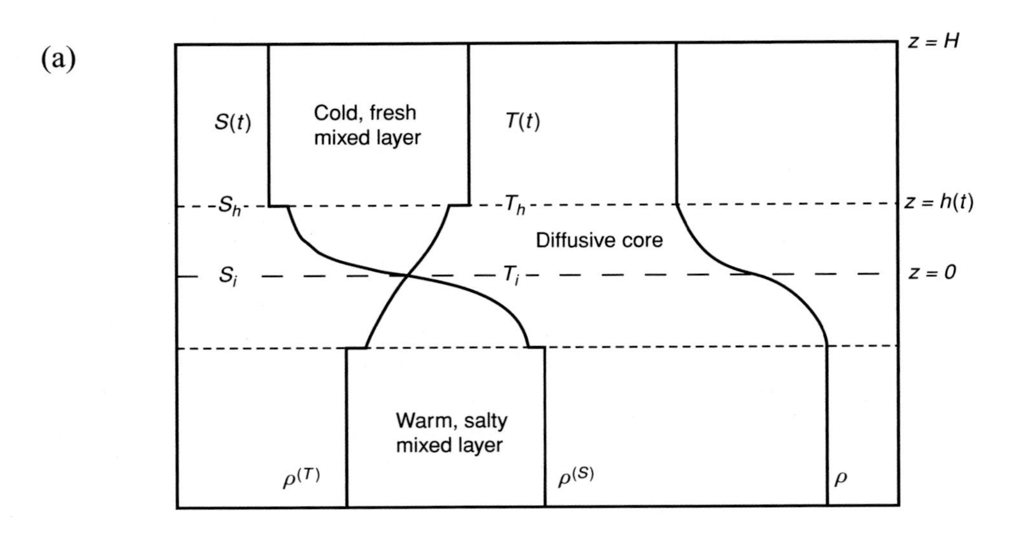

(b)

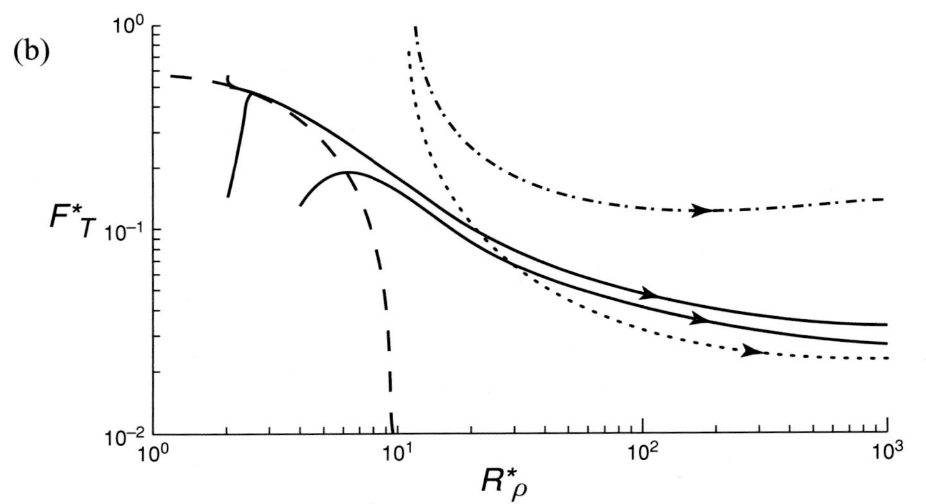

Figure 4.10 (a) Schematic diagram of a theoretical model for the two-layer diffusive system. (b) The evolutionary paths (curves with arrows) in the (R_ρ^*, F_T^*) plane predicted by the model in (a) for various initial conditions. F_T^* denotes the temperature flux normalized by the solid plane value. The long-dashed curve corresponds to the steady-state model of Linden and Shirtcliffe (1978). From Worster (2004).

(1978) model by Worster (2004), which takes these time-dependent dynamics into account, explains many features of laboratory experiments. It rationalizes the tendency of the diffusive system to approach the quasi-equilibrium state for intermediate density ratios and explains its evolutionary pattern for high R_ρ^*. The time-dependent model also draws attention to the significance of initial conditions by demonstrating their long-term influence on the system (see Fig. 4.10b). In the parameter regime where steady-state balances are expected, the Linden–Shirtcliffe–Worster model is consistent with the four-thirds flux laws (4.5), (4.6) and the interfacial thickness law (4.8).

It should be noted that an alternative model of the two-layer diffusive system exists (Fernando, 1989) in which the interface is represented by a double boundary layer structure: the outer layer is controlled by the diffusion of heat and contains a much thinner salinity layer. Fernando (1989, 1990) finds some support for the double boundary layer model in the laboratory and field measurements. However, a more systematic analysis is needed to identify the relative merits of the single- and double-layer models of diffusive interfaces.

5

The bounded layer model

The term "bounded model" in our discussion pertains to a configuration analogous to the classical Rayleigh–Bénard thermal convection problem. A fluid layer is located between rigid and slippery (i.e., stress-free) horizontal surfaces with constant prescribed values of temperature and salinity, as illustrated in Figure 5.1. This model plays an interesting role in the development of double-diffusive theory. The simplicity of its setup has led to the first major breakthroughs in the analysis of finite amplitude effects. The bounded model continues to attract considerable attention from theorists, who use it as a testing ground for the development of analytical techniques and new insights into the mechanics of double-diffusion. At the same time, the application of the model to observable double-diffusive phenomena is far from straightforward. The rigid horizontal planes – an essential component of the bounded model – are absent in most geophysical situations. In the ocean, double-diffusive regions are too separated from the bottom and sea-surface to be influenced by the vertical boundaries. Therefore, each application of the bounded model needs to be carefully and pervasively motivated. We start with the diffusive case (Fig. 5.1a), which is given priority for two reasons. One is historical: the first nonlinear treatment of the bounded model was offered for the diffusive configuration. Another consideration involves a more direct connection of the diffusive bounded model to observable structures.

5.1 Diffusive layer

It is perhaps only natural that the first successful nonlinear treatment of the bounded double-diffusive model was built on, and inspired by, the techniques and insights already developed for thermal convection. Veronis (1965, 1968) examined the role of a stabilizing salt gradient in thermal convection, analyzing linear and leading-order finite amplitude effects. Temperature and salinity values were prescribed at the vertical boundaries of the horizontally unlimited fluid layer. The solutions were

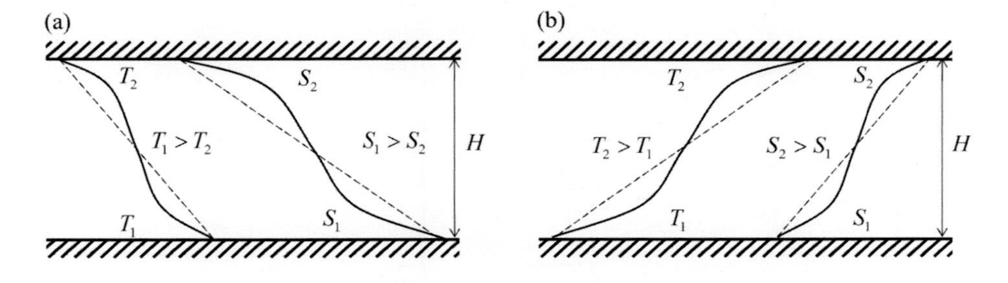

Figure 5.1 Schematic illustration of the bounded diffusive (a) and fingering (b) model. The fluid layer is vertically limited by rigid and slippery horizontal planes with prescribed temperature and salinity values.

sought as a perturbation to the purely conductive state, characterized by uniform vertical temperature and salinity gradients. Despite the similarity of the model formulation to its thermal one-component counterpart (Malkus and Veronis, 1958), the presence of a stabilizing salt gradient did not merely modify the thermal convection problem but resulted in qualitatively different, much richer and, arguably, less intuitive dynamics.

The inclusion of a salt gradient immediately leads to major changes in the linear stability properties of the bounded system. As thermal forcing ($\Delta T = T_1 - T_2$) is increased from zero, the instability initially appears in the form of oscillatory overstable perturbations. Veronis (1965) rationalized this property by offering the following physical argument. In order for direct, monotonically growing, modes to emerge first, they have to be preceded by a marginally stable steady state. Steady infinitesimal motions can occur only when the advection of the mean gradients is balanced by the diffusion of perturbations. The advection–diffusion equations for temperature and salinity (dimensional formulation) reduce to:

$$\begin{cases} w'\bar{T}_z = k_T \nabla^2 T', \\ w'\bar{S}_z = k_S \nabla^2 S'. \end{cases} \tag{5.1}$$

However, diffusion of salt is much slower than the diffusion of heat. Thus, if the background stratification is gravitationally stable ($R_\rho^* = \frac{\beta \bar{S}_z}{\alpha \bar{T}_z} > 1$) or close to neutral ($R_\rho^* \sim 1$) then, in view of (5.1), the density perturbation should be dominated by the haline component $\left(\alpha T' \sim \tau \beta S' \ll \beta S'\right)$. This means that the dynamics of such a system are controlled by the salinity distribution. The governing equations describing the evolution of the salinity field become effectively uncoupled from thermal effects. However, the background salinity stratification is stable and therefore steady infinitesimal convection is impossible; perturbations gradually decay due to molecular dissipation.

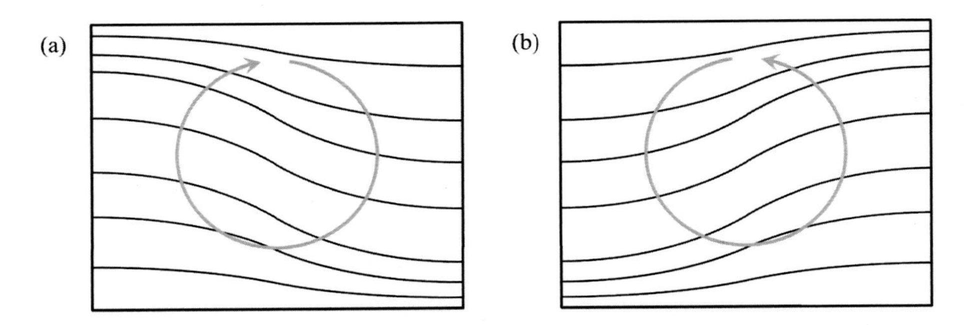

Figure 5.2 Mechanism of the oscillatory diffusive instability in the bounded model. The schematic shows the salinity surfaces for the clockwise (a) and counterclockwise (b) oscillation phases. From Veronis (1965).

This argument suggests that steady circulation in the linear model is permitted only under the rather restrictive conditions of the gravitationally unstable density distribution. Since the existence of a stable steady state is a prerequisite for the direct modes of instability, we arrive at the conclusion that, without oscillatory modes, linear destabilization of the bottom-heavy fluid is impossible. Direct modes are inherently limited in this regard. As will be seen shortly, oscillatory motions are characterized by higher dynamic flexibility – they can tap into the sources of available potential energy that are inaccessible for the direct modes.

The mechanism of overstable oscillations in the bounded bottom-heavy fluid is illustrated in Figure 5.2a,b. The evolutionary pattern of this system is characterized by periodic transitions between the clockwise (a) and counterclockwise (b) circulation patterns. Since heat diffuses much faster than salt, the isotherms tend to be more horizontal than the isohalines. Crowded curves in the upper left-hand region of the regime (a) correspond to saltier and therefore heavier fluid than in the upper right-hand region. Similarly, the fluid in the lower left-hand region is heavier than in the lower right-hand region. The counterclockwise torque produced in this manner reverses the sense of circulation, resulting in the regime indicated in (b). Reproducing the foregoing argument for the system (b), we conclude that the torque is now clockwise, causing the system to reverse the circulation pattern again, recreating the original configuration in (a). The cycle repeats over and over again. Furthermore, when thermal forcing is sufficiently strong, oscillations are not just maintained but systematically increase in amplitude. This scenario provides the physical interpretation for oscillatory diffusive instability in the bounded model. While Veronis' picture, understandably, has some similarities to that of overstable oscillations in the unbounded model (Chapter 1), it is cast in terms of more interesting, and perhaps more relevant for most applications, two-dimensional dynamics.

An analysis of nonlinear effects in the bounded diffusive model brings more surprises. For low diffusivity ratio ($\tau \to 0$), linear theory predicts instability for

$$R > \frac{Pr}{1+Pr} R_S + \frac{27}{4}\pi^4, \tag{5.2}$$

where $R = \frac{g\alpha \Delta T H^3}{\nu k_T}$ is the Rayleigh number based on the thermal component and $R_S = \frac{g\beta \Delta S H^3}{\nu k_T}$ is the salinity Rayleigh number; H is the layer thickness; $\Delta T = T_1 - T_2$ and $\Delta S = S_1 - S_2$ are the temperature and salinity variations across the layer (see Fig. 5.1a). For large Rayleigh numbers, (5.2) is equivalent to the unbounded condition (Chapter 1):

$$R_\rho^* < \frac{Pr+1}{Pr}. \tag{5.3}$$

Thus, while the bounded model permits the instability of bottom-heavy fluids, the instability range is rather narrow for any moderately large Prandtl number. For the heat–salt system ($Pr = 7$) we expect double-diffusive convection for $1 < R_\rho^* < 1.14$.

The linear result (5.3) seems to be at odds with the basic properties of diffusive convection. Laboratory experiments (Chapter 4) as well as field observations, with Arctic staircases being the prime example, offer plenty of evidence for the abundance of double-diffusive structures for density ratios as high as $R_\rho^* \sim 10$. Why then are such phenomena inconsistent with (5.3)? The contradiction is resolved by considering nonlinear effects. Before presenting the supporting analytical and numerical evidence, Veronis (1965) points out that nonlinearity is bound to destabilize the diffusive system. It is natural to expect that an active fully nonlinear circulation acts to homogenize the interior distribution of temperature and salinity. The "interior" in this context refers to a region separated from the upper and lower boundaries where temperature and salinity values are prescribed. The tendency of the circulation to mix the interior is opposed by molecular diffusion, acting to restore the stratification. This balance applies for both temperature and salinity, but diffusivity of salt is extremely low and therefore the tendency to restore interior salinity stratification is very weak. Thus, the interior salinity would be homogenized more than temperature. In this case, the interior density distribution is controlled by the (negative) temperature gradient, resulting in top-heavy stratification. The interior density distribution becomes gravitationally unstable, supporting vigorous overturning circulation. This argument – in contrast to the linear analysis – predicts that the motion in the bounded model can be maintained even if the variation in the thermal component of density is a small fraction of the corresponding salinity variation ($\alpha |\Delta T| \ll \beta |\Delta S|$).

Veronis unambiguously confirmed his physical argument through fully nonlinear simulations and an analytical model based on the highly truncated Fourier

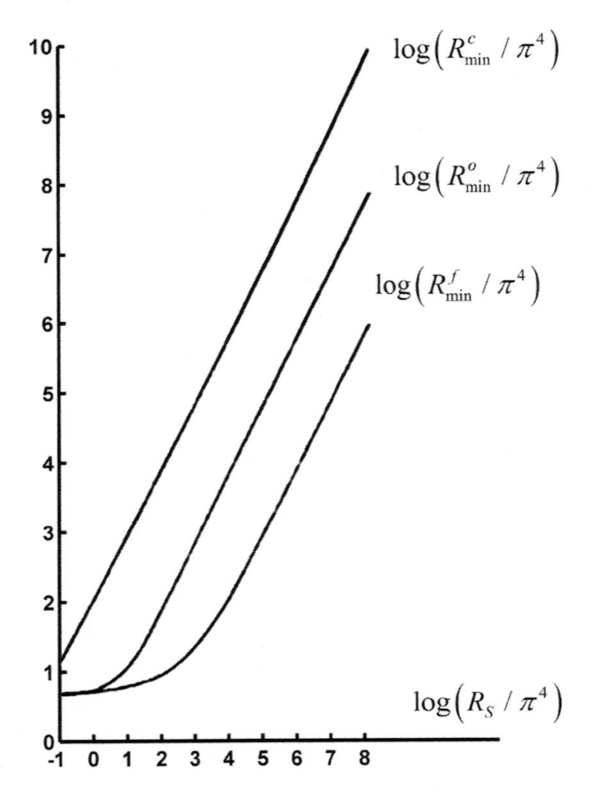

Figure 5.3 The values of three critical thermal Rayleigh numbers $R^c_{\min}$, $R^o_{\min}$ and $R^f_{\min}$ (see the text) are plotted as a function of the haline Rayleigh number R_S. From Veronis (1965).

representation of all variables. Figure 5.3 shows the values of three critical thermal Rayleigh numbers required for (i) the destabilization of the infinitesimal direct mode ($R^c_{\min}$), (ii) the infinitesimal oscillatory instability ($R^o_{\min}$) and (iii) the maintenance of finite amplitude motion ($R^f_{\min}$). All minimal thermal Rayleigh numbers are plotted as a function of the haline Rayleigh number (R_S). The Rayleigh number for linear direct instability is two orders of magnitude higher than for oscillatory modes and, perhaps an even more striking result, exceeds $R^f_{\min}$ by four orders of magnitude.

Numerical integrations reveal another counterintuitive feature of the bounded diffusive system. Although the preferred mode of linear evolution is oscillatory, the system ultimately evolves in time to a regular steady state. This fundamentally nonlinear steady state could be approached either through growth of the oscillatory modes (for $R > R^o_{\min}$) or, alternatively, by introducing a sufficiently large initial perturbation, which makes the transition to a finite amplitude steady state possible even for $R < R^o_{\min}$. The formation of the steady state for $R < R^o_{\min}$ is interesting since the *infinitesimal* steady states are possible only for much larger thermal Rayleigh numbers ($R^c_{\min} \gg R^o_{\min}$). In this regard, bounded diffusive convection is rather unusual. In the majority of instability studies (e.g., one-component convection) linear theory offers valuable insights, qualitatively applicable to the finite amplitude evolution of the system. For diffusive convection, all bets are off. The diffusive problem is principally nonlinear.

Numerical simulations in Veronis (1968) revealed that the more realistic representation of fluid (relative to the minimal model in Veronis, 1965) has a generally stabilizing effect on the system. Specific calculations were restricted to the top-heavy configuration, which takes the analysis somewhat beyond the scope of this monograph – double-diffusion in gravitationally stable configurations. However, several lessons learned from Veronis' analysis have a generic character and are likely to apply to bottom-heavy double-diffusive systems. One conclusion concerns the counter-gradient density flux. The solute gradient in the immediate vicinity of the boundary is necessarily sharper than the temperature gradient, and therefore the net tendency would be to maintain gravitationally stable boundary layers. In the interior regions, characterized by convective overturns, density flux is necessarily downward. Thus, in the boundary layers, density flux is counter-gradient and the effective diffusivity of density is negative. Quantifying this idea, Veronis (1968) goes on to predict a specific value of the boundary layer diffusivity for the nearly neutral background stratification ($R \approx R_S$):

$$k_\rho \sim -(k_T k_S)^{\frac{1}{2}}. \tag{5.4}$$

Another suggestive observation involves a systematic increase in the period of oscillations in time for the growing oscillatory modes. For a relatively narrow range of intermediate thermal Rayleigh numbers (R), the ultimate state of the system is represented by the finite amplitude oscillations, with the amplitude increasing with R. However, for sufficiently large values of R, finite amplitude oscillatory instability ultimately settles into a steady convective pattern. The tendency for the oscillation period to increase with increasing perturbation amplitude may prove to be the key for explaining the transition from linear oscillatory growth to eventual steady circulation observed in most simulations of diffusive systems.

Finally, Veronis suggests that the heat flux of the fully equilibrated diffusive system should be comparable to the corresponding heat flux observed in thermal convections for the same Rayleigh numbers. The rationale is rather straightforward. Strong finite amplitude convection would homogenize salinity in the interior zone and, since the diffusivity of salt is very low, the tendency to restore the stratification by means of molecular diffusion is weak. In this case, the inhibiting effect of the solute gradient will be reduced, and the fluid can convect nearly as much heat as it does in the absence of the solute. The available simulations for the bounded diffusive system confirm this conjecture. However, the reader is reminded that the laboratory experiments with diffusive interfaces, discussed in Chapter 4, can yield heat transports an order of magnitude larger than in corresponding thermal convection experiments. This underscores differences between the bounded and two-layer models. In both cases, heat transport is constrained by the diffusion of temperature through thin and strongly stratified boundary layers. However, in the bounded model the vertical velocity vanishes identically at the boundary and therefore molecular diffusion there is the only means for heat transfer. In the case of the two-layer configuration, however, this constraint is not nearly as severe – the interface is deformable and therefore the possibility exists for the eddy-induced transport of heat and salt.

Since Veronis' studies, the bounded diffusive model has been investigated in much greater detail a number of times. Huppert and Moore (1976) performed a comprehensive numerical analysis of the Veronis problem. Steady finite amplitude solutions were obtained and examined for linearly stable bottom-heavy stratification. Systematic exploration of the (R, R_S) parameter space made it possible to classify all nonlinear solutions into four major categories. As the thermal Rayleigh number increases, the thermohaline system undergoes transitions from (i) simple quasi-harmonic oscillations to (ii) more complicated periodic oscillations, then to (iii) aperiodic and, finally, to (iv) time-independent solutions. Somewhat counterintuitively, an increase in the Rayleigh number in the diffusive model suppresses the disordered motion. For certain parameters, two different solutions exist. Numerous instances of hysteresis were found in simulations in which an increase in Rayleigh number was followed by its decrease.

The key numerical results in Huppert and Moore (1976), particularly with regard to the dynamics of steady solutions, were rationalized by a complementary analytical study (Proctor, 1981). The rich dynamics of diffusive systems motivated numerous follow-up studies aimed to describe bifurcations and chaotic behavior observed in numerical simulations (Moore and Weiss, 1990; Knobloch *et al.*, 1992; among others). In terms of explaining diffusive dynamics from first principles, particularly promising are attempts to formulate Fourier-truncated models, which reduce the governing differential equations to a small set of ordinary differential equations (e.g., Da Costa *et al.*, 1981). These relatively simple low-order models have proven to be successful in capturing the behavior of the full system and in revealing the zero-order dynamics at play. Other applications of the bounded diffusive model include the analysis of the effects associated with non-uniformities of the background stratification (Balmforth *et al.*, 1998; Walton, 1982) and the appearance of spatially localized nonlinear states (Batiste *et al.*, 2006; Bergeon and Knobloch, 2008a,b).

An interesting approach to the analytical modeling of double-diffusive convection involves the prediction of the upper bound on the flux of the unstably stratified component using variational principles. The basic idea of this method has been introduced in pioneering studies by Malkus (1954) and Howard (1963) for thermal convection. In view of the complexities of the governing equations in fluid mechanics, the upper bound theory does not attempt to search for actual solutions but, instead, places bounds on some of their average properties. The method is based on the integral relations (power integrals) derived from, but much simpler than, the original system. Using a series of algebraic inequalities, it is usually possible to rigorously deduce upper limits for the vertical transport of the diffusing properties. The upper bound theory has been widely and successfully applied to various fluid dynamical problems (Joseph, 1976; Doering and Constantin, 1996; Kerswell, 1998). While double-diffusive convection appears to be well suited for the application of the upper bound theory, only a few such attempts (Stern, 1982; Balmforth *et al.*, 2006) have been reported.

The double-diffusive version of the upper bound theory is more challenging and intriguing than the convective case, particularly since it exhibits a remarkably rich array of dynamics. It uses the same power integrals for T and S that emerge in one-component convection but also adds a crucial mixed integral constraint, representing the interplay of temperature and salinity perturbations. Stern (1982) applied the upper bound theory to the diffusive case, whereas Balmforth *et al.* (2006) examined both diffusive and fingering regimes. The ultimate goal of any bounding exercise is to formulate the upper bound that is sufficiently close to the physically realized fluxes. Although double-diffusive theories may not yet have reached this goal, the physical insights brought about by such bounding models are already evident. Conversely, the problem of double-diffusive convection between two rigid horizontal planes offers an attractive testing ground for the development of new methods for the upper bound theory. This line of research is promising and should be pursued much further.

Before moving on to the next topic, we should note some difficulties regarding the application of the bounded diffusive model to observed phenomena. To the best of the author's knowledge, no double-diffusive configuration in nature involves setting specific temperature and salinity values at the vertical boundaries. Problems arise even in the controlled setting of laboratory experiments. Perhaps closest to the bounded model comes a setup used by Krishnamurti (2003, 2009), who maintained the vertical finger-favorable T–S differences by flushing the reservoirs located at the top and bottom of the experimental tank (Chapter 4). This laboratory configuration has not yet been used in the diffusive regime.

Nevertheless, the bounded model is likely to be relevant, albeit in an abstract and qualitative sense, to laboratory and oceanic examples of diffusive convection. A generic pattern realized in this model consists of thin boundary layers near the solid horizontal planes, controlled by the molecular dissipation of heat and salt, and a gravitationally unstable interior. The dynamics of diffusive layers in nature are similar. The key feature is the interplay between molecular diffusion acting across thin interfaces and convection in the adjacent mixed layers. Therefore, it is our belief that insights generated by studies of the bounded diffusive model (Veronis, 1965, 1968; Knobloch *et al.*, 1992; Balmforth and Biello, 1998; Balmforth *et al.*, 1998) have general applicability. The analogy between the bounded *salt-finger* model – our next topic – and realizations in nature is more tenuous, although certain dynamic similarities do exist.

5.2 Salt-finger layer

Consider the salt-finger configuration in Figure 5.1b. As previously, the bounded model prescribes temperature and salinity values at the rigid slippery vertical boundaries, although in this case both temperature and salinity increase upward ($T_2 > T_1$, $S_2 > S_1$). The aspects of fingering adequately addressed by the bounded model include the pattern selection problem and analysis of the finger-induced transport.

Planform selection

One of the first bounded finger models was developed by Straus (1972), who numerically obtained two-dimensional finite amplitude steady solutions and examined their stability with respect to three-dimensional perturbations. Computational constraints at that time prevented the author from approaching the parameter regimes that would be relevant for most applications. Nevertheless, this study raised some intriguing questions. For instance, stability analysis of the finite amplitude two-dimensional (roll-type) solutions suggested that they are stable with respect to three-dimensional perturbations as long as the horizontal wavelength is comparable to the height of the fluid layer. The stability of salt-finger rolls suggests that such two-dimensional structures should be realized in nature. The latter proposition is at odds with laboratory observations (Shirtcliffe and Turner, 1970), which clearly revealed the preference for three-dimensional patterns best described as squares (Fig. 4.5).

An attempt to identify the cause of the disagreement was made by Proctor and Holyer (1986), who treated the problem using an asymptotic expansion pivoted about the marginally unstable point. The problem of pattern selection in the

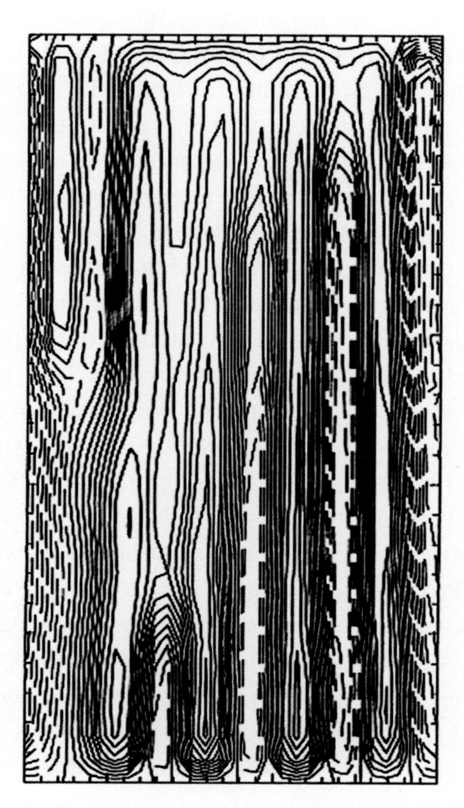

Figure 5.4 Vertical cross-section of the fully developed temperature perturbation (T') in the bounded numerical simulation (salt-finger case). The computational domain corresponds to $5 \times 5 \times 10$ fastest growing finger wavelengths. From Radko and Stern (2000).

salt-finger regime is challenging because linear instability theory (Chapter 2) provides no guidance with respect to the preferred planform. Identical growth rates can be achieved by a variety of horizontal structures and, from the perspective of linear theory, two-dimensional rolls are as likely to be realized as square cells. Apparently, the key to the planform selection lies in fundamentally nonlinear interactions between different modes present in the equilibrated state. Therefore, Proctor and Holyer (1986) incorporated the leading-order nonlinear effects by expanding in the small parameter $\frac{L}{H} \ll 1$ representing the ratio of the typical width of fingers (L) to the layer height (H). Their model assumed that fingers are vertically elongated throughout the fluid layer, thereby excluding from consideration structures of limited – comparable to the salt-finger width – vertical extent (Fig. 5.4). The weakly nonlinear analysis led the authors to conclude, in conflict with observations, that the roll-type planforms are preferred over the square cells.

The suggestion of the dynamic preference for square fingers was supported by the laboratory experiments referenced in Schmitt (1994a). In these experiments, salt fingers were generated in a triangular tank. The geometry of the container did not affect the planform: fingers remained rectangular. It should be noted that the planform selection has been discussed mostly in connection with laboratory experiments: fingers in the ocean are less accessible and more irregular. However, an interesting example was reported by Osborn (1991), who observed asymmetric salt fingers in the near-surface zone of the ocean. These fingers were characterized by narrow downward plumes separated by a weaker and broader upwelling. The spacing between adjacent fingers was consistent with the prediction of the fastest growing finger model. Schmitt demonstrated that such asymmetric structures can be obtained by superposing several fastest growing modes with differing horizontal wavenumber components. Although the selection mechanism of asymmetric structures has never been fully explained, it was hypothesized (Osborn, 1991; Schmitt, 1994a) that surface evaporation effects could have played some role.

The next attempt to solve the planform selection puzzle was made by Radko and Stern (2000). As in Proctor and Holyer (1986), this study was based on a weakly nonlinear asymptotic expansion. However, the small parameter $((R_\rho \tau)^{-1} - 1)$ in Radko and Stern measured the separation of parameters from the marginally stable point for salt fingering ($R_\rho = \tau^{-1}$). No assumption was made regarding the relative difference in vertical and horizontal scales, which made it possible to adequately represent thin boundary layers at the edges of the finger zone. These boundary layers, clearly visible in Figure 5.4, proved to be essential in the pattern selection process. Figure 5.5 shows a numerical experiment initiated by slightly perturbed rolls. The horizontal temperature pattern slowly transforms from a quasi two-dimensional structure to remarkably regular square cells. Particularly suggestive is the sequence of events – the transition from rolls to cells starts within the boundary layers and then spreads throughout the elongated fingers in the interior. The significance of boundary layers for the planform selection was confirmed by asymptotic analysis. Radko and Stern (2000) demonstrated that (i) two-dimensional rolls are unstable with respect to cross-roll perturbations and (ii) an approximation which assumes that the vertical scale of the boundary layer exceeds the salt-finger width immediately leads to the opposite (and erroneous) conclusion. The observed preference for square cells was attributed to the instability of solutions with non-square planforms, which resolved the conundrum posed by Proctor and Holyer (1986).

Vertical transport

In addition to the dynamics of the horizontal pattern selection, the bounded salt-finger model has brought some insight into the mechanisms controlling the

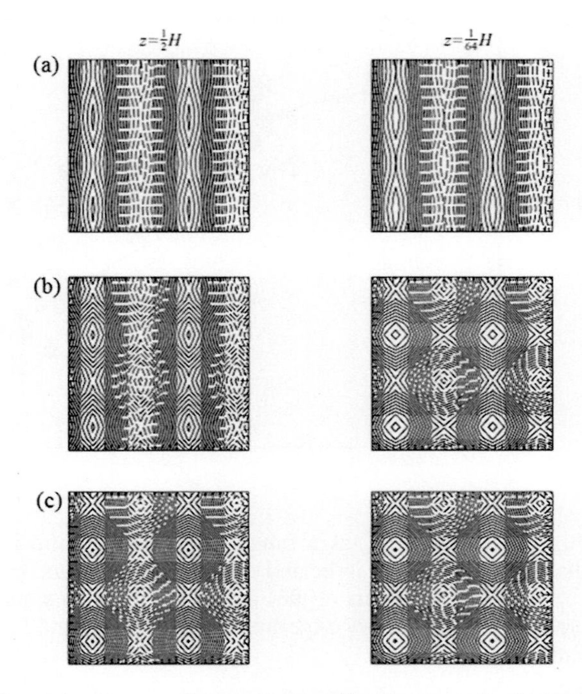

Figure 5.5 Transition from rolls to square cells in a numerical simulation. Shown are the horizontal cross-sections of temperature across the boundary layer (right) and across the interior region (left) at various stages. The time increases from (a) to (c). Transition to the square-cell planform starts with the boundary layer (b) but eventually is completed at all levels (c). From Radko and Stern (2000).

vertical *T–S* transport across the salt-finger layer. For two-layer systems (Chapter 4) dimensional arguments lead to Turner's (1967) flux law (4.6). These dimensional arguments do not necessarily apply to the bounded configuration (Fig. 5.1) since the bounded model involves an additional *independent* dimensional parameter – thickness of the salt-finger layer (H). Nevertheless, numerical simulations in Radko and Stern (2000) revealed that the bounded model is well described by the four-thirds flux law. A series of calculations in which density ratio and molecular parameters were fixed, but the height of the finger layer varied by an order of magnitude, produced a nearly uniform flux law coefficient $C'_S = \frac{\alpha F_S}{(\alpha \Delta S)^{\frac{4}{3}}}$. In an attempt to rationalize this unexpected behavior, Radko and Stern (2000) developed an analytical model for the vertical transport, predicting

$$C'_S \propto H^{-\frac{1}{10}} (\Delta S)^{-\frac{1}{30}} . \tag{5.5}$$

The exponents $-1/10$ and $-1/30$ in (5.5) correspond to extremely weak dependences on layer height and overall *T–S* variation and therefore the flux law

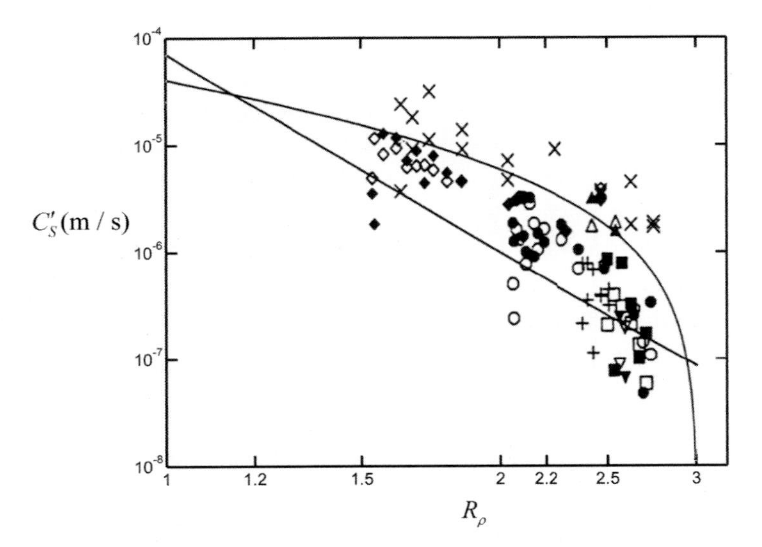

Figure 5.6 Flux law constant C'_S as a function of density ratio in the bounded model of Radko and Stern (2000) indicated by the solid curve and in the laboratory experiments of Taylor and Veronis (1996) marked by various signs. The straight line is an average of the laboratory experiments of Griffiths and Ruddick (1980). From Radko and Stern (2000).

coefficient is effectively determined by the density ratio, as in Turner's four-thirds flux law.

A perhaps even more surprising and suggestive finding of Radko and Stern (2000) is that the fluxes obtained for the bounded model were consistent with the fluxes in the laboratory two-layer experiments (Chapter 4) in which the finger zone is sandwiched between two well-mixed reservoirs. Figure 5.6 presents a plot of the flux law coefficient (5.5) from a bounded layer model (Radko and Stern, 2000) superimposed on a series of two-layer experiments (Taylor and Veronis, 1996). Both the magnitude of C'_S in experiments and its variation with the density ratio is captured by the bounded model. What makes this result particularly intriguing is that the model does not contain any significant adjustable parameters and therefore the comparison is objective. The agreement between the laboratory- and model-based estimates of C'_S in Figure 5.6 suggests that the dynamics of the finger zone in the bounded model and in two-layer experiments are similar. It is plausible that the vertical velocity of individual fingers in experiments is greatly reduced as they approach large-scale plumes in the convectively mixing layers. If so, the matching conditions at the boundary between the salt-finger zone and the mixed layers can be approximated by simple rigid lid conditions, and the bounded model can be used to explain the dynamics of two-layer experiments.

(a)

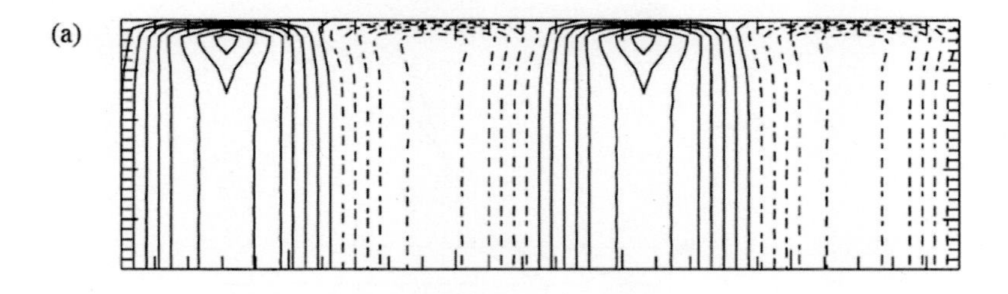

(b)

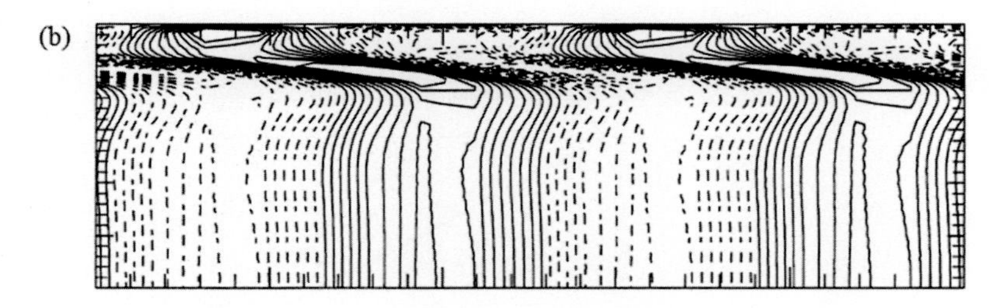

Figure 5.7 The temperature patterns in the boundary layer in two calculations with different values of H. The boundary layer Rayleigh number in (a) is $Ra = 1440$ and in (b) is $Ra = 2800$. The boundary layer for the low Ra regime is stable and laminar, while in the high Ra case it is distorted by the local instability. From Radko and Stern (2000).

Height of the finger zone

An obvious difference between the bounded model and typical two-layer laboratory experiments is related to the selection of the thickness of the finger layer (H). In the bounded model, the layer thickness is a free parameter, but in two-layer experiments it is controlled by the variations in temperature and salinity across the mixed layers (ΔT, ΔS). Still, the bounded model brings some insight into the mechanisms that could be involved in the selection of H. Radko and Stern (2000) examined a series of bounded model simulations in which H was systematically increased, and noticed that at a certain point ($H = H^*$) the flow field undergoes a dramatic transition. For relatively low values of H, temperature and salinity patterns are regular and steady. However, when H exceeds a critical value (H^*), the flow becomes disorganized and time-dependent, which is suggestive of the onset of a secondary instability. This instability has a local character and it is limited to thin conductive boundary layers forming near the vertical boundaries (Fig. 5.7).

The origin of the secondary boundary layer instabilities is clear. As the vertical velocity near the rigid boundaries is suppressed, the vertical *advective* fluxes in the boundary layers ($\alpha \overline{w'T'}$, $\beta \overline{w'S'}$) are greatly reduced relative to that in the interior, which is compensated by the increase in *conductive* transport ($k_T \alpha \bar{T}_z$, $k_S \beta \bar{S}_z$).

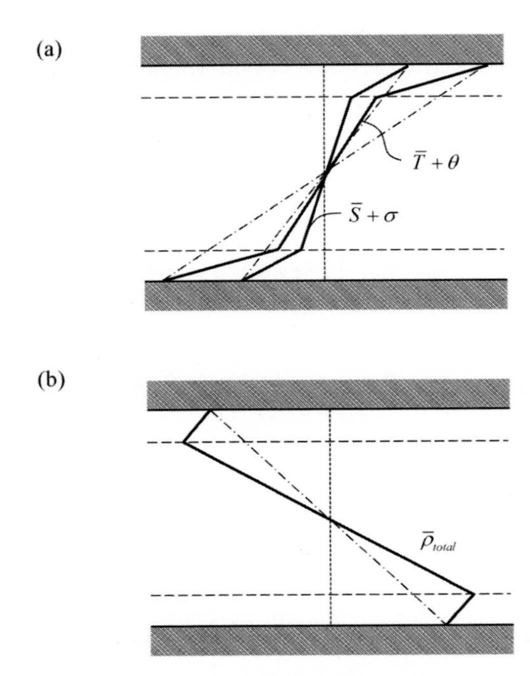

Figure 5.8 Schematic diagram of the horizontally averaged profiles of temperature and salinity (a) and density (b) in the bounded steady state. From Radko and Stern (2000).

Since vertical density fluxes due to heat and salt transport by fingers are comparable ($\gamma \sim 1$), we expect that in the boundary layers $k_T \alpha \bar{T}_z \sim k_S \beta \bar{S}_z$ and therefore $\alpha \bar{T}_z \ll \beta \bar{S}_z$. The latter inequality implies that the density stratification in the boundary layers is gravitationally unstable ($\bar{\rho}_z > 0$) in contrast to the stable interior ($\bar{\rho}_z < 0$), as shown in the schematic in Figure 5.8. A well-known criterion for the instability of the gravitationally unstable layers is based on the thermohaline Rayleigh number (*Ra*). When *Ra* exceeds a critical value on the order of a thousand, vigorous instabilities are expected to occur, which is readily confirmed by inspection of the boundary layers in Figure 5.7.

Calculations in Radko and Stern (2000) were performed for $\tau = 1/3$ – a model of the sugar–salt experiment. The order-one value of the diffusivity ratio implies comparable scales for the haline and thermal boundary layers and thereby simplifies analytical development. The local thermohaline Rayleigh number is given by

$$Ra = \frac{gh^4}{\nu}\left(\frac{\bar{S}_z}{k_S} - \frac{\bar{T}_z}{k_T}\right), \tag{5.6}$$

where *h* is the thickness of boundary layers. The boundary layer Rayleigh number was evaluated for the calculation in Figure 5.7a (regular boundary layer), resulting in $Ra = 1440$, and for the visibly unstable boundary layer in Figure 5.7b, yielding

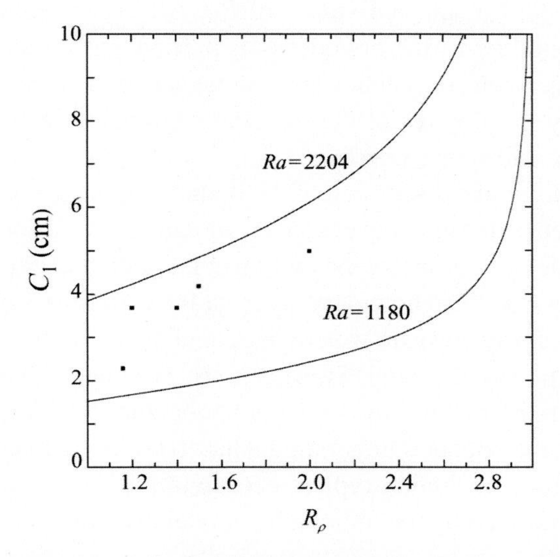

Figure 5.9 Comparison of the theoretical prediction of the coefficient C_1 in the interfacial thickness law (4.8) with the results of the laboratory experiments (Linden 1978). The analytical theory suggests that the $C_1(R_\rho)$ functions are controlled by the values of the boundary layer Rayleigh number Ra and the slope of the corresponding theoretical curves is in apparent agreement with the experimental data. From Radko and Stern (2000).

$Ra = 2800$. Analytical theory suggests that the local boundary layer Rayleigh number is controlled by the overall height of the finger layer ($Ra \propto H^{0.4}$), which explains destabilization of the boundary layers at sufficiently high H. The critical thickness (H^*) corresponding to the onset of the boundary layer instability can be estimated by requiring $Ra \sim 1000$.

Limited evidence from numerical and laboratory two-layer experiments (Chapter 4) suggests that the dynamics of such systems also undergo boundary layer instabilities and are profoundly affected by them. Numerical simulations in Shen (1989) revealed the appearance of local density inversions at the edges of the finger zone. Shen hypothesized that these inversions are essential for the breakup of fingers. Such transitional zones between convecting layers and finger interfaces are also visible in the shadowgraphs of laboratory experiments (Linden, 1978). Therefore, it is sensible to assume that the local instability of the top-heavy transitional zones is involved in the arrest of the spatial spreading of the finger zone in two-layer experiments. Radko and Stern (2000) hypothesized that the equilibrium thickness of the finger layer in two-layer experiments corresponds to the marginally unstable transitional zone ($Ra \sim 1000$). This theory finds some support in laboratory data. Figure 5.9 presents the coefficient of the thickness law (4.8) $C_1 = H(\beta \Delta S)^{\frac{1}{3}}$ deduced from measurements in Linden (1978) and predicted by the bounded model for selected values of Ra. All experimental points belong to the area corresponding to the boundary layer Rayleigh numbers in the range of $Ra \sim 1000$–2000. Such consistency lends credence to the idea of thickness control by the secondary instabilities of the top-heavy boundary layers forming near the boundaries of the salt-finger interface.

Before concluding the discussion of the bounded model, a few words of caution are in order. Based on the existing evidence, we can claim with some confidence that the bounded model is relevant for the two-layer laboratory experiments, particularly for the sugar–salt case. The boundary layers at the extremities of the finger zone in these experiments (Fig. 4.4) are well defined and appear to be dynamically similar to those in the bounded model. However, the bounded dynamics – interesting as they may be in their own right – are not applicable to all instances of fingering. The relevance of the bounded finger model has to be reexamined on a case-by-case basis. In most cases, including typical oceanographic configurations, salt fingers may be better described by the unbounded model presented in Chapter 3.

6
Collective instability

The term collective instability refers to a spontaneous excitation of gravity waves by a fully developed and initially homogeneous field of salt fingers. This effect, briefly mentioned in Chapter 3, was discovered by Stern (1969). Stern's theory was inspired by laboratory experiments with salt fingers in uniform background property gradients. In these experiments (some of which are described in Stern and Turner, 1969) large-scale waves appeared shortly after the development of salt fingers and were followed by even more dramatic transformation of the flow into a series of alternating convective cells and sharp interfaces – structures that came to be known as thermohaline staircases (Chapter 8). Ironically, the link between waves and staircases, which motivated the collective instability theory in the first place, has never been confirmed. These effects are largely independent. Nevertheless, collective instability turned out to be a very interesting phenomenon in its own right, profoundly affecting doubly diffusive fluids in a number of ways.

At least three different theoretical models (Stern, 1969; Holyer, 1981; Stern *et al.*, 2001) have been proposed to explain the generation of large-scale waves by fingers. While each model is based on distinct physics, they all lead to the same conclusion – collective instability is controlled by the Stern number:

$$A = \frac{\alpha F_T - \beta F_S}{\nu(\alpha \bar{T}_z - \beta \bar{S}_z)}. \tag{6.1}$$

An equivalent expression for A in terms of the density flux is given in (3.5). All three models are summarized and compared next (Section 6.1), followed by a more in-depth discussion of the parametric theory (Stern *et al.*, 2001) in Section 6.2.

6.1 Approaches

The first argument was proposed by Stern (1969). Stern assumed that the coupling between large-scale waves and fingers in the laboratory experiments (Stern and

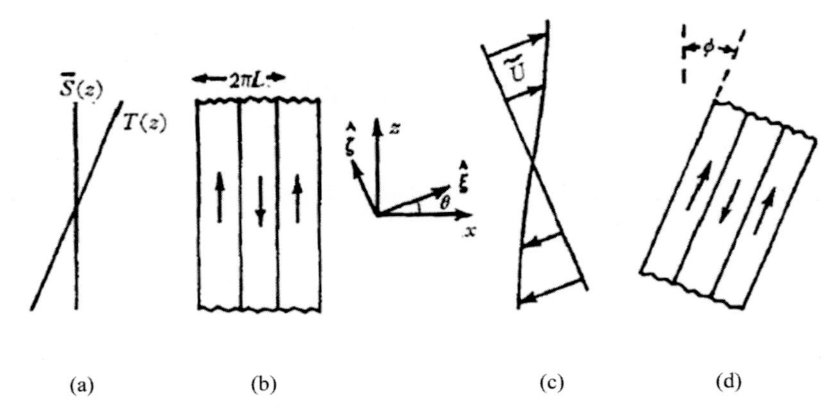

Figure 6.1 Stern's (1969) model. Stern envisioned initially vertical salt fingers in the uniform temperature and salinity gradient (a,b), which are subjected to the shear of a large-scale gravity wave (c). The wave slightly tilts the fingers (d) reducing their vertical transport of heat, salt and, ultimately, buoyancy. The buoyancy perturbation, in turn, affects the gravity wave. The feedback between the fingers and wave results in the amplification of the latter, provided that the Stern number (6.1) exceeds unity. From Stern (1969).

Turner, 1969) is due to the ability of the wave shear to tilt fingers, thereby reducing the vertical component of heat and salt fluxes, as indicated in Figure 6.1. The associated large-scale modulation of the vertical transport results in temperature and salinity convergences which, in turn, modify the buoyancy distribution in the wave. Stern argued that the buoyancy forcing induced in this manner provides a feedback mechanism for the amplification of gravity waves as long as A exceeds unity.

Stern's model was reexamined by Holyer (1981). Of particular concern was the ad hoc assumption that tilting of fingers reduces the vertical components of the temperature and salinity fluxes but not their magnitudes. Holyer (1981) offered a more mathematically rigorous, if not more physical, treatment of the collective instability using multiscale analysis. She examined stability of the two-dimensional basic state consisting of an array of steady, vertical fingers with respect to long-wavelength perturbations, as illustrated in Figure 6.2. The resulting condition for instability was $A > 1/3$, which is similar, but not identical, to Stern's (1969) criterion ($A > 1$). Holyer (1985) reproduced this calculation in three dimensions, assuming a square horizontal planform of salt fingers, and arrived at the collective instability criterion of $A > 2/3$ – even closer to Stern's original prediction. While Holyer's models avoided questionable steps in the original theory, they were replaced by other assumptions that were also not fully justified. In particular, salt

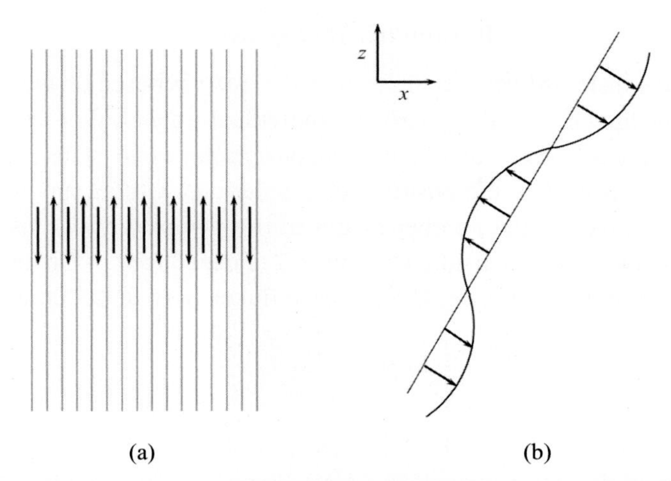

Figure 6.2 Schematic illustration of Holyer's (1984) model. Holyer analyzed stability of the marginally unstable vertical fingers (a) with respect to large-scale perturbations (b). From Holyer (1984).

fingers observed in nature or in laboratory settings are not vertical and definitely not steady. As discussed in Chapter 2, the dominant horizontal wavelength realized in a fully developed salt-finger field corresponds much better to the linearly fastest growing mode than to the steady solution used by Holyer (1981, 1985).

A third attempt to formulate a rigorous theory for the collective instability was made by Stern *et al.* (2001). The vertical temperature and salinity fluxes due to salt fingers were assumed to be controlled by the large-scale *T–S* gradients and parameterized accordingly:

$$\begin{cases} F_T = -K_T(R_\rho)\frac{\partial T}{\partial z}, \\ F_S = -K_S(R_\rho)\frac{\partial S}{\partial z}. \end{cases} \tag{6.2}$$

The flux ratio $\gamma = \frac{\alpha F_T}{\beta F_S}$ was approximated by a constant and the effects of shear on the fluxes – the essential element of the Stern (1969) model – were ignored based on compelling numerical evidence. The amplification of the gravity wave was attributed to subtle changes in the large-scale buoyancy distribution, resulting from the convergence of the vertical double-diffusive fluxes (6.2). Remarkably, linear stability analysis of the parameterized flux-gradient system recovered Stern's original instability condition: $A > 1$. Theoretical predictions of the wave growth rates were confirmed quantitatively by DNS. The very reasonable physical assumptions, its performance and utility warrant a closer look at the flux-gradient model (Stern *et al.*, 2001), which is next discussed in greater detail.

6.2 Parametric flux-gradient model

The collective instability model in Stern *et al.* (2001) is based on the Navier–Stokes equations for the large-scale (relative to individual fingers) flow components. The model assumes a horizontally homogeneous background state, which makes it possible to focus on two-dimensional (x,z) solutions without any loss of generality. In order to reduce the number of governing parameters, we use the standard double-diffusive non-dimensionalization (1.11) introduced in Chapter 1. The non-dimensional expression for double-diffusive fluxes in (6.2) becomes

$$\begin{cases} F_T = -Nu(R_\rho)\dfrac{\partial T}{\partial z}, \\ F_S = \dfrac{F_T}{\gamma(R_\rho)}, \end{cases} \tag{6.3}$$

where, as previously (Chapter 3) we assumed that the Nusselt number (Nu) and the flux ratio (γ) are controlled by the density ratio (R_ρ). The governing equations are linearized with respect to uniform vertical temperature–salinity gradients:

$$\begin{cases} \dfrac{\partial}{\partial t}T' + \dfrac{\partial \psi'}{\partial x} = -\dfrac{\partial}{\partial z}F'_T, \\ \dfrac{\partial}{\partial t}S' + \dfrac{1}{R_\rho}\dfrac{\partial \psi'}{\partial x} = -\dfrac{\partial}{\partial z}F'_S, \\ \dfrac{1}{Pr}\dfrac{\partial}{\partial t}\nabla^2\psi' = \left(\dfrac{\partial T'}{\partial x} - \dfrac{\partial S'}{\partial x}\right) + \nabla^4\psi', \end{cases} \tag{6.4}$$

where (F'_T, F'_S) in (6.4) are the linear perturbations of the temperature and salinity fluxes, and ψ' is the perturbation streamfunction, defined by $(u', w') = (-\psi'_z, \psi'_x)$. The non-dimensional expression for the temperature flux reduces to $F_T = -Nu(R_\rho)(1 + T'_z)$ and therefore

$$F'_T = -\left(\left.\frac{\partial Nu}{\partial R_\rho}\right|_{R_\rho=\bar{R}_\rho} R'_\rho + Nu(\bar{R}_\rho)T'_z\right), \tag{6.5}$$

where $\bar{R}_\rho$ is the basic density ratio and R'_ρ is the perturbation:

$$R'_\rho = \frac{(1+T'_z)}{(\bar{R}_\rho^{-1} + S'_z)} - \bar{R}_\rho \approx \bar{R}_\rho\left(T'_z - \bar{R}_\rho S'_z\right). \tag{6.6}$$

The explicit expression for F'_S is obtained by expressing the salt flux in terms of the flux ratio as in (6.3), which after linearization yields

$$F'_S = \frac{F'_T}{\gamma(\bar{R}_\rho)} + \left.\frac{\partial \gamma^{-1}}{\partial R_\rho}\right|_{R_\rho=\bar{R}_\rho} R'_\rho Nu(\bar{R}_\rho). \tag{6.7}$$

Stability properties of the resulting linear system are analyzed using normal modes:

$$\begin{pmatrix} T' \\ S' \\ \psi' \end{pmatrix} = \begin{pmatrix} \hat{T} \\ \hat{S} \\ \hat{\psi} \end{pmatrix} \exp(ikx + imz + \lambda t), \tag{6.8}$$

substitution of which into (6.4)–(6.7) yields a cubic equation for the growth rate:

$$\lambda^3 + a_2\lambda^2 + a_1\lambda + a_0 = 0. \tag{6.9}$$

The coefficients of (6.9) are represented by lengthy algebraic expressions in terms of

$$a_i = a_i\left[k, m, \bar{R}_\rho, Nu(\bar{R}_\rho), \gamma(\bar{R}_\rho), A_{Nu}, A_\gamma\right], \tag{6.10}$$

where $A_{Nu} = \bar{R}_\rho \left.\frac{\partial Nu}{\partial R_\rho}\right|_{R_\rho=\bar{R}_\rho}$ and $A_\gamma = \bar{R}_\rho \left.\frac{\partial \gamma^{-1}}{\partial R_\rho}\right|_{R_\rho=\bar{R}_\rho}$ measure the variation in fluxes with the density ratio. It is interesting to note that the variation in the flux ratio (A_γ term) has a very limited effect on the collective instability. In the original paper (Stern *et al.*, 2001), the flux ratio was assumed to be constant ($A_\gamma = 0$). In retrospect, it is clear that setting A_γ to zero was just a lucky shot that, fortunately, did not introduce any significant errors in the analysis of the collective instability. For other forms of secondary double-diffusive instabilities (Chapters 7 and 8) the variation in the flux ratio becomes essential.

To formulate an explicit condition for the collective instability on the basis of (6.9) we note that the marginal stability point corresponds to a purely imaginary growth rate

$$\lambda = i\omega, \tag{6.11}$$

where ω is real. Next, (6.11) is substituted in the growth rate equation (6.9). Taking the real and imaginary components of the result, we obtain $\omega^2 = a_1$ and $a_2\omega^2 = a_0$, respectively, or

$$a_2 a_1 = a_0. \tag{6.12}$$

Using a sequence of algebraic inequalities, Stern *et al.* (2001) demonstrated that the critical condition for instability (6.12) can only be achieved if

$$\frac{Nu(\bar{R}_\rho)}{Pr} \frac{\gamma^{-1} - 1}{1 - \bar{R}_\rho^{-1}} > 1. \tag{6.13}$$

Expressing (6.13) in terms of the dimensional temperature and salinity fluxes and gradients, we recover Stern's original criterion for instability:

$$A = \frac{\alpha F_{T\,\mathrm{dim}} - \beta F_{S\,\mathrm{dim}}}{\nu(\alpha \bar{T}_{z\,\mathrm{dim}} - \beta \bar{S}_{z\,\mathrm{dim}})} > 1. \tag{6.14}$$

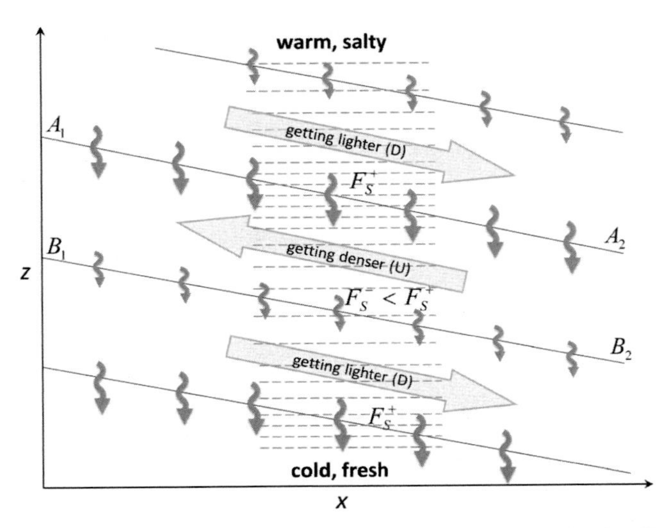

Figure 6.3 Physical mechanism of the collective instability in the Stern *et al.* (2001) parametric model. The instability is driven by the interplay between changes in stratification caused by the wave and the resulting spatial variation in the salt-finger buoyancy flux. After Stern *et al.* (2001).

6.3 Physical interpretation

The mechanism of the collective instability is physically explained in the schematic in Figure 6.3, which shows a plane internal gravity wave inclined at a small angle to the horizontal. In the upward-moving regions (U in Fig. 6.3) cold and fresh waters are advected from below across the mean isohalines and isotherms, locally decreasing temperature and salinity. Regions that are moving downward (D in Fig. 6.3) increase temperature and salinity by transporting relatively warm and salty water from above. Such a perturbation increases the vertical temperature and salinity gradients (T_z, S_z) in areas located below downward-flowing currents (D) and above the upward-moving ones (U), such as the front (A_1, A_2) in Figure 6.3. In regions immediately above the downward currents (D) – front (B_1, B_2) for instance – the gradients are reduced. The local increase (decrease) of temperature and salinity gradients results in larger (smaller) temperature and salinity fluxes. Since the salinity component of the salt-finger density flux exceeds its temperature component ($\gamma < 1$), the density flux is controlled by salinity. As a result, the density flux is distributed (Fig. 6.3) in a manner that tends to increase density in downward-flowing regions (D) and decrease it in the U-regions. Thus, the buoyancy forcing due to salt fingers has an adverse effect on the flow: forcing is downward when the velocity is upward and vice versa. Such systems are often susceptible to oscillatory "overstable" instabilities. Whether or not this mechanism leads to instability depends on the balance between the buoyancy forcing by salt fingers and viscous damping, which is quantified by the Stern number (A). If $A > 1$, viscous dissipation plays a secondary role and the wave amplifies.

It is instructive to point out a formal analogy between the mechanisms of collective instability and laminar diffusive instability (Chapter 2). The eddy diffusivity of salt due to salt fingers exceeds the eddy diffusivity of temperature, as can be ascertained by taking the ratio of the two equations in (6.2):

$$\frac{K_T}{K_S} = \frac{\gamma}{R_\rho} < 1. \tag{6.15}$$

Thus, in fully developed salt fingering, the faster diffuser is salinity and the slower diffuser is temperature. The configuration in which the stably stratified component diffuses more slowly than the unstable one leads to the oscillatory diffusive instability. This is exemplified by the laminar case of cold and fresh water above warm and salty. Thus, by appealing to the analogy between turbulent and laminar fluids, we immediately arrive at the possibility of large-scale oscillatory instability in the salt-finger case. Note that temperature and salinity switch their roles as faster and slower diffusers depending on whether we consider molecular effects at the microscale or turbulent large-scale processes. The parallel between laminar diffusive convection and turbulent salt fingering could be made more quantitative by noting that the condition for oscillatory instability in diffusive systems is determined by the Prandtl number (Chapter 2). The smaller Pr, the more unstable is the system. For the turbulent salt-finger case, the relevant Prandtl number is based on turbulent diffusivity: $Pr_{\text{turb}} = \frac{\nu}{K_T} = \frac{\nu \bar{T}_z}{F_T} \sim \frac{\nu \bar{\rho}_z}{F_\rho} = \frac{1}{A}$, which leads to the same conclusion as was rigorously derived in (6.14) – collective instability is controlled by the Stern number A.

Having examined the parametric flux-gradient theory, one might want to have a second look at the earlier attempts (Stern, 1969; Holyer, 1981, 1985) to formulate physical models of the collective instability. Focused on seemingly different driving processes, each model still managed to predict correctly the instability condition. Is there a deep reason for their success or was it largely coincidental? This question is more complicated than it appears at first. The non-dimensional parameter A is clearly an important one – it reflects the balance between the advective transfer and viscous damping; it can be thought of as the double-diffusive equivalent of the Reynolds number. Thus, it is possible that A should inevitably emerge as a critical collective instability condition in any sensible physical model, regardless of the specific assumptions. On the other hand, an in-depth analysis of the problem may reveal hidden analogies between the finger tilt (Stern, 1969), long-wave instabilities of the elevator modes (Holyer, 1981, 1985) and flux-gradient effects (Stern *et al.*, 2001), unifying different approaches to the collective instability. Borrowing one of

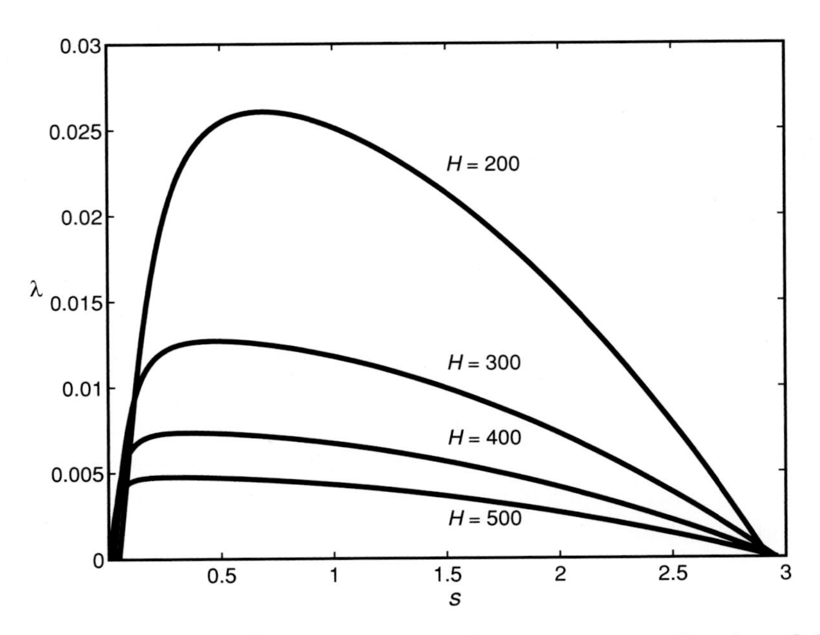

Figure 6.4 Growth rates of the collective instability waves as a function of the (non-dimensional) vertical wavelength (*H*) and the slope of the wave (*s*).

Stern's expressions, as memorable and original as his science, we can claim that each approach "has a certain degree of independent plausibility."

6.4 Specific solutions

The flux-gradient model (Stern *et al.*, 2001) not only explains the dynamics of collective instability but also affords quantitative and testable predictions for such major instability characteristics as the growth rates and dominant spatial patterns. These specific solutions, however, require knowledge of the salt-finger diffusivities of heat and salt and, in particular, of their dependence on the background density ratio. Due to computational constraints, Stern *et al.* (2001) based their analysis on diffusivities derived from two-dimensional numerical solutions. Since three-dimensional simulations for the heat–salt parameters ($\tau = 0.01$, $Pr = 7$) have recently become possible (e.g., Radko and Smith, 2012), we now reproduce several calculations in Stern *et al.* (2001) using the updated flux-gradient laws.

Simulations in Radko and Smith (2012) led to the parameterization (3.10) of the vertical temperature and salinity fluxes, which was used in (6.9) and (6.10) to evaluate the growth rate of the collective instability (λ). In Figure 6.4, λ is plotted as a function of the wave slope $\left(s = -\frac{k}{m}\right)$ for various values of the vertical wavelength

(H) and $R_\rho = 2$. Note that the system is invariant with respect to changes in the sign of the wave slope – the growth rates of the modes sloping downward in the positive x-direction (as in Fig. 6.3) are identical to those of the corresponding modes sloping upward, and therefore only positive slopes are shown in Figure 6.4. This convenient property does not apply to systems characterized by the presence of horizontal gradients of temperature and salinity (discussed in Chapter 7).

The results in Figure 6.4 indicate that the growth rates generally decrease with increasing H. The patterns of $\lambda(s)$ are similar for various values of H; they are characterized by a well-defined maximum at small but finite values of slope ($s = s_{\max}$). As H increases, this preferred maximal value of slope shifts towards lower values. The reader is reminded that the non-dimensionalization in Figure 6.4 is based on the salt-finger scale $d = \left(\frac{k_T \nu}{g\alpha|T_z|}\right)^{\frac{1}{4}} \sim 0.01$ m and $k_T \approx 1.4 \cdot 10^{-7}\,\mathrm{m^2\,s^{-1}}$. Thus, vertical wavelengths of $H = 200$–500 in Figure 6.4 correspond to the dimensional range of 2–5 m. The non-dimensional growth rates of $\lambda = 0.005$–0.025 translate to $\lambda_{\mathrm{dim}} = \frac{k_T}{d^2}\lambda = (7 - 35) \cdot 10^{-6}\,\mathrm{s^{-1}}$ with the corresponding growth period of several hours.

In Figure 6.5 we plot the (real part of) growth rates as a function of k and m in logarithmic coordinates for various density ratios. While the range of unstable wavenumbers is only weakly dependent on the density ratio, the growth rates rapidly decrease with increasing R_ρ. For $R_\rho > 4.362$, the system becomes stable with respect to the collective instability. Note also that the application of the parametric instability implies scale separation between the wave and individual fingers. Thus, the accuracy of the model becomes questionable for relatively large k and m.

6.5 Nonlinear effects

The predictions of the parametric flux-gradient theory were validated by DNS in Stern *et al.* (2001). These simulations demonstrated that salt-finger fluxes are not strongly affected by large-scale wave shear despite the significant tilt of fingers, apparent in Figure 6.6. The momentum transfer by salt fingers, not taken into account by the parametric theory, was also confirmed to be of secondary importance and the finite amplitude perturbations amplify at rates consistent with the prediction in (6.9) and (6.10). Collective instabilities generally equilibrate only after development of density overturns (Fig. 6.7). As the wave amplifies, the minimum Richardson number of the large-scale perturbation systematically decreases until the flow becomes locally susceptible to Kelvin–Helmholtz instabilities. The resulting intermittent density overturns in the regions of strong shear ($Ri < 1/4$) generate large Reynolds number convective turbulence, associated with large dissipation of heat

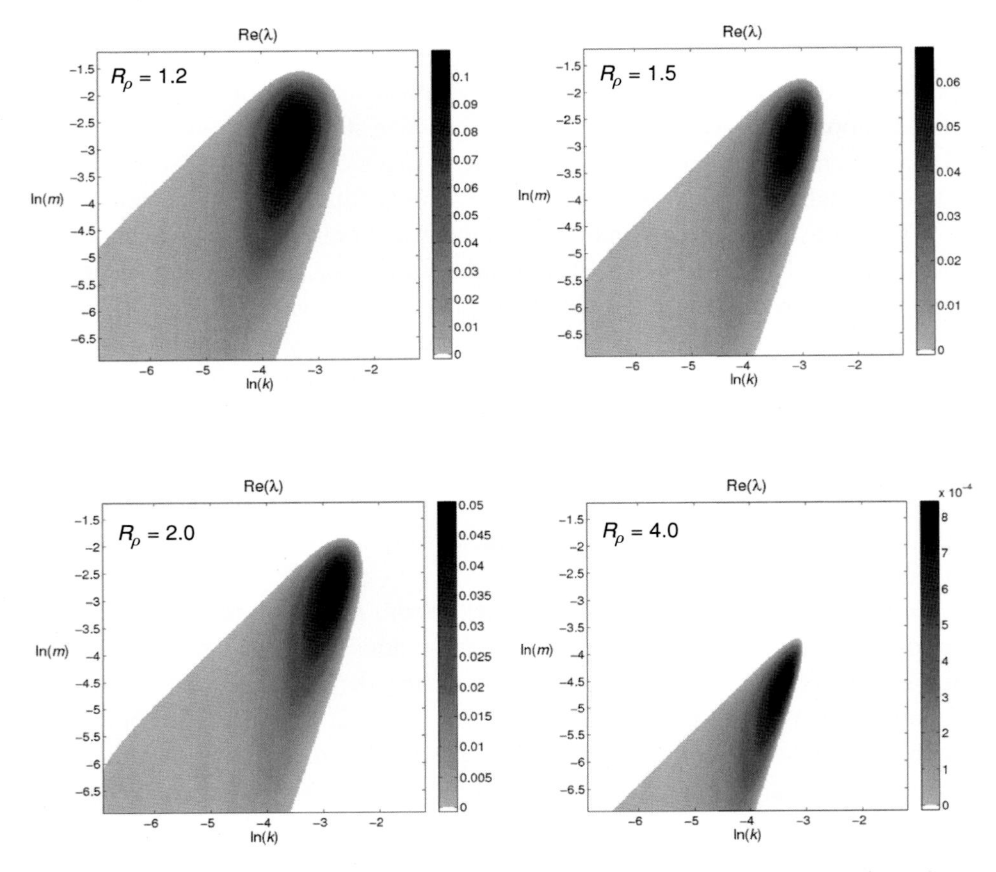

Figure 6.5 Growth rates (λ) of the collective instability waves as a function of the wavenumbers k and m for various values of the density ratio (R_ρ). Only the positive values of λ are shown. The unstable region shrinks when R_ρ is increased.

and momentum. While the connection between wave-induced overturns, turbulent patches and vertical mixing is well known (Gregg, 1987; Thorpe, 2005), such processes are usually discussed in the context of the mechanical generation of gravity waves by a combination of tidal forcing and topographic effects. The intriguing suggestion that some turbulent mixing events can be produced by the indirect action of double-diffusion (Stern *et al.*, 2001) warrants further analysis and quantification.

The nonlinear effects of collective instability have been examined more systematically by Stern and Simeonov (2002), who developed a fully nonlinear numerical parametric model of salt fingering. This model made no attempt to resolve individual salt fingers but represented their effects in terms of vertical fluxes (6.2). The parametric model was shown to be consistent with the DNS in the computationally accessible regimes. For more challenging large-scale configurations, beyond

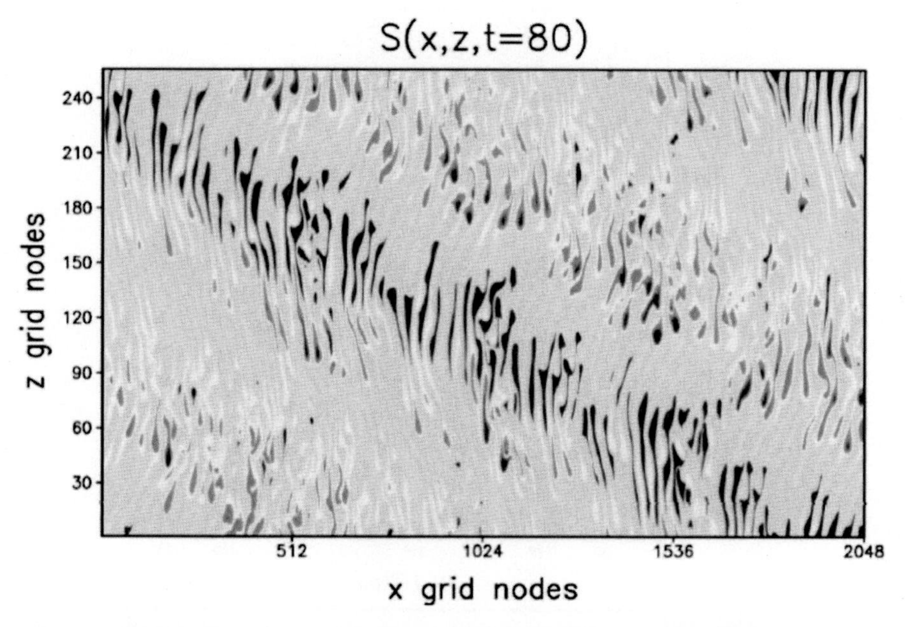

Figure 6.6 Two-dimensional direct numerical simulation showing salt fingers growing in the internal gravity wave. Presented is the departure of salinity from the uniform background gradient. Red color corresponds to high values and low values are shown in blue. From Stern *et al.* (2001). See color plates section.

the reach of DNS, the parametric model offered a unique opportunity to simulate numerically collective instability waves and the resulting overturns.

Numerical simulations in Stern and Simeonov (2002), both parametric and direct, confirmed the propensity of collective instability waves to overturn and generate localized patches of enhanced mixing. The intensity of these events was shown to be strongly dependent on the background density ratio. For large density ratios ($R_\rho > 2$), collective instability is relatively benign – it generates density overturns that are limited in size and magnitude and do not contribute significantly to the vertical transfer of properties. However, the situation changes dramatically for low density ratios ($R_\rho < 2$). In this case, the portion of the net heat flux transported by the collective instability waves is comparable to, or exceeds, the direct salt-finger flux. This result has potentially far-reaching implications for the analysis and interpretation of small-scale mixing in the ocean. So far, attempts to quantify the contribution of salt fingers to vertical mixing have been focused on the direct effects caused by the primary double-diffusive instabilities (e.g., St. Laurent and Schmitt, 1999). However, if collective instability waves are responsible for a significant fraction of net vertical transport, double-diffusive mixing would be effectively disguised as mechanically generated turbulence. To the best

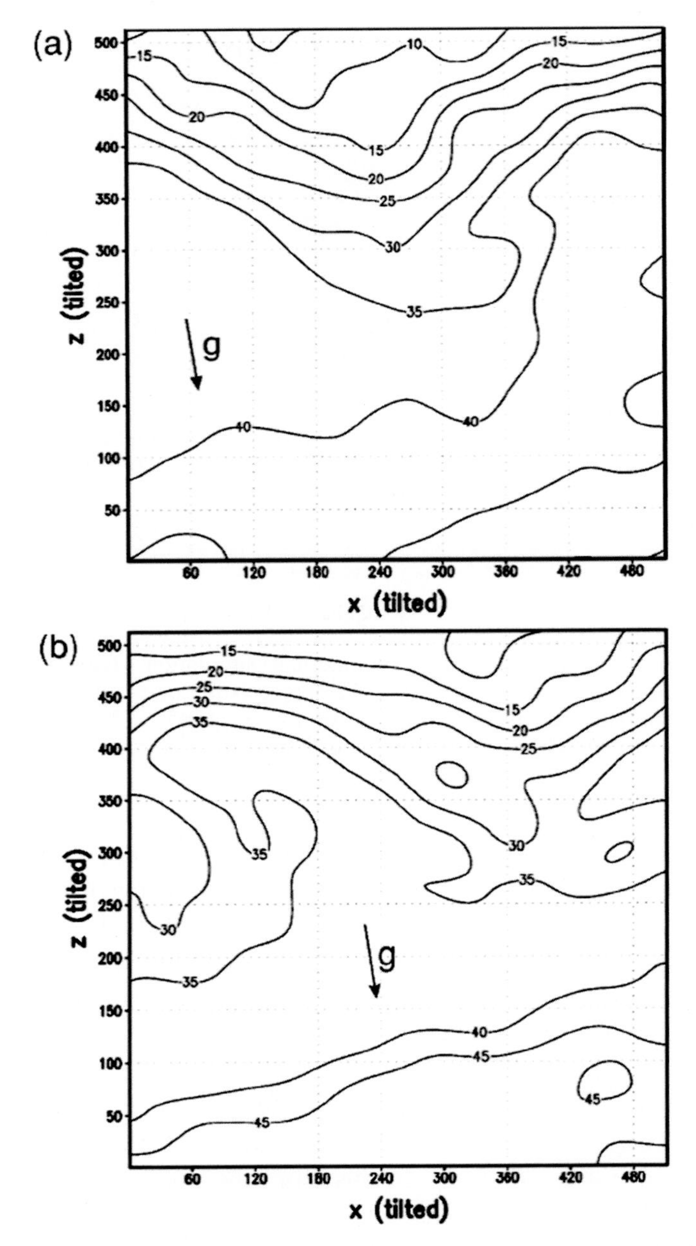

Figure 6.7 Total low-passed density distribution in the overturning collective instability wave. This direct numerical simulation was performed in the tilted coordinate system; the orientation of the gravity vector is indicated. The plots in (a) and (b) show the state shortly before and after the overturn, respectively. From Stern *et al.* (2001).

of our knowledge, this possibility has not yet been considered in the analysis of field data.

An interesting property of collective instability was identified by Radko and Stern (2011),[1] who suggested that their impact can be significantly enhanced by external large-scale shears, which are ubiquitous in the ocean. This somewhat counterintuitive conclusion was based on the stability analysis of a parallel sinusoidal shear flow in finger-favorable stratification. Salt fingers were introduced by parameterizing vertical fluxes (6.2), similar to the approach taken by Stern and Simeonov (2002). Linear stability analysis of this system revealed the existence of relatively large-scale instabilities driven by the interaction between the background shear and double-diffusion. These oscillatory thermohaline–shear modes can be interpreted as a form of the classical collective instability modes, modulated by the basic current. However, they appear to be much more effective in mixing, owing to their ability to interact with the background flow.

The importance of collective thermohaline–shear instabilities was confirmed by numerical three-dimensional simulations of the parametric model. Figure 6.8 presents the evolution of the temperature perturbation in an experiment initiated by dynamically stable shear with minimum Richardson number of $Ri = 0.5$. The background density ratio is $R_\rho = 2$, which is representative of the Atlantic thermocline. In dimensional units, the size of the computational domain in Figure 6.8 is $\sim 20\,\text{m} \times 10\,\text{m} \times 5\,\text{m}$ and the duration is equivalent to a period of a few days. Calculations of this scale are currently beyond DNS capabilities and thus are only accessible by parameterized models. This experiment confirmed that the inclusion of double-diffusive fluxes immediately destabilizes the system. Consistent with linear inferences (Radko and Stern, 2011), the instability modes that appeared first (Fig. 6.8a) strongly varied in y and z but were almost uniform in the direction of the background current (x). This pattern changed dramatically in the later stages of the experiment; the development of significant along-current temperature variation is shown in Figure 6.8b. Note that the along-current variability at this stage is particularly strong at the locations of the largest background shear (bottom, center and top of each plot in Fig. 6.8).

The along-current perturbations in Figure 6.8b were interpreted as secondary Kelvin–Helmholtz instabilities induced by the thermohaline modes in the following manner. Growing collective instability modes introduced, upon reaching finite amplitude, significant spatial variation in the vertical density gradient. The reduction of density gradient in some regions, in turn, made the flow susceptible to local

[1] This was the final scientific contribution of Melvin Stern; the paper was completed after he passed away. Despite serious health problems, Melvin maintained keen interest in the dynamics of collective instabilities and was actively involved in this project until his last days.

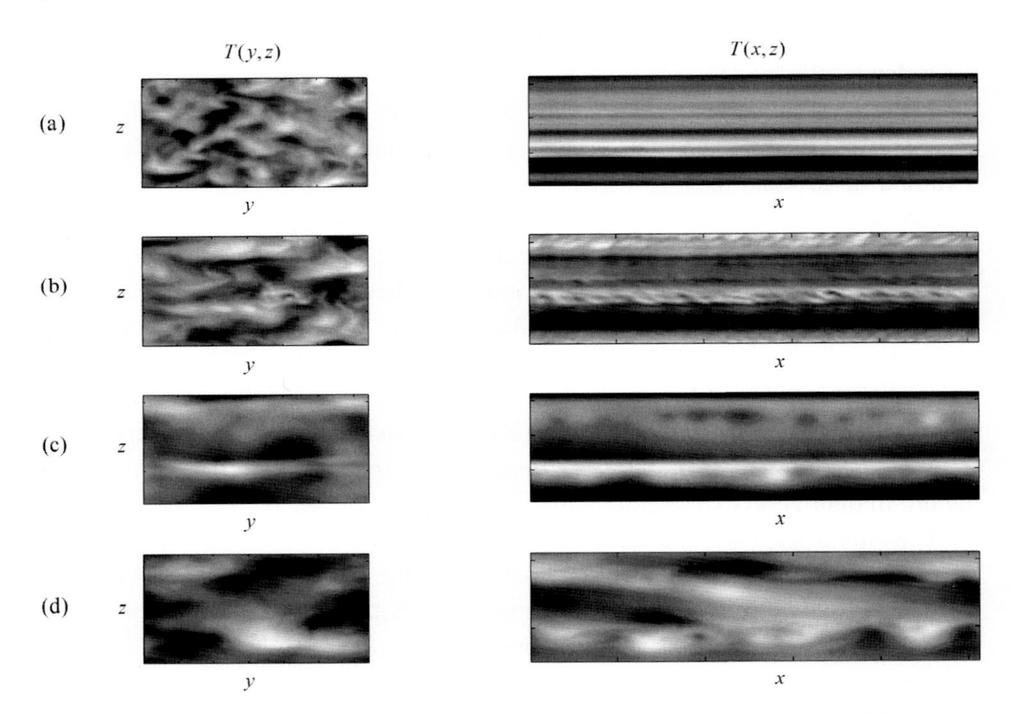

Figure 6.8 Numerical simulation of the subcritical ($Ri = 0.5$) double-diffusive shear flow. The (left) cross-current and (right) along-current vertical sections are shown at various times (increasing downward). Note rapid destabilization of the shear flow by double-diffusion. From Radko and Stern (2011).

Kelvin–Helmholtz instabilities, particularly in the zones of high background shear, resulting in intense convective mixing. However, the localization of secondary instabilities to the regions of strong shear was transient. The ensuing vertical mixing stabilized the high-shear zones and the along-current variability shifted to low-shear regions (Fig. 6.8c). In the final stage of the experiment (Fig. 6.8d) the temperature variance was distributed rather uniformly throughout the computational domain. Also apparent is the tendency for a systematic increase of typical spatial scales of the perturbation in time, associated with the sequential coalescence of relatively small eddies into larger and larger structures. Such mergers are common in turbulent stratified fluids (Balmforth *et al.*, 1998; Radko, 2007) particularly in the double-diffusive environment (Huppert, 1971; Radko, 2005; Simeonov and Stern, 2008). It should be emphasized that all the interesting turbulent dynamics illustrated in Figure 6.8 are ultimately driven by collective instabilities – in the absence of double-diffusion, the background flow would be stable.

In order to quantify the role of stable shear, the calculation in Figure 6.8 has been reproduced without the background current. The outcome was very different. The removal of the background shear decreased the probability of density overturns by an order of magnitude and substantially reduced the average mixing rates. Thus, both background current and collective instabilities appear to be essential and mutually reinforcing in terms of transition to turbulence for dynamically stable configurations. Also revealing is the comparison of the $Ri = 0.5$ experiment in Figure 6.8 with its dynamically unstable counterpart ($Ri = 0.15$). The average heat flux by the resolved scales of motion (thus excluding the direct double-diffusive transport) in the experiment with $Ri = 0.5$ was comparable to but smaller, by a factor of three, than in the dynamically unstable ($Ri = 0.15$) calculation. This similarity suggests that, with respect to the generation of density overturns and consequent mixing, in the double-diffusive thermocline there is no dramatic difference between the low Ri and high Ri regimes. This proposition can help to rationalize some observations of turbulent mixing in the main thermocline, an environment generally characterized by relatively weak dynamically stable ($Ri > 1/4$) shears (e.g., Polzin, 1996).

In summary, collective instability provides an efficient mechanism for the generation and maintenance of an active wave environment and thus can indirectly influence vertical turbulent mixing. In addition to the spontaneous generation of internal waves from random perturbations, collective instability is likely to amplify a broad band of waves generated by other causes (e.g., topographic and tidal effects), thereby facilitating density overturns and transition to turbulence. The identification of such processes in nature is challenging and more work needs to be done to quantify the large-scale impact of collective instability. Another potentially critical mechanism for irreversible mixing in the ocean is related to the tendency of double-diffusion to drive quasi-lateral intrusions in the presence of horizontal temperature and salinity gradients. The dynamics of such interleaving motions are discussed next.

7
Thermohaline intrusions

As is the case for the majority of key double-diffusive effects, the modern theory of thermohaline intrusions was launched by Melvin Stern. Using stability arguments, Stern (1967) predicted a tendency for the spontaneous generation of intrusive quasi-horizontal structures, spreading across lateral temperature and salinity fronts in a double-diffusive fluid. Stern's theory was guided by observations (Stommel and Fedorov, 1967) of "hundreds of superimposed laminae from 2–30 meters thick," coherent over large horizontal distances and readily identifiable in the vertical temperature and salinity casts in the main thermocline of the Timor Sea (Fig. 7.1). While Stommel and Fedorov (1967) could not precisely identify their origin, they hypothesized that intrusions might be "direct evidence of the more exotic types of two-diffusivity convection." One cannot help being equally impressed by the shrewd comment of Stommel and Fedorov, which seems to have been based entirely on their physical intuition, and by Stern's ability to develop a consistent intrusion model from such a subtle observational hint.

In the years following Stern's discovery, thermohaline intrusions continued to attract steady interest from oceanographers and applied mathematicians. The motivation to understand and predict intrusion properties comes from the necessity to quantify their role in the lateral mixing of oceanic water masses. A fundamental problem of physical oceanography is to explain how the differential thermodynamic forcing of the sea-surface – heating in the tropical regions and high-latitude cooling – is balanced by advective and mixing processes in the ocean interior (Wunsch and Ferrari, 2004; Song *et al.*, 2011). Ultimately, the temperature and salinity variances are dissipated by molecular diffusion acting on the microscale, with larger-scale processes playing a catalytic role in the chain of mixing events (e.g., Merryfield, 2005). While the dynamic connection between basin-scale forcing and microscale mixing has not been fully explained, plausible scenarios have been proposed and considered. Garrett (1982) suggested that an essential component of the global *T–S* cascade to small scales involves thermohaline interleaving. This

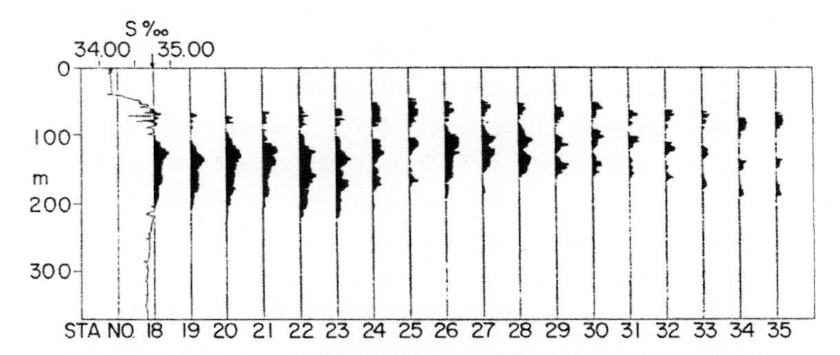

Figure 7.1 One of the first observations of thermohaline intrusions. Salinity profiles taken off Mindanao reveal multiple inversions laterally correlated between stations, taken as evidence of active interleaving of the adjacent water masses. From Stommel and Fedorov (1967).

sentiment was echoed by a number of oceanographers (Schmitt, 1994b; Ruddick and Kerr, 2003; Song *et al.*, 2011) but is in dire need of corroboration and quantification. A major predicament in this regard is caused by our limited understanding of processes controlling the intensity of intrusive mixing. Even such fundamental mixing characteristics as typical lateral eddy diffusivities due to intrusions are poorly known, with estimates reported in the literature varying by several orders of magnitude. Despite the general acceptance of its importance, no attempt has been made to incorporate interleaving into General Circulation Models, yet another sign of the challenges and uncertainties facing the community. Double-diffusion is not for weaklings; it requires the patience of a saint and the intensity of a Rottweiler. On the positive side, a host of unresolved interleaving problems opens numerous opportunities for significant advancement of this field – the opportunities that readers of our monograph are encouraged to pursue.

For pedagogical reasons, in the next section (Section 7.1) we present the linear theory of thermohaline interleaving in its most basic form: the model excludes the effects of planetary rotation, baroclinicity and ambient forcing. The most common generalizations of the stability analysis are considered in Section 7.2. While linear theory by itself cannot quantify the magnitude and transport characteristics of intrusions, it explains their spatial patterns and offers valuable physical insights. Discussion of the fundamentally nonlinear properties of intrusions follows shortly after (Section 7.3).

7.1 Linear theory

While intrusions can be produced by several mixing mechanisms (salt fingering, diffusive convection, differential diffusion, molecular diffusion or a combination thereof) we focus on finger-driven interleaving. Our starting point is the Boussinesq equations of motion, applied to the large-scale, relative to fingers, flow components. The basic state is assumed to be at rest ($\bar{\vec{v}} = 0$), whereas the temperature and salinity fields are separated into the linearly stratified basic state $(\bar{T}, \bar{S})$ and a departure (T', S') from it. In the intrusion problem, it becomes essential to include small but finite horizontal gradients, which for simplicity are assumed to be uniform and density compensated. That is, the basic density field is horizontally homogeneous. Without loss of generality, the x and y axes are oriented in the cross-front and along-front directions respectively ($\alpha\bar{T}_x = \beta\bar{S}_x > 0, \alpha\bar{T}_y = \beta\bar{S}_y = 0$). To reduce the number of governing parameters, we employ the standard double-diffusive system of non-dimensionalization (1.11) and linearize the result:

$$\begin{cases} \frac{\partial T'}{\partial t} + Gu' + w' = -\frac{\partial F_T'}{\partial z}, \\ \frac{\partial S'}{\partial t} + Gu' + \frac{w'}{R_\rho} = -\frac{\partial F_S'}{\partial z}, \\ \frac{1}{Pr}\frac{\partial}{\partial t}\vec{v}' = -\nabla p' + (T' - S')\vec{k} + \nabla^2\vec{v}', \\ \nabla \cdot \vec{v}' = 0, \end{cases} \tag{7.1}$$

where (F_T', F_S') are the perturbations of temperature and salinity fluxes due to salt fingers and $G = \bar{T}_x$ is the non-dimensional horizontal temperature gradient, representing the slope of isotherms in the basic state. Following the mainstream approach (Stern, 1967; Toole and Georgi, 1981; Walsh and Ruddick, 1995, 2000; Smyth and Ruddick, 2010) the vertical T–S transport is parameterized as a function of local gradients as in (6.2). At this point, no attempt is made to take into account eddy momentum transport since the salt-finger Reynolds stress is typically much less than molecular friction (e.g., Stern *et al.*, 2001; Krishnamurti, 2006). Note that the opposite is true for the transport of heat and salt, which is dominated by the eddy fluxes; weak molecular dissipation can be either neglected or incorporated into the eddy fluxes (F_T, F_S).

As in the case of collective instability (Chapter 6), the parametric intrusion model requires users to assume specific flux laws in (7.1). The first models of this nature (Stern, 1967; Toole and Georgi, 1981) used the eddy diffusivity of salt (K_S), which is independent of density ratio (R_ρ). The realization that double-diffusive transport is significantly R_ρ-dependent has led to improved versions (Walsh and Ruddick, 1995, 2000; Merryfield, 2000; Smyth and Ruddick, 2010) that assume various ad hoc expressions for (F_T, F_S). The following examples are based on the most recent parameterization (3.10) derived from a suite of high-resolution direct numerical simulations (Radko and Smith, 2012). As previously (cf. Chapter 6), we analyze stability using normal modes (2.1), substitution of which in (7.1) yields the cubic growth rate equation analogous to (6.9), except that its coefficients are now also affected by lateral T–S gradients (G).

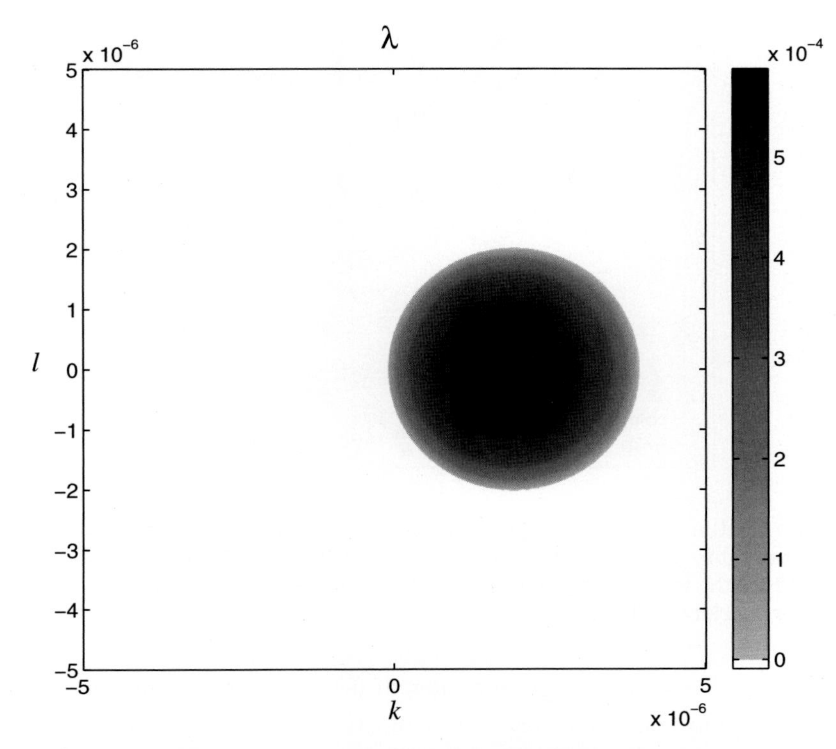

Figure 7.2 The growth rate (λ) of the salt-finger-driven intrusions as a function of (non-dimensional) horizontal wavenumbers k and l. Only the positive values of λ are shown. The largest growth rates are obtained for intrusions oriented across the front and tilted in the same sense as isotherms and isohalines.

In order to discriminate monotonically growing intrusive solutions from the oscillatory collective instability modes, we consider only real solutions of the growth rate equation. In Figure 7.2, the growth rate is plotted as a function of the horizontal wavenumbers (k, l) for a fixed value of the vertical wavenumber $m = 2\pi H^{-1}$; $H = 10^3$ (corresponding to $H_{\mathrm{dim}} \sim 10\,\mathrm{m}$) and the horizontal gradient $G = 0.001$. Two features of the growth rate pattern in Figure 7.2 are most noteworthy. First, the maximal growth rates are achieved for modes oriented across the front ($l = 0$), which is a generic property of intrusions in a non-rotating model. This property considerably simplifies the stability analysis, justifying the focus of most intrusion models on two-dimensional (x, z) dynamics. Another key observation is that positive growth rates are realized mostly for $k > 0$. This means that intrusions are spreading in a manner indicated in the schematic diagram in Figure 7.3 – warm and salty waters rise across the density surfaces, whereas the cold and fresh ones sink.

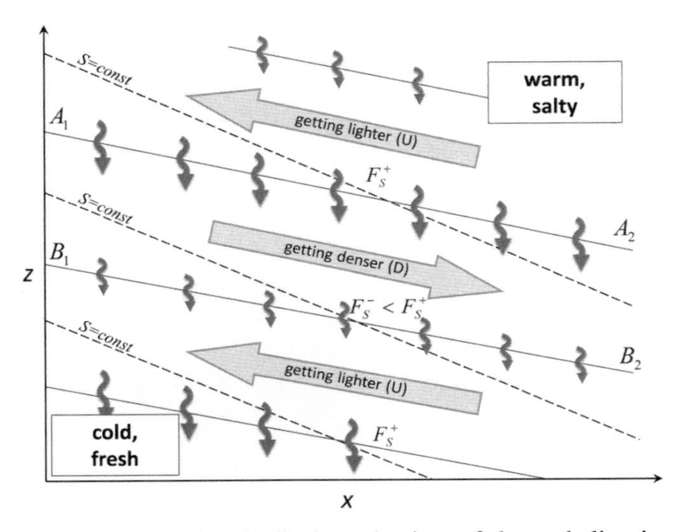

Figure 7.3 Illustration of the physical mechanism of thermohaline interleaving. The instability is driven by the positive feedback between changes in stratification caused by the interleaving and the resulting salt-finger buoyancy forcing, which further accelerates the intrusions.

In Figure 7.4, we plot the growth rates of the cross-front intrusions (recall that those are the most unstable ones) as a function of the intrusion slope $s = -\frac{k}{m}$ for various values of intrusion height. For typical oceanic conditions ($H_{\text{dim}} \sim 10\,\text{m}$, $G \sim 0.001$, $R_\rho \approx 2$) growth periods of several days are expected on the basis of linear theory. Generally, intrusions tend to grow slower than collective instability waves. This, however, does not imply that intrusions are less important. What makes intrusions ultimately more effective in modifying the background field is the persistence of their spatial pattern. Unlike collective instability waves, which periodically reverse the direction of the flow and thus partially negate the modification of the environment during an earlier phase, intrusions operate in the direct mode. They consistently transfer warm and salty fluid laterally into colder and fresher regions and vice versa, which results in strong down-gradient lateral mixing. In the vertical direction, the temperature and salinity fluxes on the intrusion scale are up-gradient (see Fig. 7.3), thereby opposing the microscale fluxes by salt fingers.

Physical interpretation

The monotonic growth of intrusions and their spatial orientation can be rationalized as follows. Consider a configuration (Fig. 7.3) in which fronts of the intrusive currents are tilted in the same sense as the temperature and salinity contours, sloping

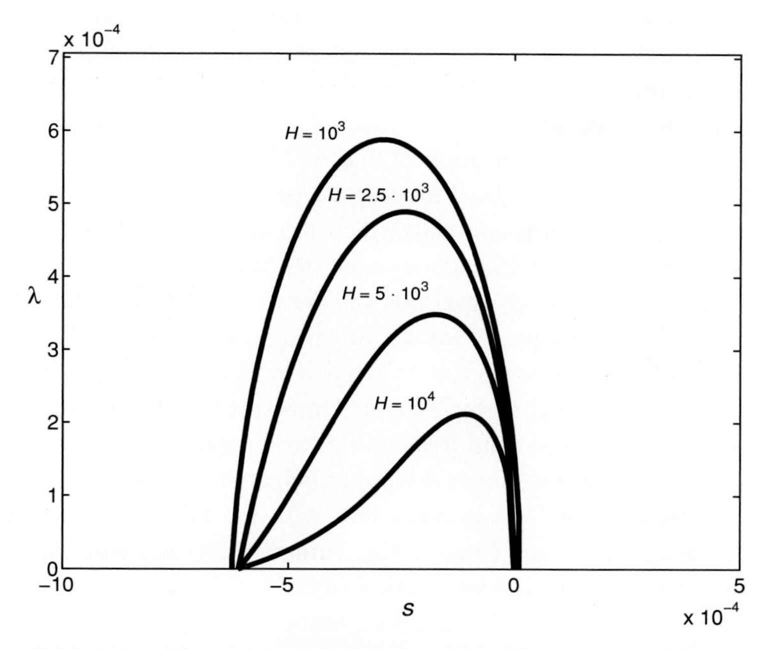

Figure 7.4 The non-dimensional growth rates (λ) of the cross-front intrusions as a function of slope for various values of intrusion thickness (*H*).

downward in the positive *x*-direction. However, the intrusions are tilted less than isotherms and isohalines. Thus, the upward-flowing intrusions (U in Fig. 7.3) advect warm and salty water across the isohalines and isotherms, resulting in a local increase of temperature and salinity. The downward intrusions (D in Fig. 7.3) reduce temperature and salinity by advecting relatively cold and fresh water. This perturbation pattern implies that the vertical temperature and salinity gradients (T_z, S_z) are enhanced in the regions below (above) the upward (downward) flowing intrusions – such as the (A_1, A_2) front shown in Figure 7.3. On the (B_1, B_2) front, located above the intrusion U and below the intrusion D, vertical temperature and salinity gradients are reduced. Higher (lower) *T–S* gradients, in turn, imply higher (lower) vertical salt-finger fluxes. Thus, the vertical *T–S* fluxes in the downward-moving intrusions converge, adding more heat and salt to the fluid. In the upward intrusions, fluid becomes fresher and colder due to salt fingers. What about density? The haline component of density flux in the salt finger environment exceeds its temperature component $(\gamma < 1)$ and therefore the pattern of density perturbation is controlled by the convergence of the salinity flux. As a result, the density of the fluid in the downward intrusions (D) continuously increases, reinforcing their initial motion. The upward intrusions (U) become lighter due to the salt-finger fluxes, and further accelerate their ascent. This positive feedback mechanism explains the monotonic growth of intrusions predicted by formal stability analysis. The negative sign of the intrusion slope is essential – an attempt to reproduce the foregoing argument for perturbations tilted with positive slope suggests suppression of such modes by the salt-finger fluxes.

Note that the physical description in Figure 7.3 pertains to finger-driven intrusions. If stratification and vertical mixing are predominantly diffusive (cold and fresh water on top of warm and salty) then the eddy density flux is dominated by its thermal, rather than haline, component: thermal stratification is the ultimate source of energy for diffusive convection and therefore $\gamma^* < 1$. Hence, the salt flux convergence argument proposed for the finger-driven intrusions (Fig. 7.3) has to be reversed. Diffusive intrusions tilted as shown in Figure 7.3 would be damped by the convergence of thermal fluxes, whereas intrusions with positive slope (warm and salty water sinking, cold and fresh rising) would grow. The spatial orientation of intrusions, diffusive and finger-driven, anticipated on the basis of qualitative physical arguments (Fig. 7.3) is consistent with the laboratory experiments on interleaving (e.g., Turner and Chen, 1974; Turner, 1978) and with the majority of oceanographic field measurements (reviewed in Section 7.6).

Two instructive comparisons should be made at this point. First, we highlight the dynamic differences and similarities between finger-driven and laminar (see Chapter 2) intrusions. While their physical mechanisms are analogous, the key difference is that for laminar fluids, temperature is the faster diffuser and salinity is the slower one. In this case, the vertical density flux is controlled by the diffusion of heat and perturbations oriented as in Figure 7.3 would be damped. Convergence of heat flux in the downward-moving regions (D) would make fluid lighter – and in the upward-moving regions denser – opposing the initial tendency and thus precluding the direct instability modes. Therefore, the preferred orientation of laminar intrusive modes is opposite to that in Figure 7.3: warm and salty intrusions sink, not rise, across density surfaces.

A second comparison can be made between intrusions (Fig. 7.3) and collective instability waves (Fig. 6.3). Both instabilities are driven by feedbacks between changes in the stratification and in the salt-finger fluxes. What makes intrusive dynamics special is the geometry of the system. In both models (Fig. 6.3 and Fig. 7.3) temperature and salinity increase upward at a given (x, y) location. However, the presence of horizontal temperature and salinity gradients, no matter how weak, makes it possible for a fluid parcel to rise and still end up in a fresher environment, which is an essential element of intrusion mechanics (Fig. 7.3). For this to occur, the parcel has to be displaced at very small angles to the horizontal – less than the slopes of isohalines. Using terminology introduced in studies of the baroclinic instability (Eady, 1949), we can claim that a necessary condition for interleaving is the presence of the "instability wedge" between the isohalines and horizontal surfaces. Displacements of parcels within this instability wedge can be reinforced (Fig. 7.3); displacements outside of it are countered by the salt-finger fluxes. Thus, spontaneous interleaving cannot occur in horizontally homogeneous systems, where the instability wedge is absent. In the horizontally stratified fluid, on the other hand, both collective and intrusive modes can be present simultaneously. The perturbations oriented within the instability wedge engage the direct intrusive dynamics. Those outside of the instability wedge experience adverse forcing and therefore are either stable or overstable; the latter produce the familiar (Chapter 6) collective instability waves.

The γ-instability

An important aspect of the parametric intrusion model concerns its sensitivity to the assumed flux-gradient laws. Despite considerable progress in representing double-diffusive transport in smooth-gradient configurations (Chapter 3), the generally accepted and observationally validated flux-gradient laws are still lacking. Does the choice of parameterization really matter in terms of predicting the general properties of intrusions? In some cases, it does. Development of a flux-gradient model for double-diffusive fluids necessarily involves parameterizations of both heat and salt diffusivities. Equivalently, one can parameterize diffusivity of one component – say, salt diffusivity $K_S(R_\rho)$ – and the flux ratio $\gamma(R_\rho)$. The model solutions are sensitive to changes in both $K_S(R_\rho)$ and $\gamma(R_\rho)$. For instance, increasing the variation in diffusivity with density ratio tends to intensify interleaving. This effect can be readily rationalized by modifying the physical interpretation of intrusive instability in Figure 7.3. Intrusions modulate not only *T–S* gradients but the density ratio as well. When K_S decreases with R_ρ, the density ratio effect strengthens the positive feedback of the salt-finger fluxes on the intrusion. Taking into account variations in diffusivity can increase the intrusion growth rates by as much as a factor of four relative to the uniform K_S model (Walsh and Ruddick, 1995).

However, it is the flux ratio model that can affect some of the most fundamental intrusion characteristics (Walsh and Ruddick, 2000). The sensitivity of intrusions to the choice of $\gamma(R_\rho)$ relation is illustrated in Figure 7.5, where we plot the growth rate of the cross-front ($l = 0$) intrusions as a function of k and m. For the calculation in Figure 7.5a, we use the parameterization (3.10). A seemingly subtle modification is made in Figure 7.5b: we retain the parameterization of $K_S(R_\rho)$ but the flux ratio is set to a constant $\gamma_0 = \gamma(\bar{R}_\rho)$, where $\bar{R}_\rho$ is the density ratio of the basic state. At first, the result of this modification appears to be most dramatic. In Figure 7.5a, the growth rates are significantly larger than in Figure 7.5b, increasing without bound with m, and the unstable modes occupy a larger fraction of the wavenumber space. Such an extreme sensitivity of the parametric model to the

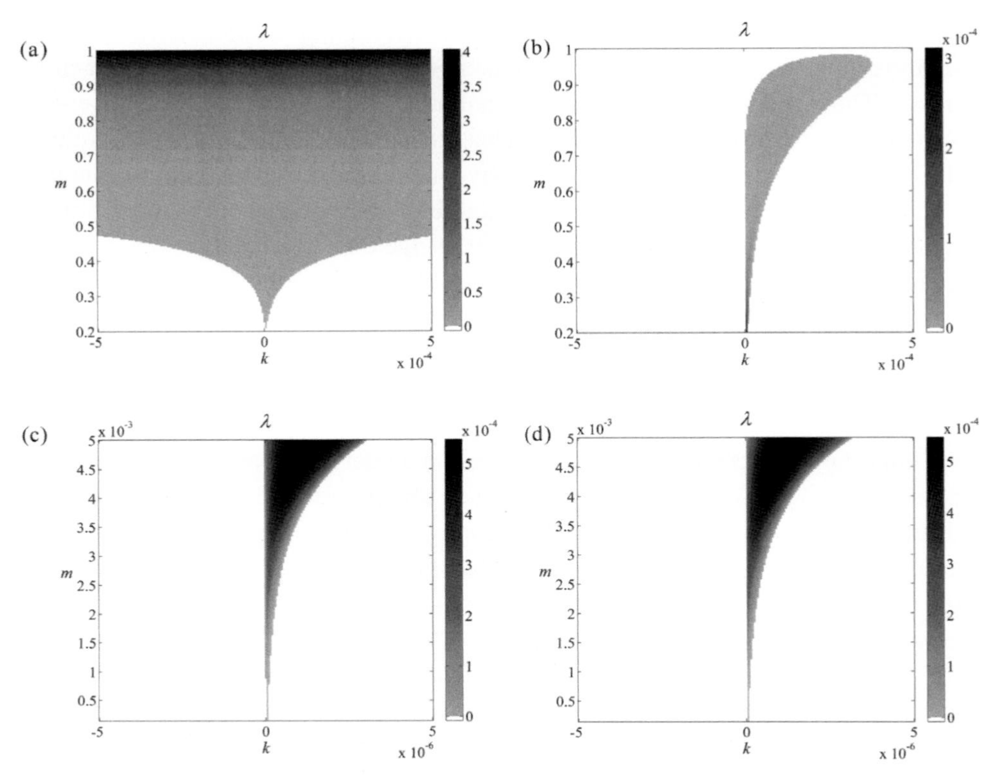

Figure 7.5 Significance of the variation in the salt-finger flux ratio for thermohaline interleaving. The growth rate is plotted as a function of the horizontal (cross-frontal) and vertical wavenumbers. The only difference between the calculations in (a) and (b) is that the former takes into account variation in flux ratio and the latter ignores it. However, the two calculations become similar when relatively small scales ($m > 5 \cdot 10^{-3}$) are excluded from consideration in (c) and (d).

assumed pattern of $\gamma(R_\rho)$ is a cause for concern, particularly since this dependence is not well constrained by lab experiments and field observations. The situation, however, does not look as bleak if we consider only the range of the parameter space occupied by relatively large-scale modes (Fig. 7.5c,d) with $m < 5 \cdot 10^{-3}$, which corresponds to a dimensional intrusion thickness of $H_{\text{dim}} \geq 10\,\text{m}$. On such scales, both models, the R_ρ-dependent one in Figure 7.5c and the uniform γ model in Figure 7.5d, are mutually consistent.

Thus, taking into account the decreasing dependence of γ on R_ρ, which is expected for relatively low density ratios, reveals new instability modes. These modes are characterized by large growth rates and relatively small spatial scales and will be referred to as the γ-instability modes. Note that a majority of theoretical intrusion models (Stern, 1967; Toole and Georgi, 1981; Merryfield, 2000; among others) assume constant flux ratio, thereby a-priori excluding the γ-instability from consideration. As a result, little is known about its dynamics and large-scale consequences, although preliminary evidence suggests that the γ-instability could have a substantial impact on the double-diffusive environment. Unlike conventional intrusive modes, the γ-instability is not contingent on the presence of lateral stratification and therefore it can potentially affect broader regions of the world ocean. The tendency of the γ-instability to operate on relatively small vertical scales complicates its representation in models and the assessment of its impact. Parametric models necessarily assume a substantial scale separation between individual salt fingers and phenomena that those models strive to represent. For relatively large-scale intrusions ($H_{\mathrm{dim}} \sim$ 10–100 m in the oceanographic context) this assumption is undoubtedly valid. Whether or not the smaller-scale γ-instability modes can be accurately described by parametric models depends on a particular configuration and therefore has to be answered on a case-by-case basis.

Preferred scales

The parametric model makes it possible to predict several key intrusion characteristics – the growth rate, slope, *T–S* amplitude ratio, and thickness – which are routinely measured in laboratory and field experiments. This opens an attractive opportunity to test and validate the model. As is conventional in instability theories, such predictions are based on the analysis of the most unstable modes. The underlying philosophy of such an approach ascribes a critical selective advantage to the modes with maximal growth rates. In the absence of strong biases in the initial amplitude, the rapidly growing perturbations are expected to dominate their slower growing counterparts during the initial phase of linear growth and thus control the flow pattern of the fully developed state as well. The application of this principle to the intrusion problem is not straightforward. Formally, the largest growth rates are attained by the small-scale γ-instability modes. The relation of these modes to the observed intrusions, which typically have much larger scales, is questionable. In order to exclude from consideration the γ-instability modes, we temporarily adopt the uniform flux ratio model as used for the calculation in Figure 7.5b,d.

The order-of-magnitude estimates of basic intrusion characteristics (thickness, growth rate and slope) can be obtained using scaling analysis of the parameterized system without detailed knowledge of the double-diffusive flux-gradient laws. For instance, Toole and Georgi (1981) suggest an explicit scaling relation for the (dimensional) intrusion thickness:

$$H_{\mathrm{TG}} \sim \left(\frac{\sqrt{\nu K_T}}{N} \frac{\bar{S}_z}{\bar{S}_x} \right)^{\frac{1}{2}}, \tag{7.2}$$

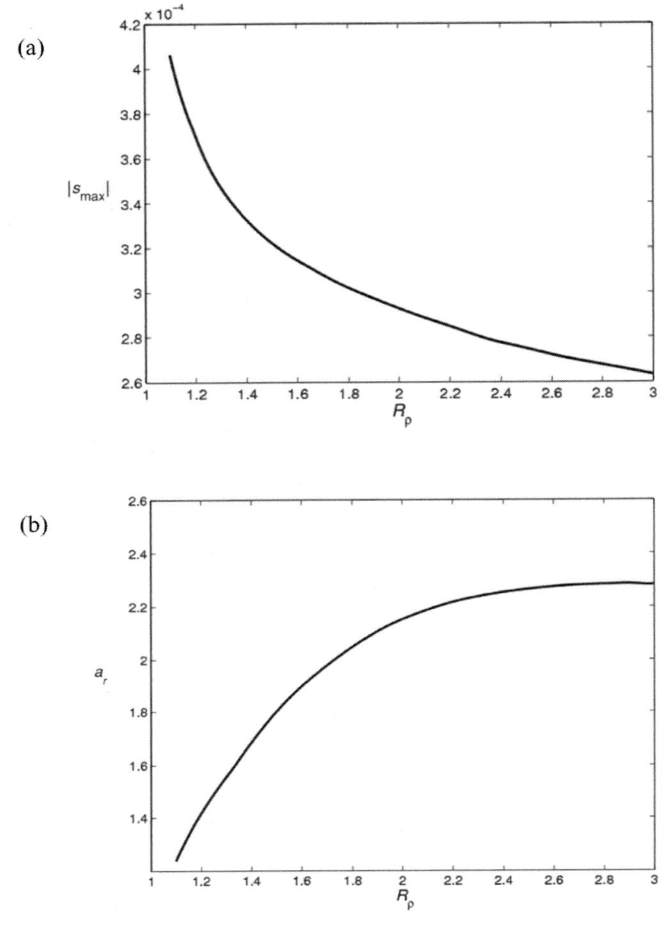

Figure 7.6 The slope (a) and the temperature/salinity amplitude ratio (b) of the most rapidly growing intrusion as a function of density ratio (R_ρ).

where $N = \sqrt{-\frac{g}{\rho}\frac{\partial \rho}{\partial z}}$ is the buoyancy frequency. The corresponding intrusion growth rate scales as $\lambda_{\text{TG}} \sim K_T/H_{\text{TG}}^2$. The molecular viscosity ν in (7.2) should be replaced by the eddy viscosity (K_M) for situations in which the latter dominates.

To be more quantitative, it becomes necessary to assume specific parameterizations of the vertical double-diffusive transport. For consistency, we shall continue to use the flux law (3.10). The experiments in Figures 7.6 and 7.7 examine the dependencies of the linear intrusion characteristics – slope, *T–S* amplitude ratio, growth rate and thickness – on the density ratio for a fixed value of the horizontal gradient ($G = 0.001$). The sensitivity to R_ρ is limited. The magnitude of the preferred slope (Fig. 7.6a) weakly decreases with density ratio, while the orientation

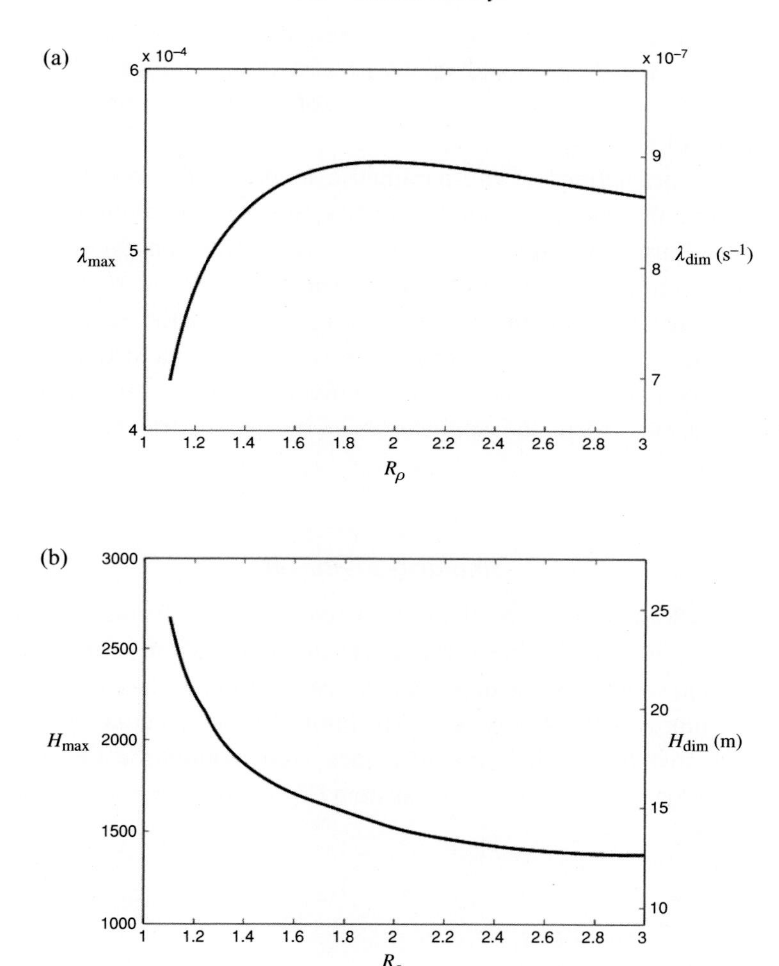

Figure 7.7 The maximal growth rate (a) and the corresponding thickness (b) of intrusions. The non-dimensional variables are indicated on the left axis. The corresponding dimensional values are for $\bar{T}_z = 0.01$ °C m^{-1}.

remains as in Figure 7.3: warm and salty intrusions move upward whereas the cold and fresh ones sink. Another observable intrusion characteristic is the amplitude ratio $a_r = \frac{\hat{T}}{\hat{S}}$, which measures the amplitude of the thermal component of density perturbation in the intrusion relative to its haline component (Fig. 7.6b). It monotonically increases with density ratio from $a_r = 1.2$ at $R_\rho = 1.1$ to $a_r = 2.2$ at $R_\rho = 2.3$. The growth rate (Fig. 7.7a) is non-monotonic: for R_ρ increasing from unity, λ_{max} first increases (somewhat counterintuitively) and then starts to gradually decrease for $R_\rho > 2$. The preferred vertical scale (Fig. 7.7b) also decreases with R_ρ. The non-dimensional wavelengths and growth rates have been converted into dimensional units using the standard double-diffusive non-dimensionalization system in (1.11). The dimensional values are also shown in Figure 7.7 (right axes) for $\bar{T}_z = 0.01\,^{\circ}\mathrm{C\,m^{-1}}$, which is representative of thermal stratification in the mid-latitude thermocline.

Overall, the predictions of the parametric model in Figures 7.6 and 7.7 are consistent with the nominal observationally inferred intrusion characteristics. The preferred intrusions slopes ($s \sim 10^{-3}$), vertical wavelengths ($H_{\max} \sim 10$ m) and amplitude ratio ($a_r \sim 2$) conform to expectations, at least in the order-of-magnitude sense. The agreement is encouraging, given the minimal nature of the model. More quantitative comparisons require the inclusion of additional ingredients, such as baroclinicity and ambient turbulence (discussed in Section 7.2). And even then, the application of the parametric model is far from straightforward.

Similarity argument

Interleaving is known to be particularly active in strong fronts, which prompts the question of how their linear characteristics depend on lateral gradients of temperature and salinity. Some dependencies can be inferred based on the structure of the governing equations, without performing detailed calculations. Merryfield (2000) noted that for small intrusion slopes – an assumption that is definitely satisfied in the ocean – the governing system (7.1) is invariant with respect to the transformation

$$\begin{aligned}
&t \to C^{-1}t,\ (x, y) \to C^{-\frac{3}{2}}(x, y),\ z \to C^{-\frac{1}{2}}z,\\
&G \to CG,\\
&(T', S') \to C^{-\frac{1}{2}}(T', S'),\\
&(u', v') \to C^{-\frac{1}{2}}(u', v'),\ w' \to C^{-\frac{3}{2}}w'.
\end{aligned} \tag{7.3}$$

So, for instance, if the horizontal temperature and salinity gradients were increased by a factor C, while the vertical gradients were kept constant, then the time scale of intrusion growth would decrease by the exactly the same factor. Furthermore, invariance (7.3) implies that the maximal growth rate of intrusions ($\lambda_{\max}$), the corresponding vertical wavenumber ($m_{\max}$) and slope ($s_{\max}$) are related to the slope of isotherms G as follows:

$$\begin{cases}
\lambda_{\max} \propto G,\\
m_{\max} \propto \sqrt{G},\\
s_{\max} \propto G.
\end{cases} \tag{7.4}$$

This elegant similarity argument effectively reduces the number of governing parameters in the intrusion problem. Stability characteristics need to be evaluated for a single value of G and (7.4) immediately extends the analysis to arbitrary G's.

It should be borne in mind that this isomorphism formally pertains only to the initial linear stage of the intrusion growth. The similarity relations (7.3) are not expected to be accurate after the appearance of convective overturns, which are characterized by flux laws that are different from those assumed for salt-finger regions (6.2). On the other hand, if these relations were to hold at least approximately in the nonlinear regime, this would imply that the equilibrium lateral fluxes due to interleaving $\left(\overline{u'T'} \propto C^{-1}\right)$ actually decrease with increasing lateral gradients $\left(\bar{T}_x \propto C\right)$. Such unusual dynamics would constitute an interesting precedent of a radical departure from Fick's law of diffusion. Conundrums of this nature reflect the challenges involved in the development of physical models of lateral intrusion-induced mixing and need to be addressed in future studies of interleaving.

Multiscale model

Recent years have witnessed significant progress in the development of the parametric model. However, a number of theoretical predictions – such as the dependence of intrusion scales on stratification, precise conditions for their occurrence, and the relative strengths of temperature and salinity signatures – are yet to be reconciled with measurements and simulations. While the parametric theory remains unquestionably the most popular means for explaining and predicting intrusion characteristics, it is prudent to explore alternatives. The heavy reliance of most interleaving models on parameterizations of microstructure, a source of great uncertainty in itself, might prove to be the Achilles' heel of contemporary intrusion theory. These concerns motivate the development of approaches based directly on the original governing equations rather than on empirical flux-gradient laws.

One of the prospects for analytical progress is related to the disparity of the salt-finger (centimeters) and intrusion (tens of meters vertically) scales. Since salt fingers operate on a distinct narrow-band range of wavelengths, problems concerning their interaction with larger scales of motion could be, but very seldom are, couched in terms of multiscale asymptotic analysis (Bensoussan *et al.*, 1978; Mei and Vernescu, 2010). A rare example of the application of multiscale methods in double-diffusion was presented by Holyer's (1981, 1985) collective instability model, discussed in Chapter 6. An analogous calculation for the intrusion problem (Radko, 2011) is briefly described below.

The first step in the multiscale model is the choice of background pattern, representing individual salt fingers, which is taken here to be harmonic in x and z:

$$T_{\mathrm{bg}} = A_T \sin(k_f x) \sin(k_f r z), \tag{7.5}$$

where A_T is the amplitude of temperature variation in salt fingers, k_f is the horizontal wavenumber of the individual fingers and r is the aspect ratio. Similar patterns are also assumed for salinity and streamfunction and the dependent variables are separated into the background field of salt fingers $\left(T_{\mathrm{bg}}, S_{\mathrm{bg}}, \psi_{\mathrm{bg}}\right)$ and a weak perturbation $\left(T', S', \psi'\right)$. Our immediate goal is to examine the linear stability of the background pattern with respect to the intrusive long-wavelength perturbations. The linearization of the governing equations, taken for simplicity to be two-dimensional Navier–Stokes, about the basic state yields:

$$\left\{\begin{aligned}
&\frac{\partial T'}{\partial t} + \frac{\partial \psi_{\mathrm{bg}}}{\partial x}\frac{\partial T'}{\partial z} + \frac{\partial \psi'}{\partial x}\frac{\partial T_{\mathrm{bg}}}{\partial z} - \frac{\partial \psi'}{\partial z}\frac{\partial T_{\mathrm{bg}}}{\partial x} - \frac{\partial \psi_{\mathrm{bg}}}{\partial z}\frac{\partial T'}{\partial x} - G\frac{\partial \psi'}{\partial z} + \frac{\partial \psi'}{\partial x} \\
&\qquad = \nabla^2 T', \\
&\frac{\partial S'}{\partial t} + \frac{\partial \psi_{\mathrm{bg}}}{\partial x}\frac{\partial S'}{\partial z} + \frac{\partial \psi'}{\partial x}\frac{\partial S_{\mathrm{bg}}}{\partial z} - \frac{\partial \psi'}{\partial z}\frac{\partial S_{\mathrm{bg}}}{\partial x} - \frac{\partial \psi_{\mathrm{bg}}}{\partial z}\frac{\partial S'}{\partial x} - G\frac{\partial \psi'}{\partial z} + \frac{1}{R_\rho}\frac{\partial \psi'}{\partial x} \\
&\qquad = \tau \nabla^2 S', \\
&\frac{\partial}{\partial t}\nabla^2 \psi' + \frac{\partial \psi_{\mathrm{bg}}}{\partial x}\frac{\nabla^2 \psi'}{\partial z} + \frac{\partial \psi'}{\partial x}\frac{\partial \nabla^2 \psi_{\mathrm{bg}}}{\partial z} - \frac{\partial \psi_{\mathrm{bg}}}{\partial z}\frac{\nabla^2 \psi'}{\partial x} - \frac{\partial \psi'}{\partial z}\frac{\partial \nabla^2 \psi_{\mathrm{bg}}}{\partial x} \\
&\qquad = Pr\left[\frac{\partial}{\partial x}(T' - S') + \nabla^4 \psi'\right].
\end{aligned}\right. \tag{7.6}$$

Next, we introduce new spatial (X, Z) and temporal (t_0) scales to describe the slow evolution of large-scale perturbations. The intrusion scales are connected to the primary (salt-finger) scales by assuming simple power laws of the form:

$$X = \varepsilon^a x,\ Z = \varepsilon^b z,\ t_0 = \varepsilon^c t, \tag{7.7}$$

where $\varepsilon \ll 1$ is a small parameter. The reason for rescaling independent variables using a single expansion parameter (ε) is a pragmatic one. The singular perturbations theory is rigorously and systematically developed only for one asymptotic parameter, and multi-parameter problems have to be solved on a case-by-case basis (Kevorkian and Cole, 1996). The most common approach requires users to focus on specific sectors in the parameter space, selected on a physical basis. For the intrusion problem, an appropriate scaling (Radko, 2011), which ensures that the model captures all essential dynamics, is given by $a = 3$, $b = 2$ and $c = 2$ in (7.7). The horizontal temperature gradient (G), another small parameter in the problem, is rescaled as $G = \varepsilon^3 G_0$. The subsequent development follows the conventional technique used in multiscale problems (e.g., Kevorkian and Cole, 1996; Mei and Vernescu, 2010): (i) (x, z, X, Z, t_0) are treated as independent variables; (ii) on short scales (x, z) we impose the same periodicity as in the basic flow; and (iii) derivatives in the governing system are replaced as follows:

$$\frac{\partial}{\partial x} \to \frac{\partial}{\partial x} + \varepsilon^3 \frac{\partial}{\partial X}, \; \frac{\partial}{\partial z} \to \frac{\partial}{\partial z} + \varepsilon^2 \frac{\partial}{\partial Z}, \; \frac{\partial}{\partial t} \to \varepsilon^2 \frac{\partial}{\partial t_0}. \tag{7.8}$$

The solution for $\left(T', S', \psi'\right)$ is obtained in terms of a series in ε:

$$\left(T', S', \psi'\right) = (T_0, S_0, \psi_0) + \varepsilon\,(T_1, S_1, \psi_1) + \varepsilon^2\,(T_2, S_2, \psi_2) + \cdots. \tag{7.9}$$

Equations (7.8) and (7.9) are substituted in (7.6) and terms of the same order in ε are collected. The resulting hierarchy of equations is sequentially solved until a closed, explicit solution is found that represents the evolution of the intrusive perturbation on large spatial and temporal scales. Stability of the modulational large-scale equations is analyzed using the normal modes proportional to $\exp(\lambda_0 t_0 + iKX + iMZ)$. The resulting growth rate equation can be expressed in terms of the original, rather than rescaled, variables. It takes the form of a fourth-order polynomial:

$$\lambda^4 + a_3\lambda^3 + a_2\lambda^2 + a_1\lambda + a_0 = 0, \tag{7.10}$$

whose coefficients a_i depend on $\left(A_T, m, \tau, R_\rho, Pr, G, s, r\right)$.

Predictions of the multiscale model (Figs. 7.8 and 7.9) are consistent with physical expectations and generally conform to the parametric model. Intrusions are oriented as in Figure 7.3 – with warm and salty (cold and fresh) water rising (sinking) across density interfaces. Typical growth rates are of the order $\lambda \sim 10^{-3}$, which corresponds to the dimensional time scale of days, and maximal intrusion slopes are of the order $s \sim 10^{-3}$.

A generic drawback of multiscale models is that the background pattern has to be specified a priori. In the intrusion problem, this implies knowledge of the amplitude of salt fingers (A_T) and their wavenumbers $\left(k_f, rk_f\right)$ in (7.5). While the horizontal wavenumber k_f can be adequately estimated on the basis of the linear model (Schmitt, 1979a), a simple theoretical prescription for the finger amplitude and the vertical wavenumber (or equivalently, the aspect ratio r) is still lacking. Nevertheless, this information is direct and often accessible in field measurements, in contrast to the flux-gradient laws used by the parametric model. Direct numerical simulations indicate that the amplitude depends on the density ratio and it is limited

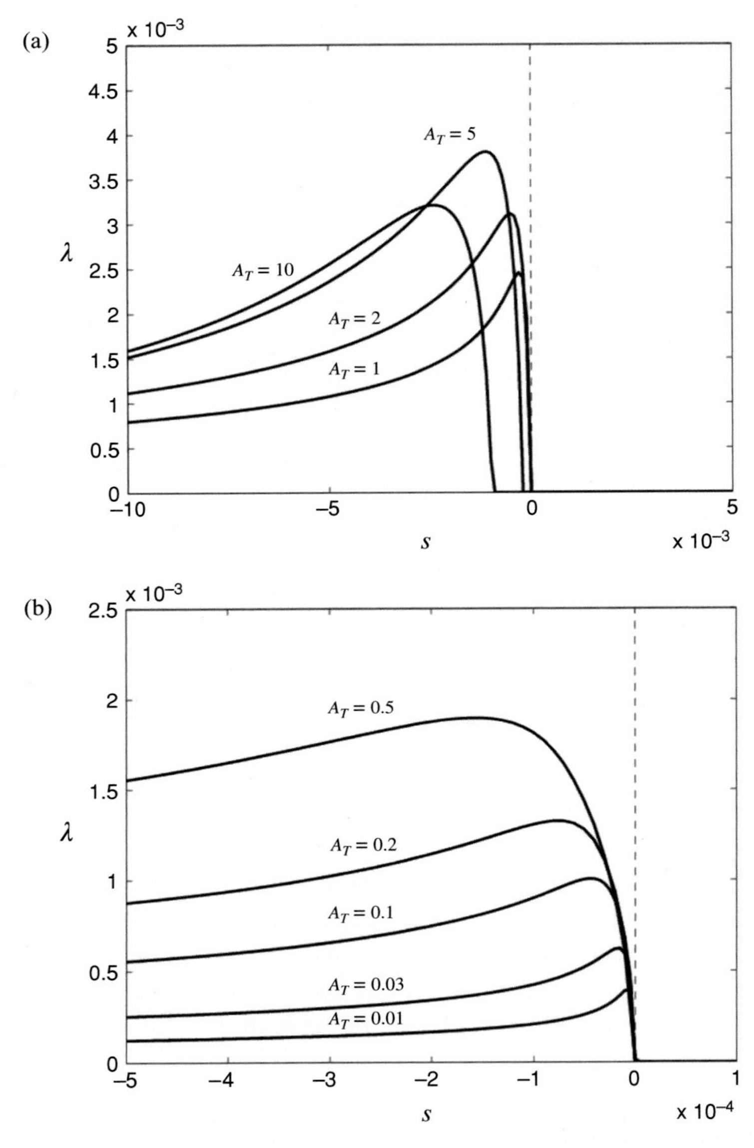

Figure 7.8 Multiscale model of interleaving. The maximal growth rate of intrusions is plotted as a function of slope (s) for various values of the salt finger amplitude (A_T) and a fixed value of finger aspect ratio ($r = 0.1$).

to the range $1 < A_T < 10$. The aspect ratio of salt fingers r is less than unity but, typically, is an order-one quantity (e.g., Taylor, 1993). The growth rates are sensitive to both A_T (Fig. 7.8) and r (Fig. 7.9), but the overall pattern of the $\lambda(k, m)$ relation appears to be structurally robust.

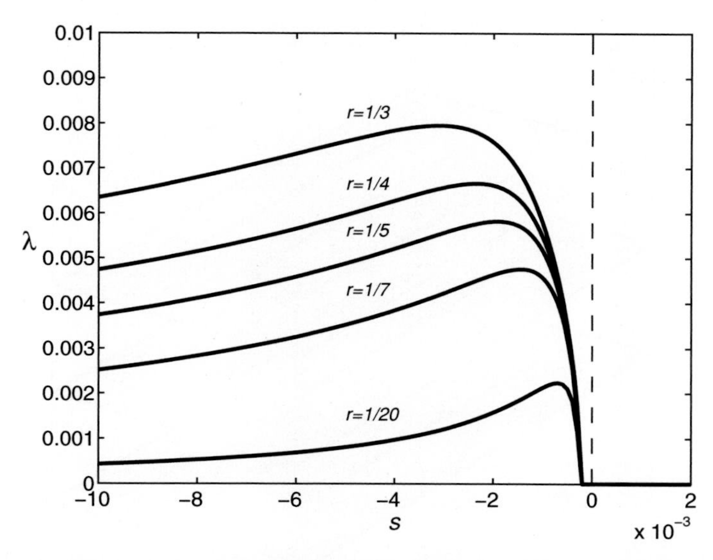

Figure 7.9 Multiscale model of interleaving. The maximal growth rate of intrusions is plotted as a function of slope (s) for various values of the finger aspect ratio (r) and a fixed value of the salt finger amplitude ($A_T = 5$).

A significant advantage of the multiscale approach, as compared to parametric modeling, is related to its transparency and, perhaps, deeper insight into the dynamics at play. The multiscale model shows explicitly how various Fourier harmonics of individual salt fingers are affected by the large-scale advection and how these finger modes constructively interact to provide a positive feedback on the intrusion. Parameterized models lose all of the connection with the dynamics of individual fingers immediately after assuming an expression for fluxes. This connection is retained by multiscale models from the first step to the last. While multiscale modeling for double-diffusion is still in a preliminary exploratory stage, its potential for addressing scale interaction problems in double-diffusion is already evident (Holyer, 1981, 1985; Radko, 2011). It is perhaps ironic that multiscale techniques have been used more frequently in configurations (Cushman-Roisin *et al.*, 1984; Gama *et al.*, 1994; Manfroi and Young, 1999; among many others) where the scale separation is not nearly as well-defined or physically justified as in the salt-finger problem.

7.2 Extensions: rotation, baroclinicity and ambient turbulence

Having examined in considerable detail the simplest configuration (Section 7.1), we now consider some oceanographically relevant extensions.

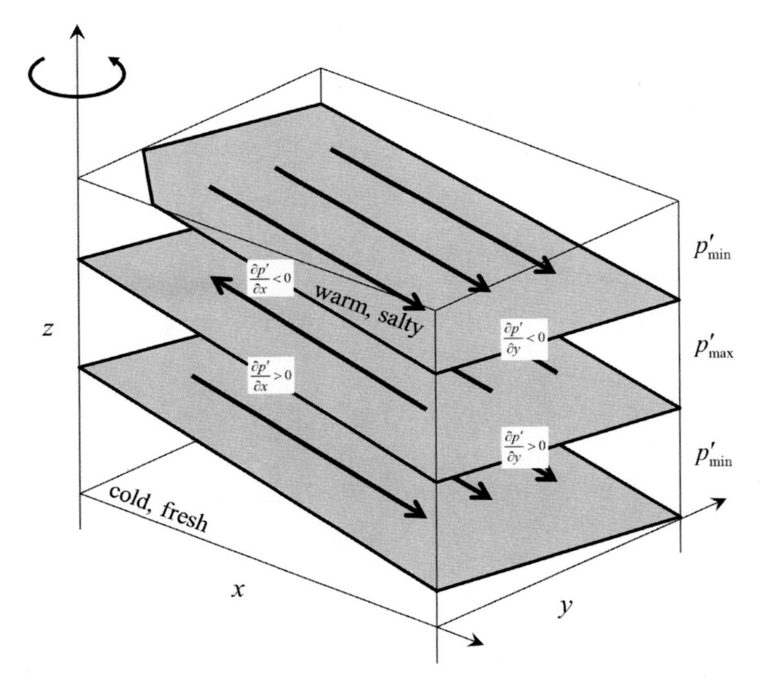

Figure 7.10 Schematic diagram illustrating the effects of rotation on finger-driven intrusions. The fastest growing intrusion develops a finite along-front slope.

Rotation

Given their slow evolutionary time scales (days), it seems likely that intrusions could be affected by planetary rotation. In reality, rotational effects in the barotropic model are rather modest. By and large, rotation does not affect the growth rate, height and the cross-front slope of the fastest growing intrusions. The most notable observable consequence of rotation is the development of a systematic tilt of intrusions in the along-front direction, as indicated in the schematic in Figure 7.10.

In order to physically interpret the rotational along-front tilt, let us have a brief look back at Figure 7.3, illustrating the essence of the intrusion mechanics, and modify this scenario by adding rotation. Planetary rotation affects circulation primarily by means of the Coriolis force. To be specific, consider dynamics realized in the Northern Hemisphere, where the Coriolis force is directed to the right of the fluid motion. Thus, intrusions advecting fluid upward (downward) in the x–z plane are forced in the positive (negative) y-direction by the Coriolis force. The next step is readjustment of the pressure distribution, which tends to balance the Coriolis force and thereby maintain the geostrophic equilibrium. A detailed calculation indicates that intrusions are fully adjusted (e.g., McDougall, 1985a) – the pressure gradient force exactly balances the Coriolis force and the along-front velocity is zero. In order to balance the Coriolis force, the along-front gradient should be positive $\left(\frac{\partial p'}{\partial y} > 0\right)$ in the upward moving intrusions and negative $\left(\frac{\partial p'}{\partial y} < 0\right)$ in the downward. How does the induced pressure distribution affect the intrusion growth? Consider first intrusions oriented as in Figure 7.10 – slightly sloping downward in the y-direction. In this case, the maximal values of the perturbation pressure are attained below the downward-moving intrusions and above the upward-moving ones. In terms of pressure distribution at the cross-front (x,z) section, this implies that in the upward-moving intrusions $\frac{\partial p'}{\partial x} > 0$. Thus, the rotationally induced pressure gradients reinforce the initial tendency: upward intrusions, moving towards the fresher side of the front, are further accelerated in the same direction. Similarly, downward intrusions are accelerated in the positive x-direction by the induced pressure gradient force. As a result, the along-front tilt in Figure 7.10 promotes the intrusion growth. On the other hand, if intrusions were tilted along the front in the opposite sense (sloping upward in y) the induced pressure gradients would act to suppress their growth. Hence, the configuration in Figure 7.10 represents the preferred orientation of intrusions, which is expected to be realized in nature as long as the selective advantage of the most rapidly growing modes is sufficient to ensure their dominance.

The analysis of intrusions driven by molecular fluxes (Kerr and Holyer, 1986) has led to similar conclusions: rotation tilts intrusions in the along-front direction but has little to no effect on other observable intrusion characteristics. Most theoretical intrusion models are concerned with the prediction of growth rates and vertical scales – quantities that control the intensity of lateral intrusion-driven mixing. Therefore, in the barotropic configuration, rotation is usually ignored for reasons of tractability. Effects of rotation, however, become more profound in the context of the baroclinic interleaving problem, which we consider next.

Baroclinicity

So far, our discussion has been focused on barotropic fronts with horizontally uniform background density. In most cases, however, horizontal density variation across oceanic fronts is significant, motivating "baroclinic" models, which take horizontal density stratification into account. Baroclinicity affects intrusions in two major ways (May and Kelley, 1997). First, there is a set of consequences directly related to the geometry of the system. The horizontal density stratification projects on the density variation along the intrusion itself, which can either intensify interleaving or suppress it. Various configurations are illustrated in Figure 7.11. When isohalines and isopycnals are inclined in the opposite sense (Fig. 7.11b), the

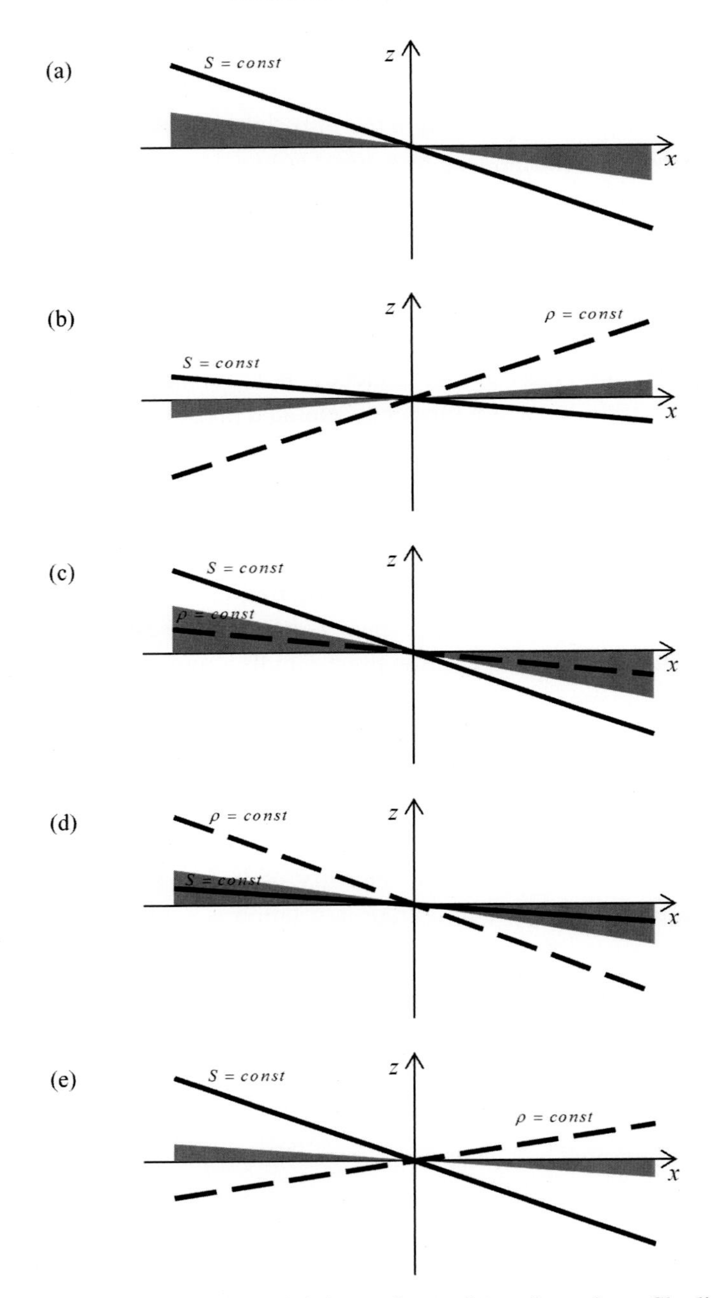

Figure 7.11 The effects of baroclinicity on finger-driven intrusions. Shading indicates the range of slopes susceptible to interleaving for various orientations of isohalines and isopycnals in (a)–(e).

density gradient along the intrusion increases relative to that in the barotropic case (Fig. 7.11a). The increase in stratification adversely affects the ability of intrusions to advect fluid across the isopycnals. As a result, intrusion growth rates decrease and the range of slopes for which unstable solutions exist reduces. On the other hand, if isohaline and isopycnal slopes are of the same sign (Fig. 7.11c), the density gradient along the intrusion is reduced. With the stabilizing influence of density stratification diminished, intrusions are able to accelerate faster and instability affects a wider range of slopes. Very different dynamics are realized when the isopycnals are inclined more steeply than isohalines (Fig. 7.11d,e). In this case, the instability region extends beyond the haline wedge $\left(-S_x/S_z < s < 0\right)$, outside of which double-diffusive effects become secondary and the dynamics become similar to the classical baroclinic instability. Predominantly baroclinic instabilities, however, are characterized by much larger vertical scales and therefore their connection with the observed ($\sim$10 m) intrusions is questionable.

The second set of baroclinic effects is associated with vertical shear. The geostrophic along-front velocity is directly linked to the cross-front density gradient by the thermal wind balance:

$$f\bar{v}_{z\,\mathrm{dim}} = -\frac{g}{\rho_0}\bar{\rho}_{x\,\mathrm{dim}}, \tag{7.11}$$

and therefore vertical shear is omnipresent at baroclinic fronts. Unlike the geometric effects (Fig. 7.11), which could enhance or suppress intrusions, shear consistently acts against intrusion growth. One such mechanism is based on the suppression of the along-flow variation in temperature and salinity by shear. Since the fastest growing intrusion modes are characterized by a finite along-front slope (Fig. 7.10) – hence significant along-front *T*/*S* variation – the tendency of shear to homogenize the flow in this direction inevitably reduces the maximal growth rate. When shear is realistically strong, the relevant fastest growing modes represent the pure cross-flow intrusions with $l \approx 0$ (e.g., Smyth, 2008). Other adverse shear-induced effects include (i) a reduction in the intensity of salt fingering caused by the tilting of fingers in the direction of shear and (ii) an enhanced probability of turbulent overturns in the high-shear environment. Both processes modify vertical diffusivities of temperature and salinity, making double-diffusion less effective in driving intrusions.

To the best of our knowledge, it has not been determined which set of baroclinic processes, geometric or shear-induced, is more significant for intrusion dynamics. The answer is likely to depend on a particular problem and background parameters – baroclinicity, stratification, and the level of ambient turbulence. It should be noted that *horizontal* shear also tends to inhibit thermohaline interleaving (Worthem *et al.*, 1983) through mechanisms that are similar to those at work in vertical (baroclinic) shear. The horizontal along-front shear homogenizes the flow in the downstream direction and thereby suppresses the fastest growing modes, which, in the presence of rotation, have a finite along-front slope.

Ambient turbulence

Within the stratified interior of the ocean, small-scale vertical mixing is accomplished by a combination of double-diffusion and mechanically generated turbulence, usually associated with overturning gravity waves (e.g., Gregg, 1987; Gargett, 1989; Thorpe, 2005). Generally, small-scale turbulence is thought to be ineffective in driving intrusions. It was shown by Stern (1967) and is apparent from the physical interpretation (Fig. 7.3) that unequal eddy diffusivities of heat and salt are essential for the spontaneous generation of interleaving motions. Fully developed turbulence, on the other hand, tends to mix temperature and salinity at equal rates. An exception is provided by differential diffusion – an incomplete turbulent mixing by weak turbulence in a strongly stratified environment. Differential diffusion is characterized by higher diffusivity of temperature and therefore it has been invoked to explain the appearance of intrusions in double-diffusively stable regions (Hebert, 1999; Gargett, 2003; Merryfield, 2002). However, the existing evidence for its general importance is, at present, too circumstantial to warrant a detailed analysis.

We proceed under a conventional assumption that eddy diffusivities of heat and salt due to mechanical turbulence are nearly equal (K_{turb}) but can differ from the turbulent eddy viscosity (A_{turb}). Another common, albeit less justified, approximation represents the net small-scale mixing as a sum of the contribution from double-diffusion and turbulence:

$$\begin{cases} K_T = K_{T\,\mathrm{dd}} + K_{\mathrm{turb}}, \\ K_S = K_{S\,\mathrm{dd}} + K_{\mathrm{turb}}, \\ K_M = K_{M\mathrm{dd}} + A_{\mathrm{turb}}, \end{cases} \tag{7.12}$$

where (K_T, K_S, K_M) are the diffusivities of temperature, salinity and momentum respectively. Subscripts dd (turb) denote the contribution from double-diffusion (turbulence). The general structure of (7.12) suggests that the impact of turbulence on interleaving is two-fold. Taking turbulence into account (i) effectively increases the net viscosity and (ii) brings the ratio of eddy diffusivities of heat and salt closer to unity. At first, it appears that both effects should have an adverse effect on interleaving. The reality, however, is not as clear-cut – the devil is in the details.

Consider first viscous effects. Double-diffusion by itself is a highly ineffective mixer of momentum. Eddy viscosity due to salt fingers is much less than the diffusivity of heat or salt; it is even less than molecular viscosity. Therefore, the presence of even a very modest amount of ambient turbulence, typical of the main thermocline, can dramatically increase the net viscosity (K_M) leaving the *T–S* diffusivities largely unchanged. Viscous damping tends to selectively suppress relatively small-scale features – the larger the viscosity, the wider is the range of scales affected by it. In this way, viscosity can exert a controlling influence on the vertical scale of intrusions. An increase in viscosity systematically shifts the range of vertical wavelengths affected by interleaving instability towards longer scales. Since intrusion models based on molecular viscosity (Fig. 7.7b) predict relatively modest vertical scales ($H_{\text{dim}} \sim$ 10–20 m) the presence of much larger intrusions in the ocean ($\sim$100 m) can be plausibly attributed to turbulent mixing. The major difficulty with this argument, of course, is that vertical eddy viscosity in the ocean is poorly constrained and spatially variable, which precludes its unequivocal corroboration. It should also be kept in mind that effects of turbulence on intrusions could be very different at barotropic and at baroclinic fronts. Turbulence can suppress interleaving entirely for strongly baroclinic conditions (Kuzmina and Zhurbas, 2000; Zhurbas and Oh, 2001), whereas in the barotropic model interleaving persists regardless of the strength of ambient turbulence, albeit intrusion growth rates are reduced (Walsh and Ruddick, 2000).

For historical reasons, it is interesting to note that the first intrusion model (Stern, 1967) did not take viscosity, molecular or eddy-induced, into account. As a result, the preferred vertical scale was absent in Stern's (1967) formulation. The growth rate of intrusive instabilities monotonically increased with decreasing height. The source of this problem was identified by Toole and Georgi (1981). They noted that the inclusion of vertical momentum flux removes the ultraviolet catastrophe in Stern's model and makes it possible to estimate the basic characteristics of intrusions by focusing on the fastest growing mode.

While the damping tendency of friction is intuitive and predictable, the possibility for viscous destabilization is anything but. Nevertheless, such a possibility is very real for interleaving motions. It was noted by McIntyre (1970) that a distinct class of instabilities can be induced by differences in eddy viscosity and diffusivity. If the eddy Prandtl number $\left(Pr_e = \frac{K_M}{K_T}\right)$ exceeds unity, the baroclinic flow becomes susceptible to instabilities that are similar to the classical thermohaline intrusions in pattern but dynamically different. Also characterized by vertically periodic structure and inclined at small angles, McIntyre modes are not dependent on unequal diffusivities of heat and salt.

The essence of the McIntyre effect resides in the ability of elevated friction to upset the equilibrium balances of stable frictionless motions. Consider for instance the configuration in Figure 7.12. Displacements within the baroclinic instability

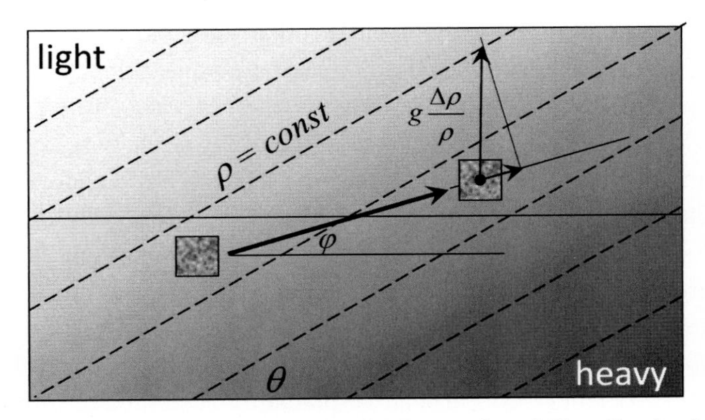

Figure 7.12 Schematic diagram of the McIntyre instability. In the baroclinic model, displacement of a fluid element upward within the wedge of instability places it in denser surroundings. The resulting upward buoyancy force tends to reinforce the initial displacement, thereby supporting the instability.

wedge ($0 < \varphi < \theta$) bring lighter fluid upward. This, by itself, suggests instability: the upward buoyancy force reinforces the initial perturbation. However, in the ideal (frictionless, non-diffusive) fluid, this tendency is countered by the Coriolis force, which consistently acts to the right – we are still in the Northern Hemisphere – and ultimately reverses the direction of the initial displacement. The result is a stable periodic oscillation known as the inertia-gravity wave. Now, let us reintroduce viscosity, high enough to dramatically reduce the perturbation velocity. The density perturbation, we assume, remains of the same magnitude as before. In this regime, the Coriolis effect is reduced to such low levels that it is no longer able to counteract the destabilizing buoyancy forcing. The perturbation monotonically grows in time.

The McIntyre instability operates at the intermediate range of vertical scales ($\sim$10–100 m), which makes it difficult to distinguish it from thermohaline intrusions on the basis of oceanographic observations. The situation is further complicated by the very likely possibility that both effects are simultaneously present and interact in a non-trivial manner. The ocean is at least triply diffusive. The rates of heat, salt and momentum dissipation are unequal. Their relative significance varies in space and time, and so do intrusion properties. For instance, Ruddick (1992) analyzed intrusions operating on the lower flank of meddy Sharon and established their predominantly thermohaline origin. Intrusions operating on the Faroe Front – another well-studied (Hallock, 1985) example of interleaving – are better described by the McIntyre model.

Smyth and Ruddick (2010) systematically analyzed the linear parametric intrusion model in terms of its sensitivity to the level of ambient turbulence. The key

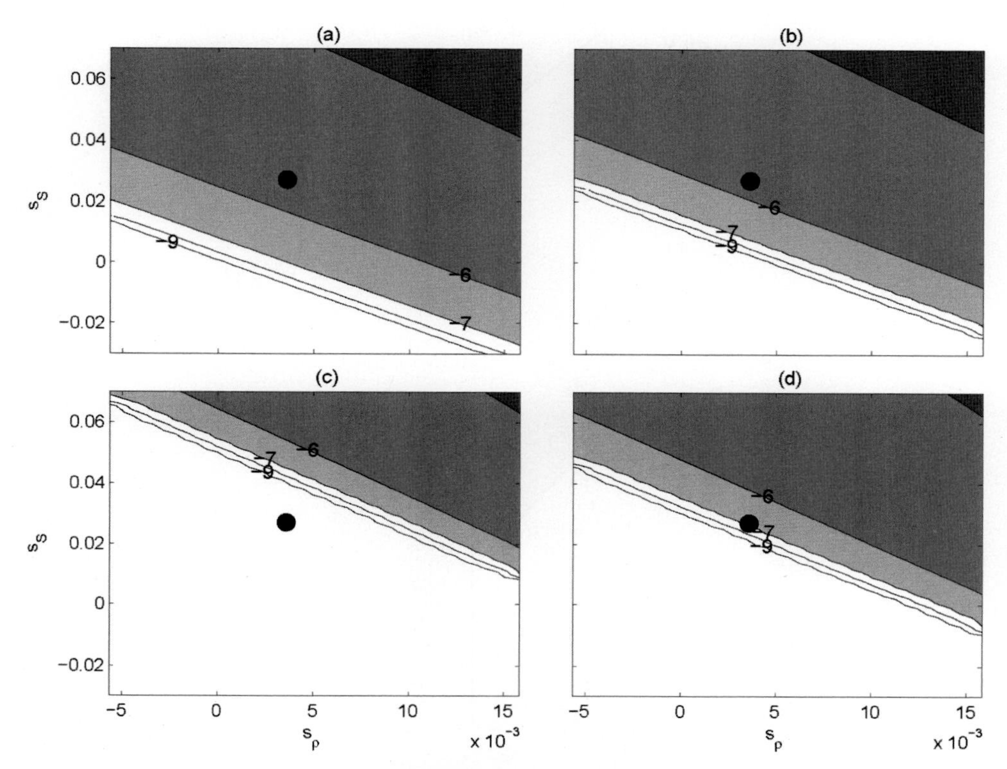

Figure 7.13 Common logarithm of the growth rate of the fastest growing interleaving mode as a function of isohaline and isopycnal slopes (darker shading corresponds to larger growth rates). Diffusivities are (a) double-diffusive only; (b) double-diffusive and molecular; (c) double-diffusive, molecular and turbulent with $Pr_e = 1$; and (d) same as (c) except $Pr_e = 20$. Large filled circles indicate the isopycnal and isohaline slopes characteristic of the lower flank of meddy Sharon. From Smyth and Ruddick (2010).

results are summarized in Figure 7.13. Adding turbulent mixing with the turbulent Prandtl number of unity to the double-diffusive model substantially reduces growth rates and shrinks the unstable region in the (s_S, s_ρ) parameter space. This effect is partially reversed when the Prandtl number is increased significantly above unity (Fig. 7.13d), which apparently reflects McIntyre dynamics at play. An increase in turbulent diffusivity (K_{turb}) generally tends to stabilize the system. Figure 7.14a shows the uniform retreat of the intrusion-favorable region following an increase in $\mu = \frac{K_{\text{turb}}}{K_{S\,\text{dd}}}$. The exception (of course there is an exception – it is double-diffusion!) is the strongly turbulent regime in which $\mu \gg 1$. In this case, the dynamics become essentially McIntyre-driven and the instability is determined by the orientation of isopycnals; the isohaline slope is largely irrelevant. The result is the tilting of the stability boundary in the (s_S, s_ρ) space towards the line $s_\rho = const$ with increasing μ

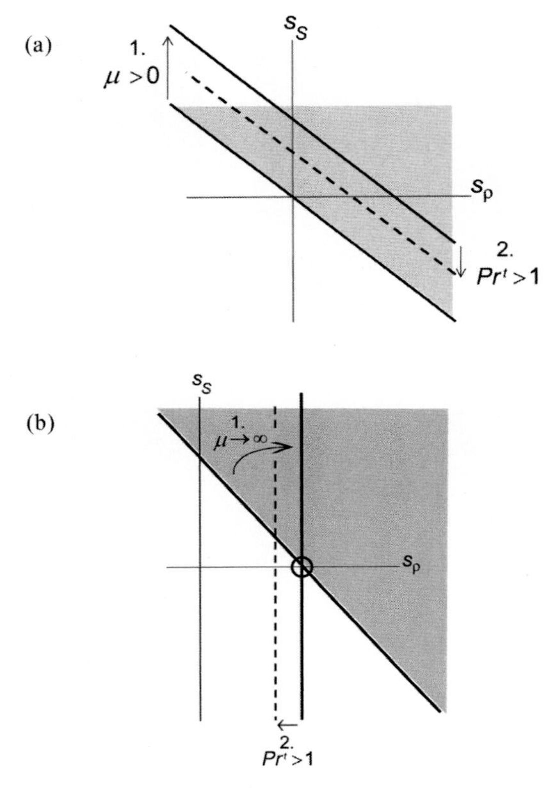

Figure 7.14 (a) Schematic showing the lowest-order effects on the stability boundary due to the imposition of weak turbulent thermal diffusivity (arrow 1) and an increase in the turbulent Prandtl number (arrow 2). Shading indicates instability. (b) Schematic for the "strong turbulence" limit, showing the tilting of the stability boundary as $\mu \to \infty$ (arrow 1) and the expansion of the unstable region as the turbulent Prandtl number is increased above unity (arrow 2). From Smyth and Ruddick (2010).

(Fig. 7.14b). This change is stabilizing if the slopes of isohalines and isopycnals are of the same sign. If not, the increase in turbulent diffusivity destabilizes the system.

7.3 Nonlinear effects

As suggestive and interesting as linear theories are, they are severely limited by (i) their principal inability to predict the amplitude of fully developed intrusions and (ii) an uncertainty in the relevance of inferred characteristics for the nonlinear stage. Problems with nonlinear analysis are exacerbated by the tendency of intrusions to transform profoundly during their evolution. This tendency is related to the subcritical nature of intrusive instabilities.

With respect to nonlinear effects, most instabilities can be separated into two distinct categories. Instabilities where nonlinearity tends to suppress linear growth are known as supercritical. Unstable supercritical modes can equilibrate at relatively low amplitudes. In subcritical instabilities, on the other hand, nonlinearity reinforces the linear growth and therefore their equilibration generally occurs after the system evolves into a new state, dynamically different from the original configuration. Secondary large-scale double-diffusive instabilities generally tend to be subcritical and intrusive instabilities are no exception. The existing evidence – laboratory, observational and numerical – is consistent with the subcritical character of interleaving. The growth of salt-finger intrusions usually persists until they develop inversions in the *T–S* profiles, where salt fingering is partially supplanted by diffusive and convective mixing.

In such circumstances, standard analytical techniques of weakly nonlinear theory, which builds upon linear results, are of limited use. Only a few analytical results have been reported for fully nonlinear intrusions. McDougall (1985b) formulated a set of necessary conditions for the existence of steady-state solutions. Intrusions in his model were represented by a series of vertically homogeneous mixed layers separated by thin high-gradient interfaces with alternating diffusive and salt-finger favorable stratification. Tractability was achieved by assuming that the *T–S* fluxes are determined by the salinity variation across the interfaces rather than by local gradients. Kerr (1992, 2000) examined secondary instabilities of finite amplitude intrusions, albeit molecularly driven, and concluded that the equilibrium states are unstable with respect to secondary two- and three-dimensional perturbations – an important suggestion offering a first glimpse into the complexity of the equilibrium intrusion dynamics.

In view of the difficulties inherent in the analytical modeling of nonlinear intrusions, more promising (and definitely more popular) are models based primarily on numerical methods. Let us review a couple of the most common categories. First, there is the class of one-dimensional intrusion models, which extend the parametric theory (Section 7.1) into the nonlinear regime (Walsh and Ruddick, 1998; Merryfield, 2000; Mueller *et al.*, 2007). Such models assume that large-scale properties of intrusions are uniform along the intrusion fronts and therefore

$$T = T(\eta, t), \tag{7.13}$$

where the variable η represents the coordinate normal to the intrusion fronts; analogous expressions are assumed for S and ψ. For such patterns, the advective terms in the governing Navier–Stokes equations vanish identically. Nonlinearities now appear only in the expressions for fluxes (F_T, F_S) in (7.1). The parameterization

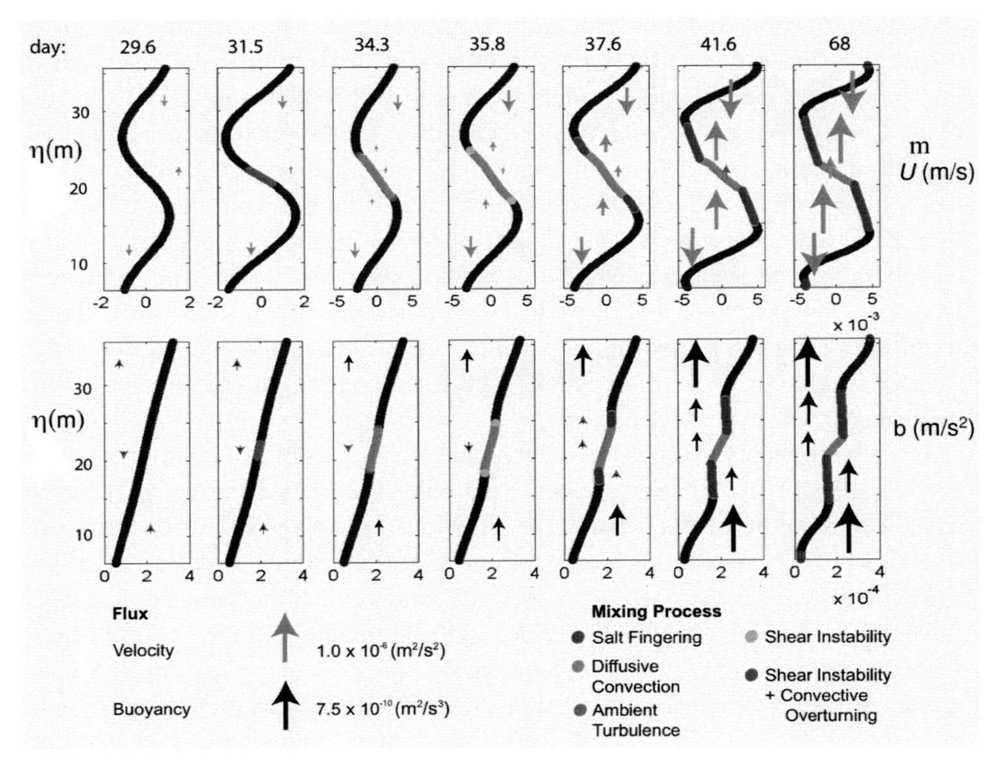

Figure 7.15 Evolution of velocity (top panel) and buoyancy (bottom panel) profiles. Arrows indicate across-intrusion fluxes. From Mueller *et al.* (2007).

of mixing necessarily involves assuming distinct flux-gradient laws for regions controlled by salt fingering, diffusive convection and convective or shear-driven turbulence.

The apparent advantage of one-dimensional models lies in their efficiency. After substitution of (7.13) and the corresponding expressions for S and ψ in the governing equations, the problem reduces to integration of a system of partial differential equations in (η, t). The simplicity of such models makes it possible to explore the parameter space in considerable detail, something that is still computationally prohibitive for direct numerical simulations (DNS). Figure 7.15 represents a typical evolution of the buoyancy field in the one-dimensional parametric model. This calculation (Mueller *et al.*, 2007) was initiated by a small-amplitude harmonic perturbation; the slope and the vertical scale were chosen in a way that maximizes the linear growth rate. In time, the system evolved into a layered configuration consisting of alternating diffusive and salt-finger interfaces separated by well-mixed layers – the "conventional" intrusion pattern. As was pointed out by Merryfield (2000), other scenarios are also possible, including

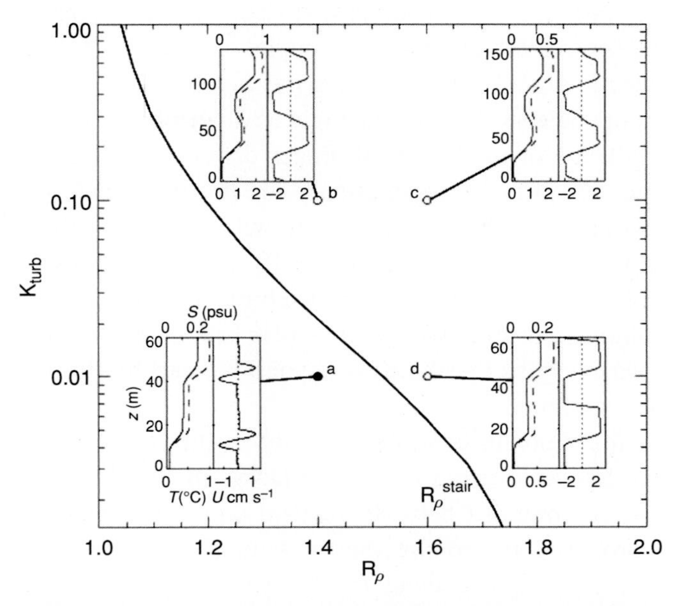

Figure 7.16 Equilibria of intrusions as a function of the density ratio (R_ρ) and turbulent diffusivity (K_{turb}). Insets show profiles of S (solid curves in the left panels), T (dashed curves), and horizontal velocity u for equilibrated intrusions as functions of z. The heavy solid curve separates the low-R_ρ region characterized by staircase patterns from the high-R_ρ region of conventional intrusions (see the inserts). From Merryfield (2000).

the "staircase" configuration characterized by the absence of diffusive interfaces. This regime is realized for relatively low density ratios. Figure 7.16 attempts to map the regions occupied by conventional intrusions and staircase solutions in the $(R_\rho, K_{\text{turb}})$ parameter space. While the parametric model can be extended to two and three dimensions, attempts of this nature are rather rare. Merryfield (2000) reports two-dimensional parametric simulations of intrusions which, for computational reasons, have not been carried on into the strongly nonlinear regime characterized by vigorous density overturns. However, no fundamental obstacles exist for multi-dimensional parametric simulations of fully nonlinear interleaving. On the contrary, the successful implementation of three-dimensional parametric modeling for the collective instability problem (Radko and Stern, 2011) suggests that such simulations can and should be profitably explored for interleaving studies as well.

While the significance of the conceptual insights generated by one-dimensional parametric models is apparent, their ability to produce accurate quantitative predictions is questionable. The limitations of one-dimensional models include (i) their reliance on mixing parameterizations, (ii) their inability to represent the along-intrusion variation of properties and (iii) the a-priori prescription of the intrusion slope (s). At this point I can imagine a frustrated reader complaining, "Why won't they just use DNS? Let's stop the guesswork and produce something reliable!" It is not that easy. Despite the rapid advancement of computational capabilities, the challenge of modeling salt-finger driven intrusions directly, without invoking various simplifying assumptions, has not been met. The fundamental problem is associated with their small slopes, which are typically on the order of $s \sim 10^{-3}$. The vertical scale of intrusions ($\sim$10 m) does not present a major computational problem in itself, but the associated horizontal scale $L_x \sim H/s \sim 10\,\text{km}$ does. The requirement to simultaneously resolve salt fingers ($\sim$1 cm) and horizontal intrusions wavelengths ($\sim$10 km) is beyond the capabilities of even the most advanced super-computers.

Fortunately, the problem of scale disparity in the intrusion problem can be, to some extent, bypassed using the "tilted DNS" model. In what appears to be a very sensible compromise between realism and convenience, Simeonov and Stern (2007) proposed integrating the governing equations in the tilted coordinate system (ξ, η) aligned along the intrusion fronts (see the schematic in Fig. 7.17a). Doubly periodic boundary conditions are assumed for the perturbation fields in ξ and η. The tilted DNS configuration makes it possible to simulate intrusions of arbitrary slope using a modest computational domain ($L_\xi \sim L_\eta$). Of course, there is a price to pay: the intrusion slope throughout the simulation is given by the initially prescribed value. Thus, the possibility that the slope could change in time is a-priori excluded. On the positive side, mixing is now simulated explicitly and not parameterized. This advantage becomes critical in the fully nonlinear stages of simulation. The equilibrium state is characterized by complicated interactions between various components – large-scale shear, salt fingers, diffusive interface, convective turbulence. These interactions are very difficult, if not impossible, to accurately represent in the parametric model.

Figures 7.18–7.20 (Hebert, 2011) give us a glimpse into the remarkably rich phenomenology of interleaving. All experiments were performed using the tilted DNS model and initiated by a small amplitude harmonic (normal mode) perturbation in T, S and ψ. For each value of $\left(R_\rho, G\right)$, Hebert (2011) considered a range of vertical scales and slopes comparable with, but not necessarily identical to, the fastest growing values. The most common outcome of such experiments is the quasi-equilibrium state consisting of diffusive and salt-finger interfaces separated by a well-mixed convective zone (Fig. 7.18). The simulation in Figure 7.19 is different. As the intrusion grows, its along-front velocity reaches magnitudes sufficient to trigger secondary Kelvin–Helmholtz instability, manifested in the formation of mixing billows, which profoundly affect the flow pattern and its dynamics.

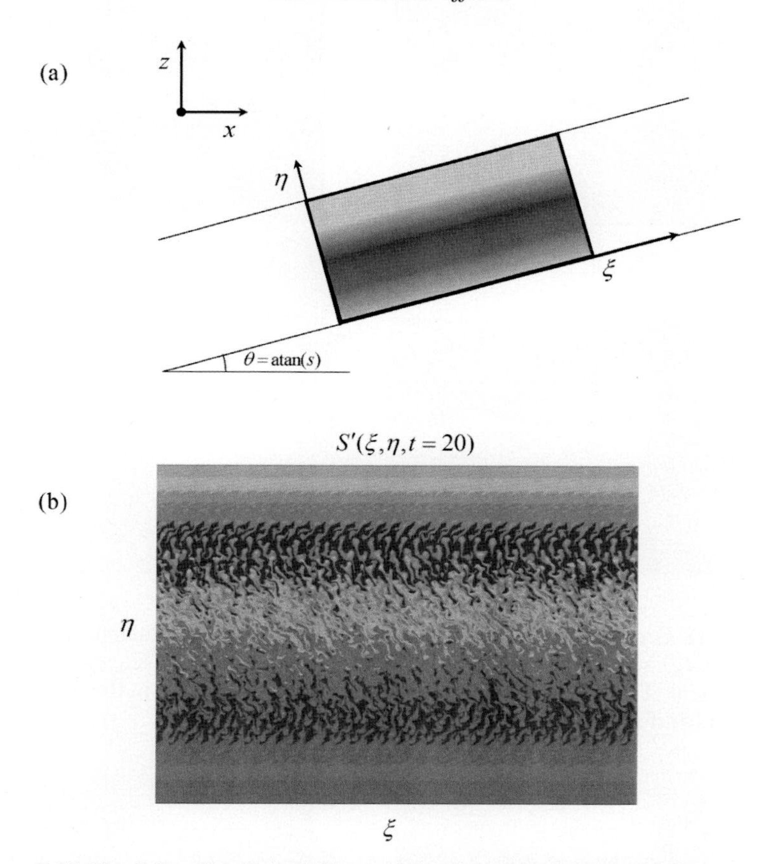

Figure 7.17 The “tilted box” modeling of intrusions. (a) Schematic diagram of the computational domain. The new coordinates are oriented along (ξ) and normal to the intrusion fronts (η). (b) Initial stage of linear intrusion growth in DNS. Presented is the departure of salinity from the background gradient. From Simeonov and Stern (2007).

The experiment in Figure 7.20 results in the staircase configuration, characterized by the absence of diffusive interfaces. Tilted DNS simulations reveal another effect that would have been impossible to identify with, and difficult to incorporate in, the one-dimensional models – the spontaneous generation of gravity waves. Figure 7.21 shows the evolution of the diffusive interface in time, the most pronounced feature of which is the cyclic amplification and breaking of waves, propagating along the intrusion fronts.

One of the evolutionary features of interleaving, which has not been fully explored by DNS, involves the sequential mergers of intrusions. Such mergers are often observed in laboratory experiments (Thorpe *et al.*, 1969), in the solutions of the one-dimensional parametric model (Walsh and Ruddick, 1998) and in the

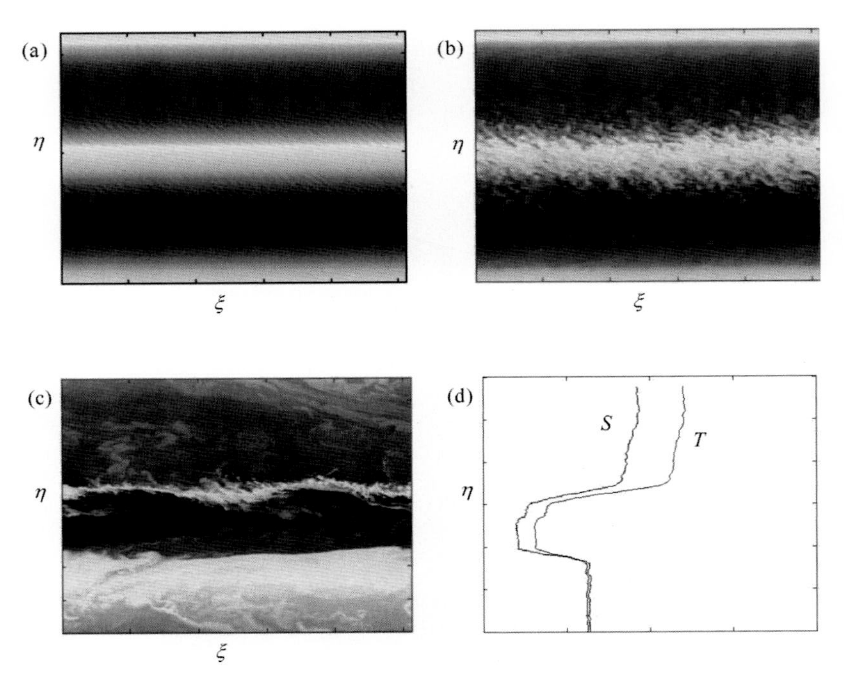

Figure 7.18 Growth and equilibration of "conventional" intrusions in the tilted DNS. (a)–(c) show the temperature perturbation at various stages of intrusion development. The vertical profiles of temperature and salinity in the final state are shown in (d). From Hebert (2011).

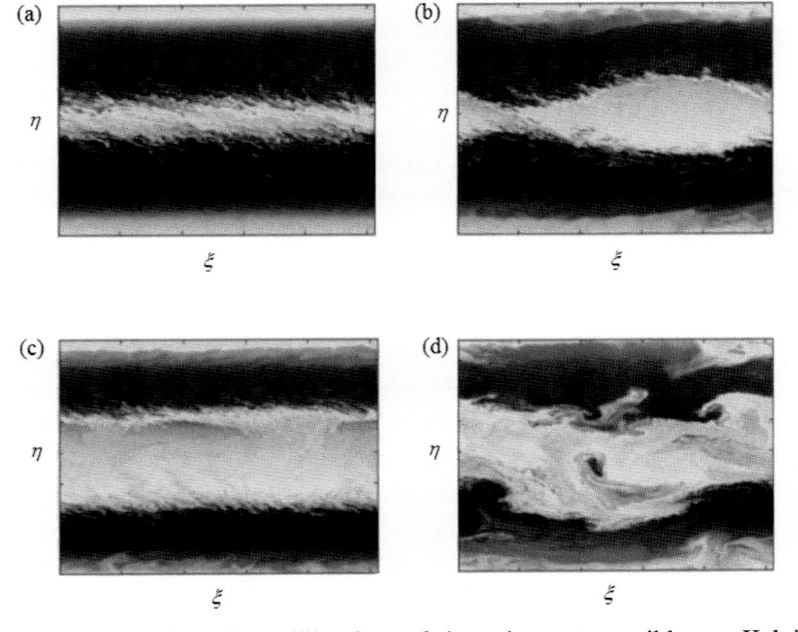

Figure 7.19 Growth and equilibration of intrusions susceptible to Kelvin–Helmholtz instabilities. The temperature perturbation is shown at various stages of intrusion development in (a)–(d). Red color corresponds to high values and low values are shown in blue. From Hebert (2011). See color plates section.

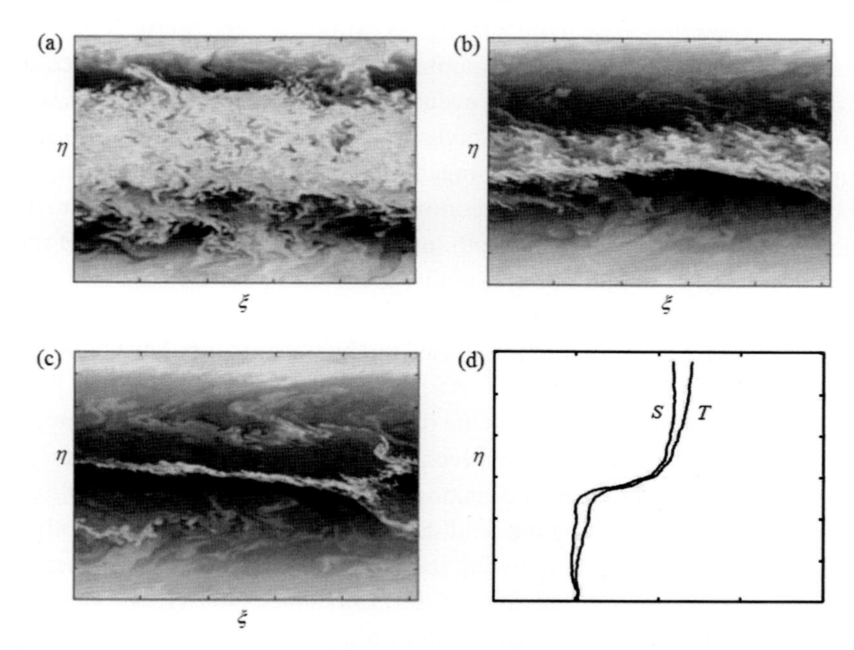

Figure 7.20 An experiment resulting in the staircase-type intrusion. The temperature perturbation is shown at various stages of intrusion development in (a)–(c). The vertical profiles of temperature and salinity in the final state are shown in (d). From Hebert (2011).

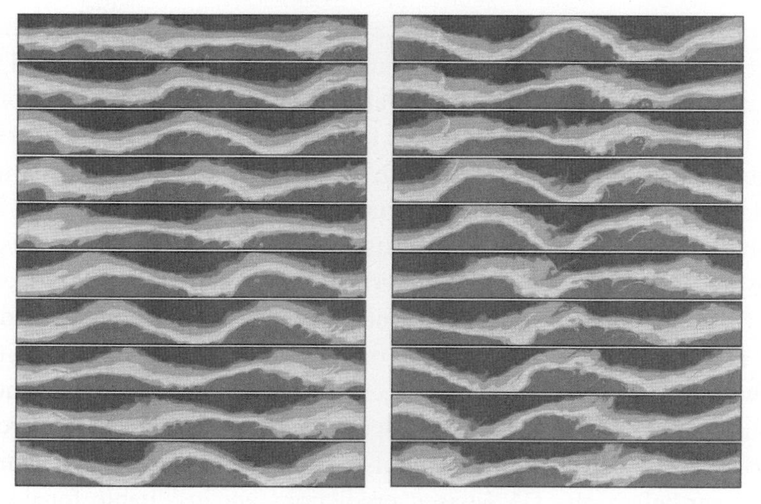

Figure 7.21 Generation of interfacial waves across the diffusive interface in the equilibrated intrusion (a tilted DNS experiment). Presented is the total salinity field in the immediate vicinity of the interface. Time is increasing downward along the left panels and then continues at the top right panel. From Simeonov and Stern (2007).

DNS of intrusions driven by molecular fluxes (Simeonov and Stern, 2008). Merging appears to be a generic property of double-diffusive structures, not limited to interleaving. For instance, layer-merging events are common in thermohaline staircases and play a prominent role in the establishment of a final quasi-equilibrium state (Huppert, 1971; Radko, 2005). For finger-driven intrusions in DNS, mergers have only been briefly mentioned by Simeonov and Stern (2007), although the merging phenomenon deserves further in-depth investigation. The significance of merging, should it prove to be ubiquitous in doubly diffusive fluids, lies in the appearance of structures with dimensions significantly exceeding the scale of the fastest growing modes. In this case, most predictions based on the linear theory have to be critically reevaluated.

An interesting question concerns the direction and magnitude of the net vertical heat flux in the equilibrium finger-driven intrusions. As discussed earlier (Section 7.1), large-scale advection by intrusions results in up-gradient heat flux: warm water is advected upward and the cold downward. Salt fingers, on the other hand, transfer heat down gradient, opposing the large-scale intrusive component. Based on tilted DNS, Simeonov and Stern (2007) conclude that (i) the net heat flux could be either upward or downward, depending on the background parameters; (ii) its large-scale and finger components are comparable in magnitude; and (iii) the presence of intrusions reduces the magnitude of the heat flux. Properties of the salt flux are markedly different. The net salt flux is dominated by the contribution from fingers and therefore is directed down gradient. Somewhat counterintuitively, the development of intrusions actually increases the net salt flux. This happens because the large-scale gradients in the salt-finger zone of a fully developed intrusion are elevated relative to the background value and so are the finger fluxes of salt. The adverse large-scale intrusion salt flux is too weak to compensate for the increased finger flux and the net salt transport with intrusions is higher than without.

The lateral heat and salt fluxes diagnosed by Simeonov and Stern (2007) are down gradient and correspond to diffusivities of $K_L \sim 2\,\mathrm{m^2\,s^{-1}}$. This value is likely to underestimate the typical interleaving diffusivities in the ocean, given that (i) for computational reasons, Simeonov and Stern employed a diffusivity ratio of $\tau = \frac{1}{6}$, significantly higher than the heat–salt value of $\sim$0.01; (ii) simulations were two-dimensional, which generally under-represents the intensity of double-diffusive processes; and (iii) calculations were limited to relatively strong fronts with $a = \frac{\bar{S}_x}{\bar{S}_z} > 0.0075$ (the lateral eddy diffusivities increase with decreasing a). Nevertheless, the DNS-based predictions are invaluable for constraining the plausible range of diffusivities – the estimates reported in the literature go as high as $K_L \sim 10^3\,\mathrm{m^2\,s^{-1}}$ – and placing a lower bound on this hard-to-quantify number.

Ultimately, the lateral fluxes of heat and salt are set by the equilibrium large-scale velocity $U_{\max}$. Simeonov and Stern (2007) find that the numerical data are adequately described by the relation

$$U_{\max} = CNH, \tag{7.14}$$

where $N = \sqrt{-\frac{g}{\rho}\frac{\partial \rho}{\partial z}}$ is the buoyancy frequency, H is the intrusion thickness and $C \approx 0.14$ is the non-dimensional coefficient calibrated using a series of tilted DNS. We shall return to the discussion of (7.14) in the context of laboratory experiments and bounded models (Section 7.4).

7.4 Laterally bounded fronts

All simulations and theoretical intrusion models described so far (Sections 7.1–7.3) consider effectively unbounded fronts. The usage of the term unbounded in double-diffusive literature is somewhat loose. It is generally understood that "unbounded" implies several related conditions: (i) the front width significantly exceeds the typical intrusion wavelength, in which case, periodic lateral boundary conditions are adequate; (ii) the intrusions are controlled by the background T–S gradients; and (iii) all that happens outside the front hardly matters in terms of its effect on the intrusion characteristics. Such conditions may exist in relatively wide fronts but not in the sharp ones. Therefore, a distinct branch of intrusion theory is focused specifically on finite-width thermohaline fronts. There are a few obstacles along this path. Most notably, the simple and convenient periodic boundary conditions used for the unbounded model no longer apply. Boundary conditions are often replaced by the requirement of no flux across the frontal margins ($u = 0$ at $x = 0, L_x$) but even these should be viewed, at best, as a plausible approximation of the conditions realized in nature. On the positive side, the finite-width case can be easily modeled by laboratory experiments and is more amenable to numerical treatments.

The first rigorous, aside from minor untidiness with respect to the lateral boundary conditions, quantitative analysis of the bounded system was performed by Thorpe *et al.* (1969). Using a combination of linear stability arguments and supporting laboratory experiments, Thorpe *et al.* examined the stability of a fluid layer with vertical salinity and horizontal temperature gradients. The diffusion of properties in this configuration was due to molecular (rather than eddy-driven) fluxes, which simplified the analysis and permitted the unambiguous comparison with experiments. In contrast to the unbounded case, which is always unstable,

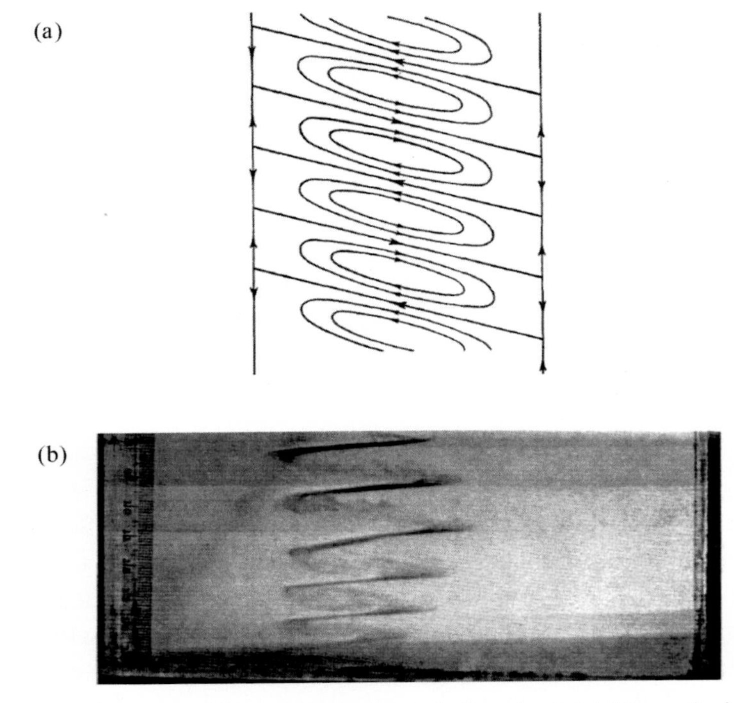

Figure 7.22 Thorpe *et al.*'s model of a thermohaline front. (a) Theoretical configuration; (b) corresponding laboratory experiments. From Thorpe *et al.* (1969).

bounded systems can be stable provided that the horizontal gradients are sufficiently weak. Thorpe *et al.* predicted marginal instability conditions and the corresponding circulation patterns, taking the form of slightly tilted intrusive structures (Fig. 7.22a).

Overall, theory and experiments agreed well. However, the experiments revealed one salient effect that was not captured by linear theory – the amalgamation of adjacent intrusions. While theory predicted the appearance of a series of cells with an alternating circulation pattern (Fig. 7.22a), the observed cells had the same sense of rotation, with the motion up (down) the heated (cold) boundary. The coalescence of adjacent cells takes place in the very early stages of development and therefore the detection and analysis of primary instabilities in experiments is difficult. Thangam *et al.* (1982) and Kerr (1990) trace the origin of the amalgamation to fundamentally nonlinear interactions that enhance the convection cells characterized by fluid rising near the heated wall and sinking at a certain distance from it. The initial effort to explain molecularly driven interleaving on fronts of finite width stimulated numerous extensions of Thorpe *et al.*'s model, of which it is important to mention the extension to a rotating system (Kerr, 1995). In the bounded configuration, the

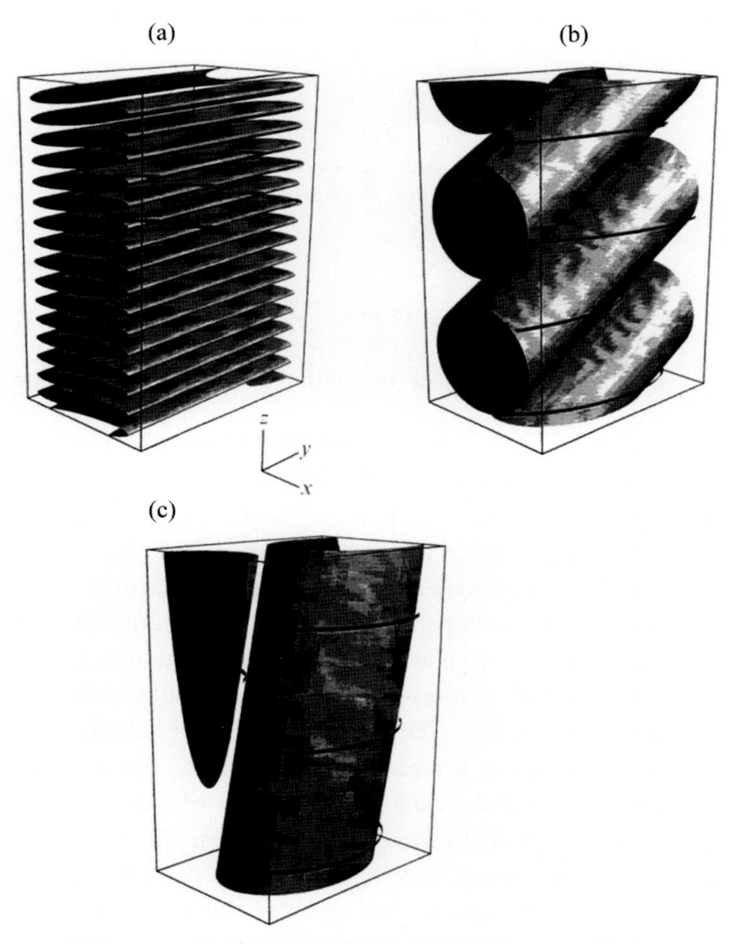

Figure 7.23 Effects of rotation on interleaving at finite-width fronts. Increase of rotation rate in (a)–(c) increases the vertical scales and slopes of intrusions. From Kerr (1995).

effects of rotation are more profound and have a generally destabilizing effect on interleaving. An increase in the rotation rates steepens intrusions in both along- and cross-front directions, as illustrated in Figure 7.23.

A counterpart to the Thorpe *et al.* (1969) analysis for salt-finger-driven intrusions was proposed by Niino (1986). Oddly enough, Niino's analysis, in contrast to Thorpe *et al.*'s model, reveals the unconditional instability of finite-width thermohaline fronts with respect to interleaving instabilities. A direct analogy between molecularly and salt-finger-driven intrusions has been generally accepted and heavily exploited in the interleaving theory, making Niino's counter-example particularly intriguing. Among numerous dynamical insights, the Niino (1986) study reveals the significance of two key non-dimensional parameters: the stability parameter

$$R_{\text{Niino}} = \frac{\left[g(1-\gamma)\beta \Delta S_{\frac{1}{2}} \right]^6}{K_S K_M a^2 N^{10}} \tag{7.15}$$

and the stratification parameter

$$\mu_{\text{Niino}} = g(1-\gamma)\frac{\beta \bar{S}_z}{N^2}. \tag{7.16}$$

Here, $\Delta S_{\frac{1}{2}}$ is the half-difference of the salinity variation across the front, a is its half-width and $N = \sqrt{-\frac{g}{\rho}\frac{\partial \rho}{\partial z}}$ is the buoyancy frequency. The values of R_{Niino} and μ_{Niino} determine whether the front could be considered wide [for $R_{\text{Niino}} > 2 \cdot 10^5 (1 + \mu_{\text{Niino}})^{4.9}$] or narrow [for $R_{\text{Niino}} < 40(1 + \mu_{\text{Niino}})^{5.4}$].

Niino's model assumed uniform eddy diffusivity of salt ($K_S = const$). Its long-awaited generalization, which took into account dependence of the salt-finger fluxes on the density ratio, was developed by Simeonov and Stern (2004). Using linear stability analysis and supporting numerical simulations, these authors demonstrated that finite-width thermohaline fronts can be linearly stable, provided that the lateral gradients are sufficiently weak. The critical cross-front salinity variation (ΔS_{cr}) required for instability rapidly decreases with decreasing density ratio. This result marks the second reversal of the view on the stability of weak fronts: from Thorpe *et al.* (1969) to Niino (1986) to Simeonov and Stern (2004). The disagreement between these studies is instructive. It underscores the urgent need to develop accurate physically based parameterizations of salt-finger transport.

No discussion of interleaving would be complete without reference to the Ruddick and Turner (1979) experiment. The laboratory configuration, ingenious and simple at the same time, is shown in Figure 7.24a. The lab tank was initially divided by a removable dam and both sides were filled, using the double-bucket method, with stably stratified aqueous solutes: sugar solute on one side and salt solute on the other. As in most laboratory experiments, sugar stratification (slower diffuser) represented salinity in the ocean and salt stratification (faster diffuser) – the oceanic temperature distribution. Great care was taken to ensure identical density stratification in both compartments. After the removal of the barrier, active interleaving ensued (Fig. 7.24b).

The main focus of the analysis in Ruddick and Turner (1979) was on the vertical scale of the intrusions. These authors used imaginative physical arguments based on the energetics of a thermohaline system to show that intrusion thickness is bounded from above and from below by scales proportional to $\frac{(1-\gamma)g\beta\Delta S}{N^2}$; here

(a)

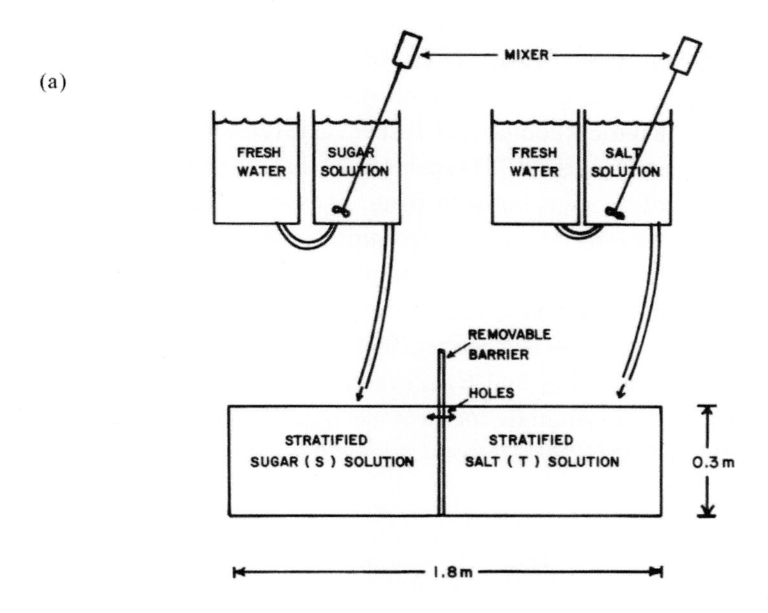

(b)

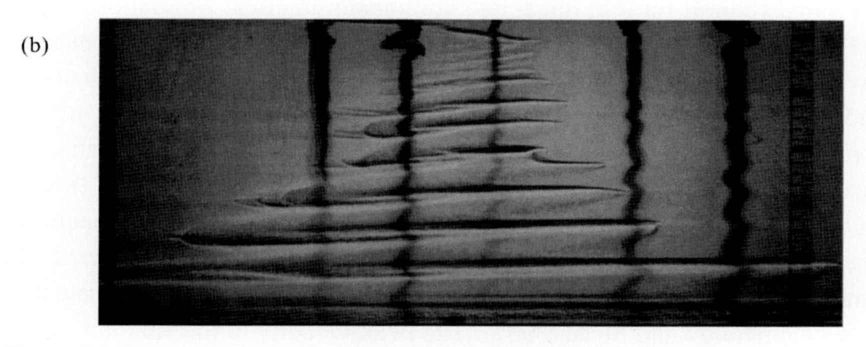

Figure 7.24 The experimental setup (a) and the resulting flow pattern (b) in Ruddick and Turner's experiments. From Ruddick *et al.* (1999).

ΔS is the lateral variation in salinity at a given level. This suggestion was validated by experiments. The best fit to the laboratory data made it possible to pin down the proportionality constant:

$$H_{\mathrm{RT}} \approx \frac{3}{2}\frac{(1-\gamma)g\beta\Delta S}{N^2}. \tag{7.17}$$

The monotonic increase in intrusion thickness towards the bottom of the tank in Figure 7.24b, where ΔS is the largest (Fig. 7.24b), is consistent with the proposed dependence of H_{RT} on ΔS. Scaling (7.17) also appears in several earlier models of finite-width interleaving (Thorpe *et al.*, 1969; Chen *et al.*, 1971). In the order-of-magnitude sense, it is equivalent to the so-called Chen scale – the height through which a heated fluid element must rise in order to become neutrally buoyant in a given density gradient (Chen *et al.*, 1971; Chen, 1974). It is also occasionally referred to as the "natural" length scale in intrusion literature (Turner, 1979).

It should be noted that the Ruddick and Turner scale (H_{RT}) is fundamentally different from the Toole and Georgi (1981) scale (H_{TG}) that emerges in the unbounded model (7.2). Such a difference is not surprising. In essence, Ruddick and Turner's experiment models intrusions in a very sharp, initially discontinuous front. In terms of frontal width (L), their experiment and the unbounded models (e.g., Toole and Georgi, 1981) represent the opposite ends of the whole spectrum of intrusion models. Niino (1986) and Simeonov and Stern (2004) demonstrated, in different ways, that the two extreme values, H_{RT} and H_{TG}, can be recovered from the finite-width model by considering the asymptotic limits of narrow ($L \to 0$) and wide ($L \to \infty$) fronts. Simeonov and Stern (2004) introduced yet another scale that is relevant for fronts of intermediate width:

$$H_{\mathrm{SS}} \sim \left(\frac{K_T L}{N}\right)^{\frac{1}{3}}, \tag{7.18}$$

and found support for it in linear and nonlinear numerical simulations.

There are many other major differences and intriguing similarities between interleaving on sharp and wide fronts. For instance, the unbounded intrusion models tend to produce step-like vertical *T–S* profiles with a large fraction of the intrusion occupied by well-mixed convective regions (e.g., Fig. 7.18). Such convective regions are notably absent in the Ruddick and Turner experiments (Fig. 7.24b). On the other hand, the propagation velocity of intrusions in experiments, as measured by the spreading of intrusion noses, is adequately represented by (7.14), the same relation that describes the equilibrium velocity of intrusions in the unbounded model (Simeonov and Stern, 2007). The peak velocity in the Ruddick and Turner experiments, realized at the center of the front, exceeds the nose extension velocity by a constant factor of 3.7 and therefore also conforms to (7.14). The values of the non-dimensional coefficient C are different: $C = 0.005$ for intrusion noses in the lab and $C = 0.14$ in Simeonov and Stern's simulations. However, the difference in coefficients can be readily attributed to the different molecular properties of the diffusing substances: $(Pr, \tau) = (10^3, \frac{1}{3})$ in the sugar–salt case and $(Pr, \tau) = (7, \frac{1}{6})$ in Simeonov and Stern (2007).

Intrusions in sharp fronts are inclined in the same sense as in the unbounded model, with the intrusion slope of the same sign as the slope of T and S surfaces. In a nice example of a sequel that lives up in quality to the original, Ruddick *et al.* (1999) examined mechanisms setting the geometry of intrusions in the Ruddick and Turner configuration. The essence of the Ruddick *et al.* theory lies in the hypothesis of continuous geostrophic adjustment. This hypothesis assumes that the fluid elements in the finger region, made more (less) buoyant by the salt-finger fluxes, continuously rise (descend) to the ambient density level that matches their new density. As a result, the zero-order density distribution could be taken as horizontally uniform; the circulation in such a configuration is driven by a small perturbation to the basic adjusted state. Ruddick *et al.* (1999) demonstrate that all key experimental measurements are consistent with the theoretical model, a-posteriori validating the adjustment hypothesis. Theory in Ruddick *et al.* (1999) attributes the orientation of intrusions in sharp-front experiments to the dominance of the salt-finger transport. On the other hand, if the system is controlled by the diffusive buoyancy fluxes, the slope of intrusions is expected to be of opposite sign. Both inferences find support in oceanographic field measurements (Ruddick, 1992). With regard to the significance of the hypothesis of continuous hydrostatic adjustment, it should be mentioned that its relevance is likely to extend far beyond the intrusion problem, with numerous applications to be found in other slowly evolving buoyancy-driven systems.

The next step in the laboratory study of intrusions was made by Krishnamurti (2006). The initial state in these experiments consisted of density-compensated gradients of T and S modeled by salt and sugar solutes. The smooth horizontal gradients in Krishnamurti's setup replaced the discontinuous initial variation of T and S in Ruddick and Turner's experiment (Fig. 7.24). This change made the experimental configuration markedly more realistic – "realism" is defined in this context by dynamic similarity to typical oceanic conditions. By varying the vertical T–S gradients, Krishnamurti was able to explore interleaving in various types of background stratification: salt-finger favorable, diffusive and doubly stable. In all cases, intrusions were pronounced and visibly different from those observed by Ruddick and Turner (1979).

The most obvious difference is the presence, in Krishnamurti's experiments, of extended regions occupied by convection (Fig. 7.25). These convecting layers play a critical role in intrusion dynamics. Particle image velocimetry (PIV) has made it possible to analyze details of the velocity field, including direct calculation of the Reynolds stress. Experiments show that momentum transfer in the salt-finger zone is minimal, far less than molecular friction and thus completely negligible for most intents and purposes. Note that the same conclusion has been reached in numerical studies (e.g., Stern *et al.*, 2001), leaving little doubt regarding its validity. The momentum balance is very different in the convective regions, where momentum transfer is significant and generally up-gradient, acting against molecular friction. Thus, it is the convective Reynolds stress that maintains the intrusions against viscous dissipation. A corollary of this observation is that the dynamics of intrusions in distributed fronts can be substantially different from those operating across sharp fronts (Ruddick and Turner, 1979) where convective zones are virtually absent.

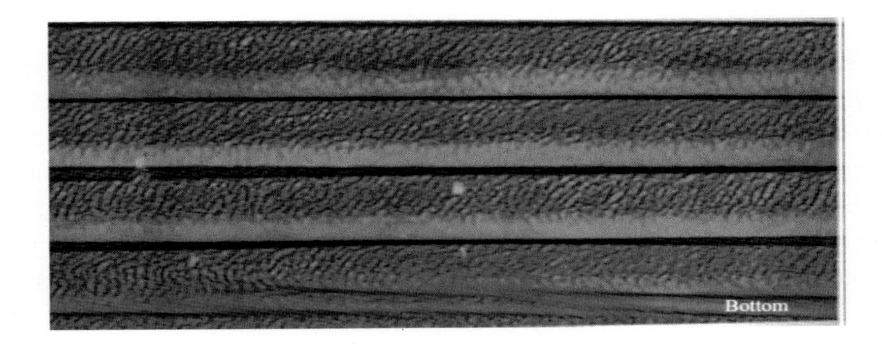

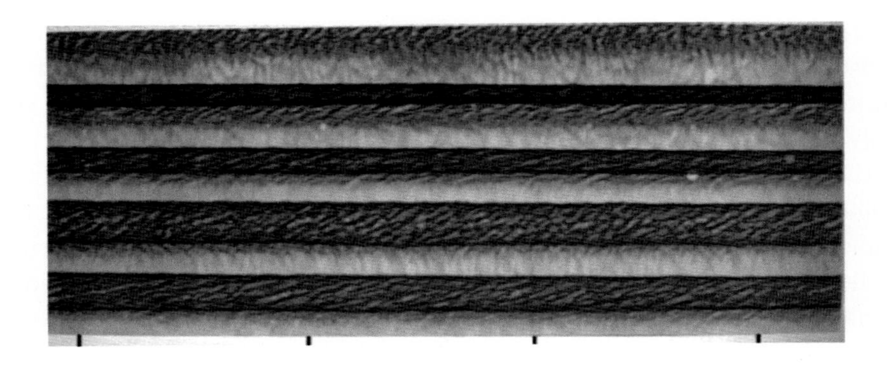

Figure 7.25 Laboratory experiments on interleaving in finite horizontal gradients. Active interleaving is observed in all configurations, including salt-finger favorable stratification (top) diffusive (center) and doubly stable (bottom). From Krishnamurti (2009).

Another distinct feature of finite-width fronts concerns the evolution of the intrusion slope. Initially, intrusions in Krishnamurti's experiments with finger-favorable stratification tilted towards the high T and S region. This is opposite to the orientation of intrusions in the Ruddick and Turner experiment, or, for that matter, in the unbounded models (Section 7.1). The "incorrect" initial inclination may be attributed to molecularly driven interleaving (Holyer, 1983). Shortly afterward, intrusions became nearly horizontal. The "correct" tilt, corresponding to the slope of intrusions of the same sign as the slope of isotherms, was observed when intrusions entered a stable region without horizontal background gradients. Such a variation of the intrusion slope in time – not only in the magnitude but even in sign – points to a potentially serious problem for the analysis of intrusions in field measurements. If the slope of thermohaline intrusions varies dramatically even in the controlled setting of laboratory experiments, then the prospects for reliable interpretation of oceanic intrusions based on the statistics of their slope are limited. Attempts of this nature (e.g., Beal, 2007) have to be critically evaluated since the situation in the ocean is confounded by the simultaneous presence of numerous external processes: gravity waves, mesoscale variability, horizontal and vertical shears.

7.5 Sidewall heating experiments

A certain group of interleaving studies can be combined under the heading of sidewall heating (cooling) experiments. The underlying theoretical concept is one of a semi-infinite fluid forced by the temperature, and possibly salinity, flux at the boundary. In the context of laboratory or numerical experiments, this implies that the processes under consideration are sufficiently localized to the region where forcing is applied and the system is largely unaffected by the presence of distant boundaries. An argument can be made that sidewall heating experiments come closer to representing oceanic intrusions. There are no walls in the ocean interior and having only one dynamically significant boundary is, perhaps, better than having two (as in the experiments of Section 7.4). Some of the experiments by Thorpe *et al.* (1969) discussed earlier were performed in sufficiently wide containers and therefore can be placed into the sidewall heating category.

A more focused investigation was performed by Tanny and Tsinober (1988). These authors used the classical configuration (Thorpe *et al.*, 1969). The working fluid, initially isothermal but vertically salt-stratified, was heated from a sidewall. However, the experimental parameters were chosen to ensure that intrusions forming near the heated boundary do not spread laterally across the whole laboratory tank but remain localized near the heated boundary throughout most of the experiment. A typical sequence of events in such experiments is illustrated in Figure 7.26. The first stage is the generation of small-scale cells with order one aspect ratio. In time, cells extend horizontally and sequentially merge, forming larger and more developed intrusions. Traces of dye in Figure 7.26 also indicate that the extent of penetration of shearing motion in the tank interior exceeds that of warm fluid that originated near the wall. The relevant scale of the intrusive layers is consistently represented by the Chen scale:

$$H_C = \frac{g\alpha\Delta T}{N^2} = \frac{\alpha\Delta T}{\beta\left|\bar{S}_z\right|}, \qquad (7.19)$$

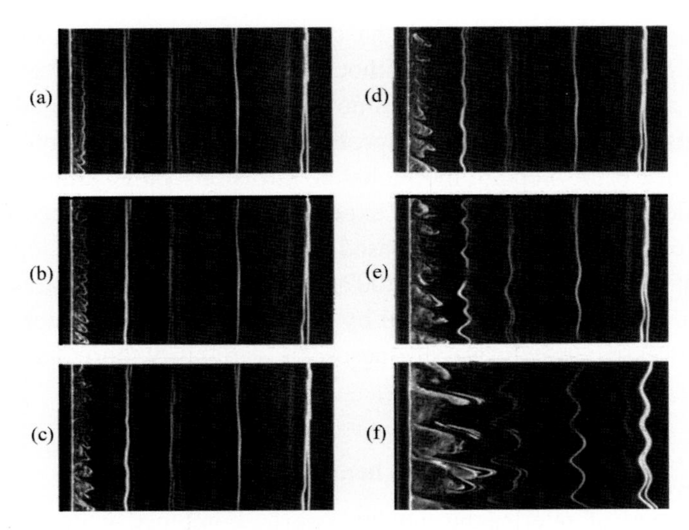

Figure 7.26 Sidewall heating experiments at various stages in (a)–(f). Note the appearance of initially small-scale slanted structures near the heated wall in (a) and (b) and their subsequent merger into the fully developed intrusions in (e) and (f). From Tanny and Tsinober (1988).

and various sidewall heating experiments suggest a relatively well-constrained range of intrusion thickness (h):

$$0.6H_C < h < H_C. \tag{7.20}$$

The intrusive instability in the sidewall heating experiments occurs when the Rayleigh number $R = \frac{g\alpha\Delta T H_C^3}{\nu k_T}$ based on the Chen scale exceeds a critical value ($R_{cr} \sim 1.5 \cdot 10^4$). Curiously enough, combining this observation with (7.19), we conclude that the vertical scale of the initial instabilities is

$$h_I = C\left(\frac{k_T \nu}{g\beta S_z}\right)^{\frac{1}{4}}, \tag{7.21}$$

which is analogous to the nominal double-diffusive scale d introduced in Chapter 1. The prefactor in (7.21) deduced from the laboratory experiments is $C \approx 29.1$. A more detailed analysis by Kerr (1989) suggests that the most relevant stability parameter for the sidewall heating problem is given by

$$Q = \frac{(1-\tau)^6 g(\alpha\Delta T)^6}{\nu k_S l^2 \left|\beta \bar{S}_z\right|^5}, \tag{7.22}$$

where l is the horizontal length scale, initially set by molecular diffusion of temperature from the heated boundary. Note the structural similarity of Kerr's parameter to the stability criterion (7.15) for the salt-finger driven intrusions in bounded fronts (Niino, 1986).

If heating is strong and persistent, the sidewall temperature increases far beyond the critical value for instability and the natural scale of intrusions (7.19) increases accordingly. In this regime, the final layer thickness is expected to significantly exceed its initial value (h_I). The merging events illustrated in Figure 7.26 represent an efficient physical mechanism for the interleaving system to conform to this requirement. The mergers are initiated in the immediate vicinity of the heated wall and then spread into the interior along the intrusion fronts. A detailed inspection of mergers in the sidewall heating experiment reveals that they take one of two forms: either (i) the high-gradient interface separating intrusions weakens and eventually disappears, resulting in the coalescence of the adjacent cells, or (ii) the interface drifts vertically causing one of the intrusions to extend vertically at the expense of the adjacent one, which shrinks and ultimately disappears. Such evolutionary patterns – referred to as *B*-mergers and *H*-mergers respectively (Radko, 2007) – are common in various layered systems. Double-diffusive periodic structures just happen to be particularly susceptible to coarsening through mergers.

The "iceblock" configuration (Huppert and Turner, 1980) was designed to investigate the role of double-diffusive processes in the melting of icebergs, calved from Arctic and Antarctic glaciers. One of the goals of this study was to assess the feasibility of towing icebergs towards the coasts and harvesting the fresh meltwater from them. Who said that double-diffusers are not concerned by societal needs? In Huppert and Turner's experiment, the iceblock was vertically inserted into salt-stratified water at rest. Despite different geometry and boundary conditions (in addition to thermal forcing, melting ice also produces the associated freshwater flux) dynamic similarities between the iceblock and sidewall heating experiments are apparent. Shortly after insertion of the iceblock, the flow pattern became dominated by the laterally spreading intrusions (Fig. 7.27, top panel). The intrusions tilt slightly upward, away from the iceblock, which is consistent with the dominance of fluxes across sharp diffusive interfaces in the strongly diffusive background stratification. Interleaving persisted even after the removal of the iceblock (Fig. 7.27, bottom panel) although the intrusions became more horizontal and voluminous. What is particularly striking is the extent to which quantitative inferences from the sidewall heating studies apply to the iceblock experiment. For instance, the thickness of intrusions measured by Huppert and Turner falls within the sidewall heating range (7.20). The laboratory-based predictions find support in field measurements taken in the vicinity of Antarctic glaciers (Jacobs *et al.*, 1981). Such agreement is yet another indication of the very robust physics of interleaving.

During the past three decades, numerous other extensions of the sidewall heating problem have been considered. Quite a few questions have been answered, many

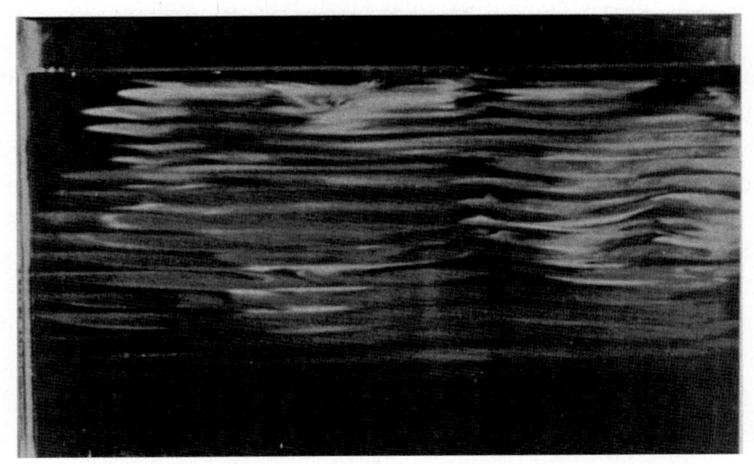

Figure 7.27 Melting iceblock experiment. Dye has been frozen into the ice to visualize ensuing interleaving motions. The top panel shows the state after inserting the iceblock and the bottom panel after its removal from the tank. From Huppert and Turner (1980).

more remain. Our intention is not to review here all aspects of sidewall heating but rather to convey a sense of the interesting and rich dynamics of the problem. Still, some key developments should be highlighted:

(i) A series of papers examined intrusions developing from localized sources, such as heating at a point (Tsinober *et al.*, 1983; Belyaev and Chashechkin, 1989; among others) or injection of fluid with anomalous properties (e.g., Turner, 1978; Nagasaka *et al.*, 1995);

(ii) Narusawa and Suzukawa (1981) and Schladow *et al.* (1992) imposed a fixed heat flux, instead of the temperature difference in classical experiments, and examined ramifications of this change for stability of the system;
(iii) Huppert *et al.* (1984) performed a series of sidewall heating experiments with various substances and discovered that intrusions are largely insensitive to molecular characteristics (Pr and τ);
(iv) Chereskin and Linden (1986) explored the (limited) effects of rotation on sidewall heated intrusions;
(v) Chen and Chen (1997) studied salt fingers developing in the convection cells and emphasized the importance of incorporating finger-driven fluxes into theoretical models of sidewall heated intrusions;
(vi) Dijkstra and Kranenborg (1998) and Kranenborg and Dijkstra (1998) reproduced key features of laboratory experiments using two-dimensional numerical simulations;
(vii) Chan *et al.* (2002) used laboratory experiments to examine the three-dimensional structure of intrusions associated with the appearance of secondary horizontal recirculation cells;
(viii) Malki-Epshtein *et al.* (2004) examined the effects of a distant wall on the propagation of intrusions and quantified conditions for which the tank length appears as an important variable.

Perhaps the most important message that we can take from this extensive body of literature is that interleaving in sidewall heating experiments is not particularly sensitive to the specifics of each setup. The general evolutionary patterns of intrusions, their geometry and mechanics, exhibit strong similarities across various experimental configurations and parameter regimes. Of course, quantitative differences exist. They are interesting and should be investigated further. However, it is comforting to know that major insights into intrusion dynamics are likely to be transferable between different systems. In the next section we shall discuss to what extent these insights might be relevant for interleaving in the ocean.

7.6 Oceanographic observations

Given all the effort invested in theoretical, laboratory and numerical modeling of interleaving, given the elegant and generally consistent physical picture that emerges, it is incredibly tempting to apply our hard-won understanding directly to all intrusive structures in the ocean. But is it reasonable to assume that the dynamics of oceanic intrusions are, in most cases, governed by the same physical principles as intrusions in the laboratory and theoretical models? Remarkably, after almost a half-century of theorizing, modeling and observing, we are still in doubt. The lack of definitive answers cannot be attributed to the rarity of the phenomenon. On the contrary, intrusions are ubiquitous; almost any temperature and salinity profile in the world ocean contains inversions indicative of lateral thermohaline intrusions. It is also not because of insufficient effort. A decade-old review of observations (Ruddick and Richards, 2003) already referred to 156 predominantly field-based studies of interleaving. The principal difficulty is that there are a number of processes – mesoscale variability, external shears, internal waves, to name a few – that can potentially produce intrusions in the ocean or at least substantially affect their evolution. These factors are a-priori excluded in the majority of idealized studies. We know that double-diffusion is responsible for interleaving in Ruddick and Turner's experiments; there is nothing else to blame. Nature is more diverse and complicated.

Of course, there are several well-known examples of interleaving for which observational evidence of their double-diffusive origin is undeniable and the dynamics are adequately captured by conventional theories. We shall present such cases first. These success stories will be followed by more controversial observations, which seem to contradict, at least in some respects, the basic double-diffusive theory of interleaving; alternative explanations will be discussed. The section ends with the list of considerations that we believe should be taken into account to reconcile oceanographic measurements with theoretical models.

Success stories

While interleaving is widespread in the world ocean, much observational effort for its detection is focused on regions characterized by strong lateral gradients of temperature and salinity. There are several reasons for that. Since interleaving is driven by lateral gradients of temperature and salinity, it is natural to expect it to be more active in areas where the gradients are large. At the same time, property contrasts between adjacent intrusions are higher if interleaving occurs at a boundary between distinct water masses, which helps to identify and analyze interleaving in field measurements. Finally, the critical large-scale consequences of interleaving may be related to its ability to diffuse sharp lateral gradients (Garrett, 1982; Schmitt, 1994b). It is generally accepted that the ocean is actively stirred by mesoscale (10–100 km) eddies driven by baroclinic instability. Mesoscale variability, however, does not result directly in irreversible mixing but, rather, acts as a catalyst by transporting tracers towards high-gradient fronts where mixing is finalized by some irreversible diabatic processes. Interleaving may represent this elusive "missing link" between adiabatic mesoscale stirring and the ultimate destruction of T–S variability by molecular dissipation.

The largest lateral gradients in the ocean are generally found in (i) quasi-permanent frontal regions separating global-scale water masses and (ii) isolated coherent vortices, which are able to transport a substantial volume of trapped fluid into regions with distinct ambient temperature and salinity properties. A clear example of frontal interleaving, supplemented by its comprehensive analysis, is given by Joyce *et al.*'s (1978) observations of the polar front of the Antarctic Circumpolar Current in the Drake Passage. Within the front, water masses were observed to intrude with characteristic vertical scales of 50–100 m. Joyce *et al.* argue that at least three features of the observed intrusions are consistent with double-diffusive dynamics: the vertical buoyancy flux is up-gradient; the intrusions tilt in the direction expected for salt-finger driven interleaving; and the pattern of vertical stratification conforms to the structure of typical thermohaline intrusions. A number of studies (Joyce, 1976; Williams, 1981; Schmitt and Georgi, 1982) document intrusive fine-structure across the Gulf Stream and North Atlantic Current and provide evidence for their salt-finger driven dynamics.

Some of the most spectacular examples of interleaving come from the Arctic regions where, due to the relatively low energy environment, double-diffusion could be a dominant mechanism for vertical and, through interleaving, horizontal mixing. An abundance of well-defined intrusions, extending laterally for hundreds of kilometers, is a characteristic feature of the Arctic environment. In the Arctic, intrusions have been observed in all types of background stratification: salt-finger favorable, diffusive and double-diffusively stable. While density stratification is stable, vertical temperature and salinity profiles are frequently characterized by zigzag patterns with numerous inversions – the tell-tale sign of thermohaline interleaving. Representative examples are shown in Figure 7.28 (measurements in the Laptev Sea, Rudels *et al.*, 2009). The instantly recognizable signatures of interleaving also appear in the *T–S* diagram (Fig. 7.28, bottom right): properties vary in a coherent manner consistent with their partial density-compensation. The very sharp features of all profiles in Figure 7.28 are also indicative of active double-diffusive processes; small-scale turbulence is generally characterized by the opposite smoothing tendency. Estimates of lateral diffusivities associated with intrusions are in the range of 100–1000 m s^{-2}, significantly exceeding that for mid-latitude interleaving. The ability of intrusions to profoundly affect high-latitude circulation has been emphasized by several studies (Perkin and Lewis, 1984; Walsh and Carmack, 2002; Rudels *et al.*, 2009). Of particular importance for the Arctic climate is the role of interleaving in mixing water masses of Pacific (colder and fresher) and Atlantic (warmer and saltier) origin (Carmack *et al.*, 1997; Rudels *et al.*, 1999).

A number of field studies report active interleaving in isolated coherent vortices. The most convincing example of this type is presented by meddy Sharon, one of

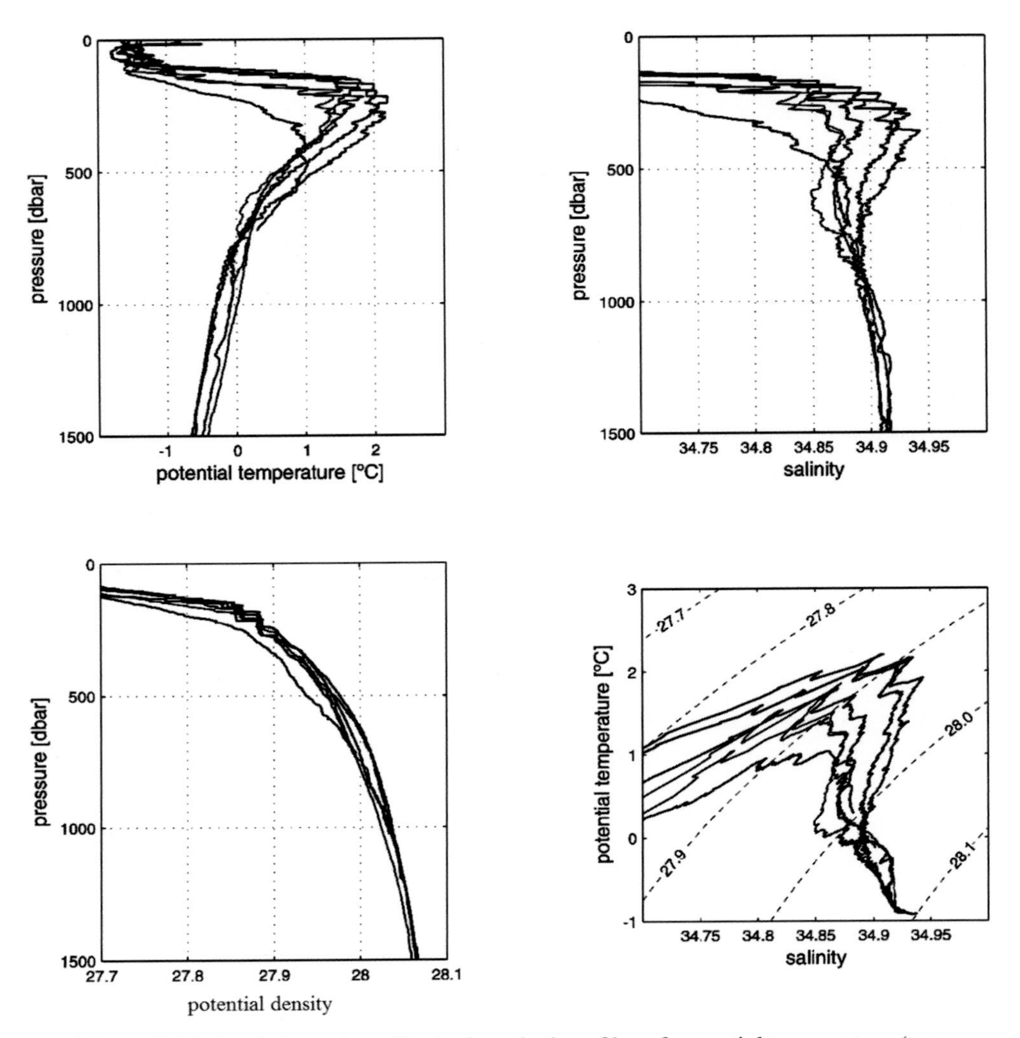

Figure 7.28 Arctic intrusions. Typical vertical profiles of potential temperature (top left), salinity (top right), potential density (bottom left) and the temperature-salinity diagram (bottom right) all exhibit numerous density-compensated inversions. From Rudels *et al.* (2009).

the intra-thermocline lenses of Mediterranean origin, which was systematically observed in the North Atlantic from 1984 to 1986. Sharon died a horrible death – she was eaten alive by thermohaline intrusions. The gradual erosion of heat, salt and velocity signatures was well documented and has been the subject of extensive analyses (Armi *et al.*, 1988, 1989; Hebert *et al.*, 1990; Ruddick, 1992, Ruddick *et al.*, 2010; May and Kelley, 2002). Stratification in the upper (lower) part of the meddy was diffusively (salt-finger) favorable, as shown in Figure 7.29a. Vertical diffusion

(a)

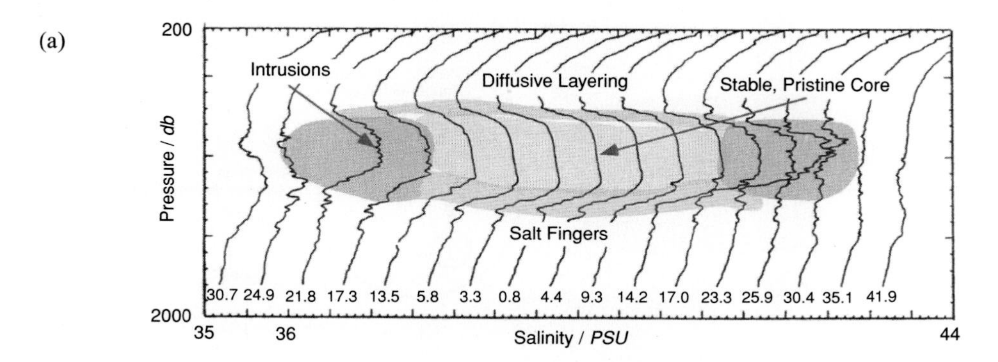

(b) (c)

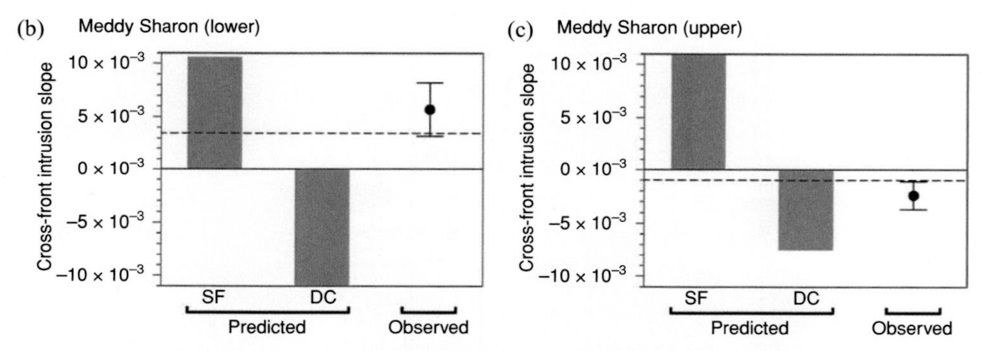

Figure 7.29 Observations of meddy Sharon. (a) Salinity structure (from Ruddick *et al.*, 2010). The range of intrusion slopes predicted on the basis of interleaving theories along with the observed values for the lower (b) and upper (c) parts of the meddy (from May and Kelley, 2002).

by itself was clearly insufficient to account for the observed erosion rates. The exclusion of lateral exchanges in the meddy heat/salt budget would lead to decay time scales of about twenty years, an order of magnitude underestimate (Hebert, 1988). This mismatch led to the suggestion that thermohaline intrusions, active and abundant at the meddy front (Fig. 7.29a), could be the primary mixing agent. The overall structure of intrusions and their slopes (Fig. 7.29b) were consistent with their double-diffusive origin in both diffusive and salt-finger regions. Predominantly baroclinic mechanisms (McIntyre, 1970) have been unambiguously ruled out (Ruddick, 1992). The vertical scales of intrusions and their advective velocities were in agreement with the predictions based on the sharp front model (Ruddick and Turner, 1979; Ruddick *et al.*, 1999).

An even more definitive calculation was made by Ruddick *et al.* (2010). Building on an earlier proposal (Joyce, 1977), Ruddick *et al.* utilized measurements of thermal microstructure to quantify interleaving-driven lateral mixing. The model

(a)

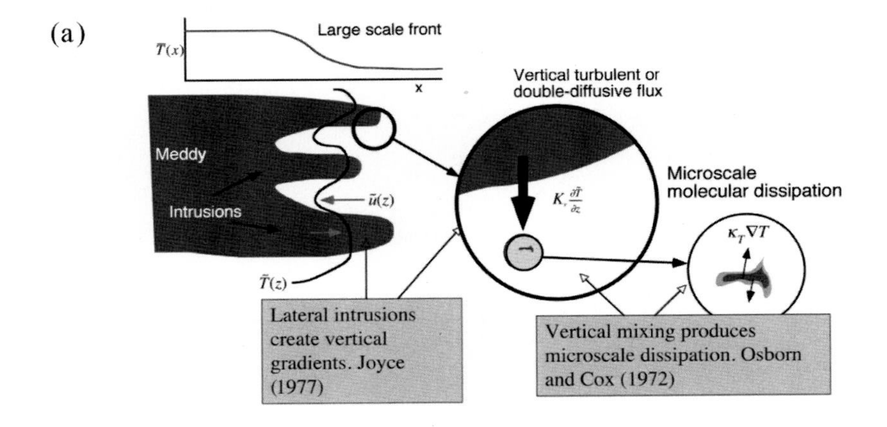

(b)

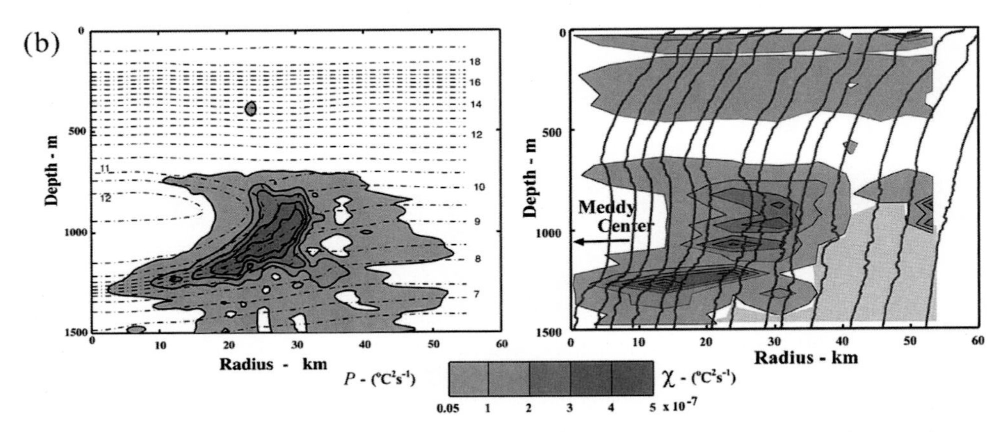

Figure 7.30 (a) Schematic illustration of the link between the production of thermal variance by the lateral interleaving motions and its ultimate dissipation by the microstructure. (b) Comparison of the production terms based on the erosion rate of meddy Sharon (left) with the microstructure dissipation (right). From Ruddick *et al.* (2010).

was based on the temperature variance equation, which under assumptions of statistically steady state and homogeneous turbulence reduces to the balance between advective variance production and molecular dissipation:

$$P = -2\overline{\vec{v}'T'} \cdot \nabla \bar{T} = 2k_T \overline{|\nabla T'|^2}, \tag{7.23}$$

where $\bar{T}$ represents the large-scale temperature distribution on scales greatly exceeding the dimensions of individual intrusions – see the schematic in Figure 7.30a. Joyce (1977) assumed that in regions of active interleaving the production term is dominated by the horizontal component associated with interleaving. Temperature variance is ultimately dissipated on the microscale (~cm), which reduces (7.23) to

$$P \approx -2\overline{u_i T_i}\frac{\partial \bar{T}}{\partial x} \approx 6k_T\overline{\left(\frac{\partial T_m}{\partial z}\right)^2}, \tag{7.24}$$

where subscripts i and m pertain to the intrusion and microstructure scales respectively. Note that (7.24) also assumes isotropy of microstructure, $\overline{\left(\frac{\partial T_m}{\partial x}\right)^2} \sim \overline{\left(\frac{\partial T_m}{\partial y}\right)^2} \sim \overline{\left(\frac{\partial T_m}{\partial z}\right)^2}$, a questionable but rather conventional step in observational mixing studies (Osborn and Cox, 1972).

Equation (7.24) makes it possible to evaluate the lateral production term $P = -2\overline{u_i T_i}\frac{\partial \bar{T}}{\partial x}$ from high-resolution microstructure measurements taken during the 1985 survey of meddy Sharon. The outward heat transport was estimated directly from the erosion rates of the meddy. These two completely independent estimates agreed nicely: not only typical values of P but their spatial distribution matched (Fig. 7.30b). This agreement is significant. First of all, it leaves no doubt that the erosion of the meddy should be attributed to thermohaline interleaving. It also validates key assumptions of Joyce's (1977) interleaving model: (i) the production of thermal variance in strong fronts is dominated by its horizontal component and (ii) the production–dissipation balance is satisfied even in the local sense.

Several attempts have been made to analyze interleaving in other coherent vortices. The results are encouraging but not as clear-cut as for Sharon. Schmitt *et al.* (1986) reported observations of intrusions at the edge of a warm-core Gulf Stream ring. Their fine- and micro-structure measurements revealed clear signatures of double-diffusive mixing. Intrusions were found to slope in the diffusive sense, which is consistent with microstructure data showing that diffusive interfaces in this ring were more unstable and had higher turbulence levels than nearby fingering regions. More quantitative comparisons were precluded by the complex three-dimensional structure of the intrusive features associated with wrapping of "streamers" of shelf, slope and Gulf Stream waters around the ring. The situation was further complicated by vertically sheared advection and high levels of mechanically generated turbulence.

Observations of thermohaline structures in the ocean have been invigorated in recent years by a new and highly promising technology – multichannel seismic imaging. Seismic imaging, which has been used by geophysicists for decades in the analysis of the Earth's subsurface, can also be used to display oceanic finestructure in unprecedented detail (Holbrook *et al.*, 2003). In a seismic survey, sound is sent from a ship-towed source, reflected by water masses with significant

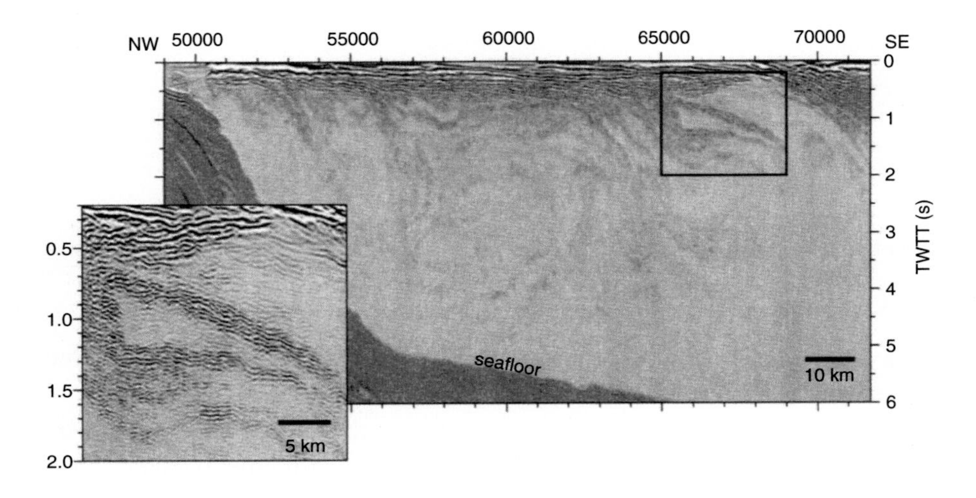

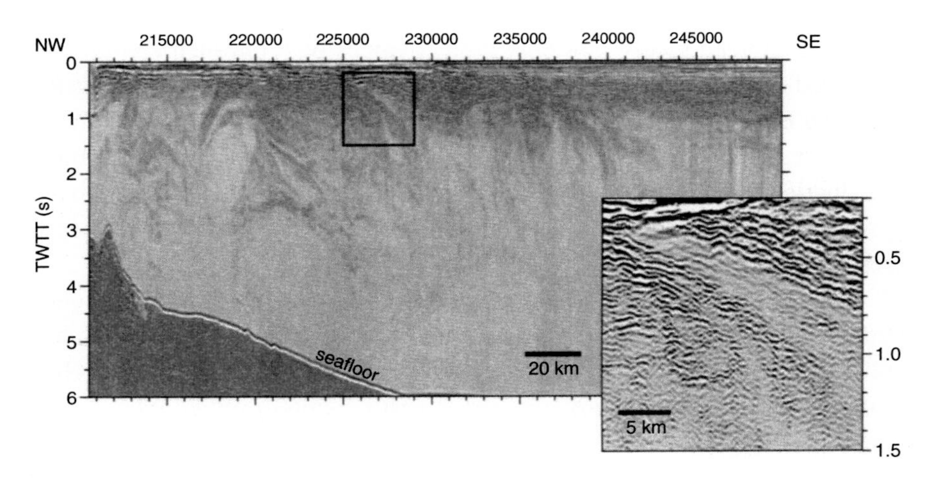

Figure 7.31 Seismic images off the coast of Newfoundland. Insets show an enlarged image of coherent slabs penetrating to ~1000 m depth, indicated by the corresponding boxed areas. From Holbrook *et al.* (2003).

temperature and salinity anomalies, and received by an array of towed hydrophones. The acoustic signal is then processed to map the thermohaline structure to depths of 1 km or more. The resolution attained by seismic imaging is on the order of 10 m in the vertical and approximately 100 m horizontally, which makes this technique perfectly suited for representation of secondary double-diffusive structures – intrusions and staircases. The acoustic impedance is affected by temperature and, to a lesser extent, by salinity. Therefore, seismic images could be interpreted as scaled maps of the temperature gradient (Ruddick *et al.*, 2009). Figure 7.31

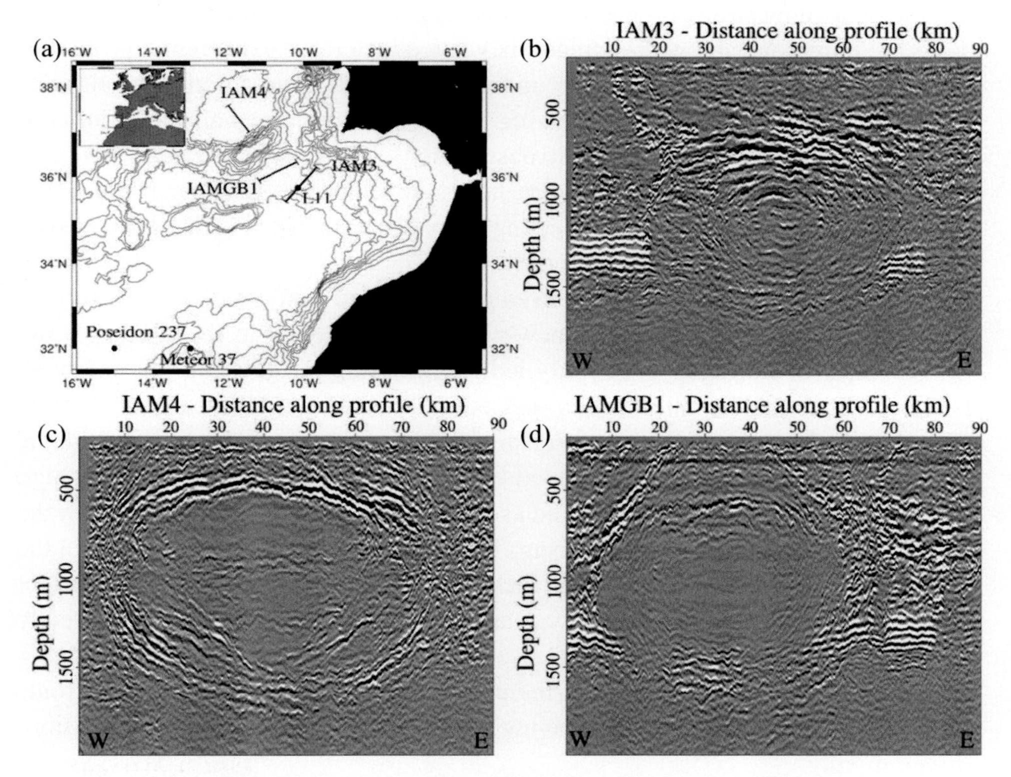

Figure 7.32 Thermohaline interleaving in meddies. (a) Map showing the geographical location of the study zone. Lines indicate the position of the multichannel seismic (MCS) profiles taken during the Iberian–Atlantic Margin (IAM) survey. MCS images of profiles (b) IAM3 on August 31, (c) IAM4 on August 29 and (d) IAMGB1 on September 7. From Biescas *et al.* (2008).

presents some of the first seismic images (Holbrook *et al.*, 2003). Clearly visible in these images is a series of well-defined laterally coherent intrusions that occupy a significant fraction of the upper ocean. The observational location at the front separating Gulf Stream from colder and fresher Labrador waters, as well as the characteristic finescale patterns, is consistent with the double-diffusive origin of interleaving (Ruddick, 2003).

Another promising location for the seismic analysis of intrusions is at the peripheries of meddies. Strong temperature and salinity variations between their interior and surrounding waters create perfect conditions for double-diffusive interleaving, as was illustrated earlier in the example of meddy Sharon. At the same time, temperature contrasts are associated with high reflectivity, which makes seismic images of meddies (Biescas *et al.*, 2008) most spectacular. The relatively homogeneous central parts of meddies (Fig. 7.32) are only weakly reflective. These core regions are surrounded by strong reflectivity bands suggestive of active interleaving. Intrusions are numerous and well-defined; they spread laterally, exchanging properties between the meddy interior and surrounding waters. The proposition of the ultimate demise of meddies through interleaving mechanisms becomes all the more convincing in view of these new seismic observations. The images in Figure 7.32 also reveal systematic differences between layers developed at the upper and lower meddy boundaries, apparently reflecting their different types (diffusively driven above and finger-driven below the core).

Song *et al.* (2011) performed an even more detailed analysis of seismic images of water masses in and around a newly discovered meddy in the Iberian Basin off the coast of Portugal. This study led to intriguing insights into the interplay between thermohaline interleaving and eddy stirring in the cascade of thermal variance to microscale dissipation. Seismic observations revealed that thermohaline intrusions are concentrated in isolated bands, associated with anomalously high mixing. These bands were interpreted as "spiral arms" of water, removed from the meddy periphery by mesoscale stirring. The structure of finescale intrusions in the spiral arms was fully consistent with their double-diffusive origin and inconsistent with direct mesoscale stirring. It was argued that the partnership between eddy stirring and thermohaline intrusions is an essential element in a chain of mixing processes in the Iberian Basin. Stirring produces strong lateral fronts; these fronts become sites of spontaneous interleaving and are ultimately dissipated by intrusion-driven mixing.

Admittedly, most current applications of seismic oceanography are inherently descriptive. However, there is significant interest in developing this technology into a source of quantitative information (e.g., Paramo and Holbrook, 2005; Sheen *et al.*, 2009) that can match conventional data in terms of accuracy and unambiguous interpretation. Regardless of the outcome of such efforts, the significance of the insights that seismic imaging provides into the finescale stage of mixing is already evident. These images convey a powerful sense of the omnipresence and intensity of interleaving – that ominous tangle of squirming serpents jumbled together in the snake-pit of the ocean. This sense is blunted in more conventional oceanographic observations.

Alternative arguments

As we have seen in the foregoing examples, double-diffusion can produce strong intrusive motions provided that it is the dominant mixing agent at a given location. In such cases, interleaving is active; it profoundly affects water-mass distribution and circulation patterns on much larger scales. However, it is not clear what happens if double-diffusion is absent, weak or dominated by some other mixing process present at the same location. Can we still expect interleaving of substantial magnitude and characteristic intrusive-like structure? The answer to this question is "maybe."

The most intriguing in this regard are observations of intrusions in double-diffusively stable regions – the regions where vertical background *T–S* gradients support neither salt fingering nor diffusive convection. The initiation of intrusions in such conditions requires mechanisms that are not represented in conventional intrusion theories (Stern, 1967; Toole and Georgi, 1981). And if the origin of intrusions is non-thermohaline, it becomes questionable whether double-diffusion is essential for their maintenance. One possibility is that the first stage of interleaving is the development of molecularly driven intrusions (Chapter 2), which can grow in lateral gradients regardless of vertical stratification (Holyer, 1983; Stern, 2003; Thompson and Veronis, 2005). When the amplitude of intrusions is large enough to create inversions in vertical *T–S* profiles, double-diffusive mixing takes over. The vertical scale of the fastest growing molecularly driven intrusions is much less than the scale of intrusions typically observed in the ocean. However, shortly after formation, molecularly driven intrusions start to merge sequentially, which can increase their thickness to the observed levels (Simeonov and Stern, 2008).

An alternative scenario was proposed by Hebert (1999), who attributed interleaving in stable regions to the action of differential diffusion. In the case of incomplete turbulent mixing (supposedly caused by overturning gravity waves), eddy diffusivity of heat can exceed that of salt. Since molecular *T–S* diffusivities are different ($k_T \gg k_S$), some turbulent events of limited duration would effectively mix temperature but not salinity. If effective diffusivities of heat and salt are different, spontaneous interleaving is once again possible, which can be shown by straightforward reinterpretation of the classical theory (Stern, 1967). Merryfield (2002) applied the intrusion model to observations of the Upper Polar Deep Water in the Arctic (Anderson *et al.*, 1994) and found that the differential diffusion model is not inconsistent with observations. Kuzmina *et al.* (2011) reexamined the differential diffusion mechanism of interleaving using data from the Eurasian Basin and suggested that it remains a viable explanation for the observed intrusions in doubly stable regions. This study emphasized the significance of baroclinic effects in controlling the intrusion slopes and intensity of interleaving in the differential mixing model. The lateral diffusivities in doubly stable regions were estimated to be in the range of 17–57 $m^2\ s^{-1}$. Intrusion-driven mixing of such strength constitutes an important mechanism for water-mass transformation, particularly in view of weak internal wave activity in the Arctic Ocean.

Both scenarios, molecular interleaving and differential diffusion, could still be broadly described as double-diffusive. The difference in molecular diffusivities

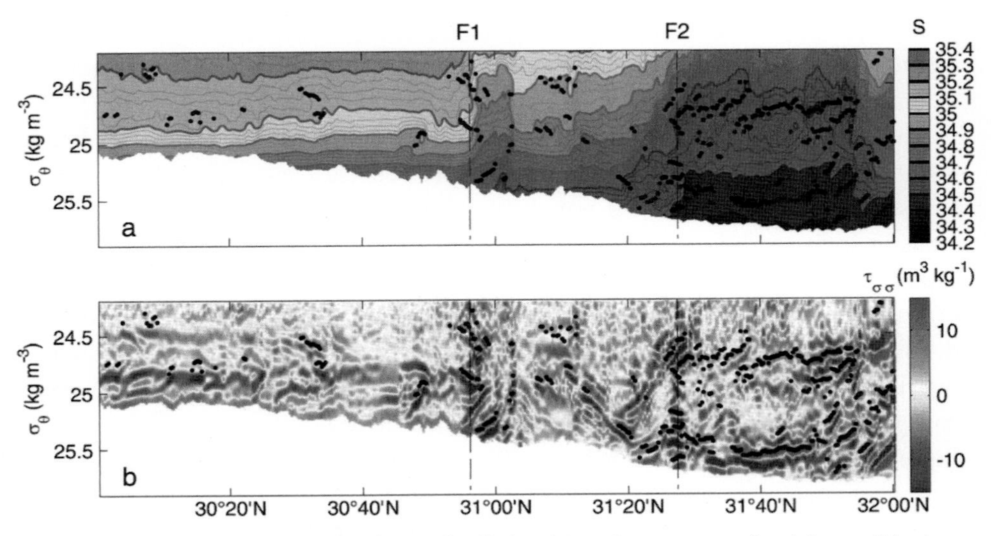

Figure 7.33 The vertical sections of salinity (a) and curvature of spiciness (b) at the North Pacific Subtropical Frontal Zone. The spiciness diagnostics reveal the presence of numerous intrusions that have very limited expression in terms of salinity. From Shcherbina *et al.* (2009). See color plates section.

of heat and salt is essential. However, completely different possibilities have been considered. For instance, Shcherbina *et al.* (2009) examined thermohaline intrusions in the North Pacific Subtropical Frontal Zone (NPSFZ). This region is only mildly susceptible to salt fingering ($R_\rho > 2$). Interleaving is also less active or, at least, less visible. Generally, fully developed interleaving is easily detectable by inversions of temperature and salinity (e.g., Fig. 7.28). Detection of intrusions at the NPSFZ required more sophisticated diagnostics, based on a water-mass characteristic called spiciness. Spiciness (τ) can be defined through its differential as

$$d\tau = \rho(\alpha dT + \beta dS). \tag{7.25}$$

In terms of thermal and haline components (αT and βS), spiciness is orthogonal to density, which makes it a rather sensitive indicator of variations in water-mass composition. The second derivative of spiciness with respect to potential density ($\tau_{\sigma\sigma}$) accentuates such variations even more. Numerous intrusive structures that are too weak to have a visible impression on the salinity distribution (Fig. 7.33a) are clearly revealed by the pattern of $\tau_{\sigma\sigma}$ (Fig. 7.33b).

The analysis of intrusion patterns by Shcherbina *et al.* was inconclusive – the range of the observed intrusion slopes was much broader than could be expected on the basis of classical theories; slopes of both signs were observed. Shcherbina *et al.* (2009) hypothesized that intrusions at this location are not self-driven but produced by "passive" mechanisms, associated with stirring of thermohaline gradients by mesoscale eddies. This argument hinges on two fundamental properties of rotating stratified turbulence – inverse transfer of energy and density towards large vertical scales and direct transfer of *T–S* variability to relatively small scales. As a result, mesoscale eddies are expected to be ineffective in creating density variability on intrusion scales ($\sim$10 m). Temperature and salinity, on the other hand, could vary significantly, albeit in the coherent density-compensated manner (Ferrari and Polzin, 2005; Smith and Ferrari, 2009). While the idea is certainly interesting, it is clear that much more detailed analyses, theoretical and observational, are required to quantify the role of passive processes in the dynamics of interleaving under various thermohaline conditions. Song *et al.* (2011) argued against the "passive interleaving" mechanism by noting that the slopes of most intrusive features observed in seismic images of oceanic fine-structure are much flatter than expected for interleaving driven directly by mesoscale stirring.

Another version of the passive interleaving hypothesis invokes internal waves as the ultimate driver of intrusive motions. In its simplest form, the idea was originally suggested by Georgi (1978). Georgi envisioned slow inertial waves with nearly horizontal velocity that varies vertically in a periodic manner. The resulting vertically sheared motion would differentially advect properties across lateral thermohaline fronts creating a *T–S* imprint that is oscillatory in the vertical and is horizontally extended – something that does resemble intrusions. This hypothesis in its raw form is easy to refute. First of all, internal waves periodically reverse displacements. The lifetime of intrusions is uncertain but it definitely exceeds the period of near-inertial waves ($\sim$1 day). In addition, Joyce *et al.* (1978) ruled out internal waves as a source of interleaving by comparing the amplitudes of temperature and velocity in the observed intrusions. This ratio was very different, by at least two orders of magnitude, from what could be expected for wave-driven perturbations. Still, Georgi's suggestion cannot be dismissed entirely. Even if waves do not cause interleaving directly, it does not necessarily mean they are not important.

The thirty-year pause in the debate on this topic was recently interrupted by the laboratory study of Griffiths and Bidokhti (2008). In their experiments, two buoyant plumes were released into the opposite ends of a long channel filled with water. The immediate consequence of such forcing was the generation of internal waves, which was followed by the appearance of lateral intrusions, clearly visible in Figure 7.34. The experiments were performed in both single-component (Fig. 7.34a) and two-component (Fig 7.34b) stratifications. Remarkably, the outcomes of these experiments were very similar; salt fingers appearing in the double-diffusive case (Fig. 7.34b) had almost no effect on the magnitude and

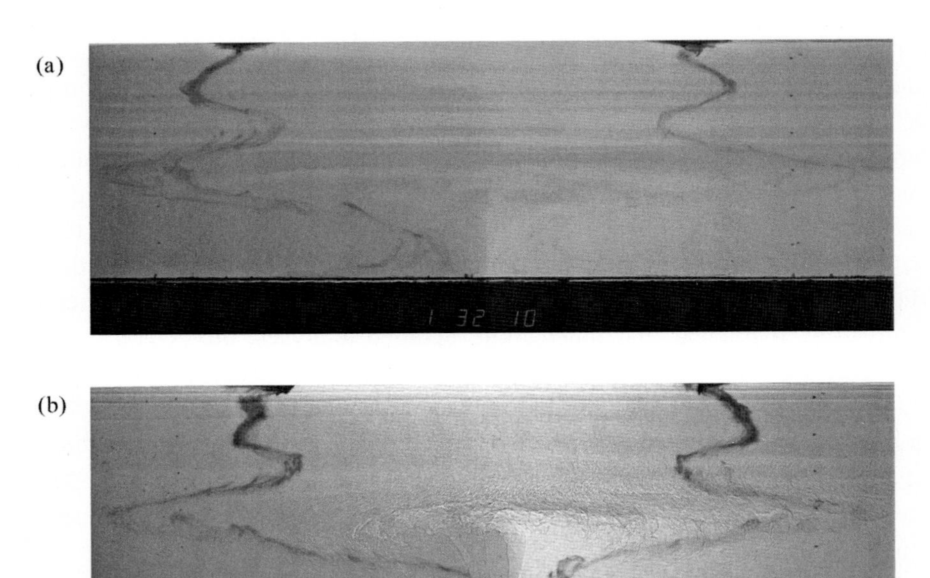

Figure 7.34 Intrusions generated by releasing buoyant plumes into a long channel filled with one-component (top panel) and double-diffusive (bottom panel) fluids. From Griffiths and Bidokhti (2008).

pattern of intrusions. The authors attributed interleaving to the action of internal waves. This mechanism undoubtedly involves nonlinear interactions between various modes of internal waves. However, the precise mechanics of such interaction, as well as its relevance to interleaving in the ocean, remain unclear.

Our final example of an ambiguous and poorly understood set of observations (those are kind of fun, aren't they?) concerns equatorial interleaving. A number of studies (Toole, 1981; Richards and Banks, 2002; Lee and Richards, 2004) have reported observations of intrusions in the equatorial Pacific thermocline. The interleaving signatures there are unmistakable. Characterized by vertical scales of a few tens of meters and laterally coherent over hundreds of kilometers, these intrusions are as formidable as one can find in the world ocean. The lateral heat transfer by equatorial interleaving can substantially affect the temperature distribution in the tropical thermocline and the speed of the equatorial undercurrent (Pezzi and Richards, 2003). Given their potential large-scale impact, it is imperative to identify the mechanisms for establishment and maintenance of equatorial intrusions. In addition to the general oceanographic interest, identification of their origin is essential for the development of physically based parameterizations of interleaving in large-scale numerical and theoretical models. The biggest problem here is that, in addition to thermohaline mechanisms, equatorial intrusions could also be generated by inertial instabilities. Edwards and Richards (1999, 2004) argue that both processes produce similar *T–S* signatures, which makes it difficult to discriminate between them on the basis of observations. But wait, things get even more complicated. In certain parameter regimes, inertial and thermohaline modes interact, resulting in the mixed thermohaline–inertial instability (Edwards and Richards, 1999). The presence of such mixed modes may preclude identification of a single dominant player in equatorial interleaving. In addition to the Pacific, vertically periodic layered structures have also been observed in the equatorial Atlantic, where they also could be associated with inertial instabilities, thermohaline processes or some combination thereof (D'Orgeville *et al.*, 2004). The problem of equatorial interleaving is complicated and remains largely unresolved.

Complications

There is no doubt that our understanding of interleaving and the general interest in this field have grown substantially in recent years. Continuous developments of the classical theory – inclusion of baroclinicity, effects of background turbulence, and more physical parameterizations of vertical transport – start to unveil the remarkably rich dynamics of thermohaline intrusions. Simulations have become more reliable and realistic. The observational database of interleaving accumulated in numerous field programs is already impressive, covering various forms of interleaving and geographic locations. These positive trends can only accelerate in the future as the significance of interleaving becomes more and more evident.

What seems to be lagging at the moment is our ability to link the interleaving theory with field measurements. Some problems are temporary. As we get wiser and better understand the relevant dynamics, we will find new ways to diagnose and interpret oceanographic observations. Other barriers could be more fundamental and related to the limited predictability of any turbulent environment. In view of these challenges, I would like to close this chapter with an (incomplete) list of considerations that should be taken into account in order to bring intrusion theories into agreement with observations:

(i) Intrusion characteristics are affected by the (less predictable) small-scale turbulence due to overturning gravity waves in the ocean; the turbulent component is spatially inhomogeneous and often controlled by external factors, such as topography and tidal forcing.
(ii) The instantaneous measurements of intrusions may reflect conditions that existed in the past history of a thermohaline front.

(iii) Secondary sub-harmonic instabilities of intrusions can lead to merging events, systematically increasing their vertical scale.
(iv) Intrusive perturbations could be initiated by processes unrelated to double-diffusion but maintained and enhanced by fundamentally thermohaline processes. In this case, the selective advantage of the most rapidly growing modes may be insufficient to ensure their dominance in the fully developed state.
(v) Significant uncertainties still exist in the formulation of the flux-gradient laws – the key element of parametric models. The effects of background shear, intermittent turbulence and internal waves on the salt-finger transport are typically not taken into account and can be substantial.

8

Thermohaline staircases

Oceanographers tend to get a bit emotional when discussing thermohaline staircases. The literature on this subject is sprinkled with colorful epithets – staircases are "dramatic," "spectacular," "striking" and "easy to admire." The reason why otherwise restrained and business-like folks suddenly become so passionate and poetic is clear from just a brief look at the structure of staircases. Thermohaline staircases (Fig. 8.1) consist of remarkably regular homogeneous layers in vertical temperature and salinity profiles. Although these mixed layers are tens of meters deep, they are created and maintained by double-diffusive processes operating on a centimeter scale. It is the ocean falling on its knees at the first sight of an army of merciless salt fingers; it is a complete surrender of large scales to the tiniest ones. Why would Nature, usually predisposed to turbulent and disorganized geophysical flows, create something so strangely precise and elegant?

It is no surprise that such an intriguing phenomenon has attracted continuous attention from observationalists and theoreticians alike. The first observations of staircases in the ocean were reported in the late 1960s (Tait and Howe, 1968; Cooper and Stommel, 1968) and were followed by a stream of similar measurements from various locations. The link between staircases and double-diffusion was recognized almost immediately and was widely and painlessly accepted by the oceanographic community. During the same period, staircases were reproduced in the laboratory (Turner, 1967; Stern and Turner, 1969), which stimulated new ideas on their dynamics. Almost there, right? It only remained to identify the exact mechanics of staircase generation, beyond the vague "salt fingers did it" level. Double-diffusers never made this final step. More than forty years later, we are still debating the relevance of at least half a dozen different layering mechanisms. Staircase dynamics is a subject filled with mysteries and speculations.

Our discussion of staircases starts with some key oceanographic observations (Section 8.1) and is followed by a summary of different hypotheses for their origin (Section 8.2). Our intentions are two-fold. First, we wish to offer readers an unbiased view on the subject. It is also important to realize that the evolution

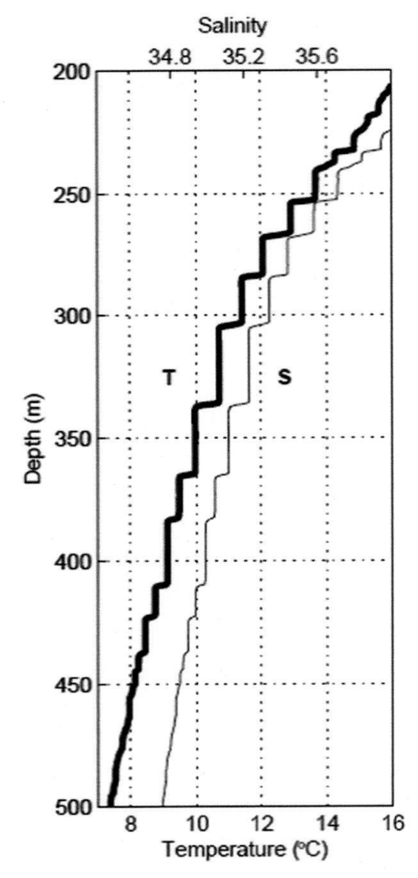

Figure 8.1 Typical profiles of potential temperature and salinity in the tropical Atlantic staircase taken during the SFTRE program. From Schmitt *et al.* (2005).

of complex systems in the ocean is usually governed by a combination of several processes. Very rarely is there only one explanation for anything that is remotely complicated in fluid dynamics and there may well be elements of truth in each staircase theory. The most recent model, and the author's favorite – the instability of the flux-gradient laws – is discussed in greater detail afterward (Section 8.3). We end the chapter by discussing the salient evolutionary features of fully developed staircases in Section 8.4.

8.1 Observations

Salt-finger staircases

The majority of observations of salt-finger staircases have come from three locations: the western tropical Atlantic (Schmitt *et al.*, 1987, 2005), Tyrrhenian Sea

(Zodiatis and Gasparini, 1996), and the Mediterranean outflow in the northeast Atlantic (Tait and Howe, 1968, 1971; Magnell, 1976). The common feature of these regions is the anomalously low values of density ratio, which appears to be both necessary and sufficient for staircase formation. No staircases have been reported for $R_\rho > 2$ and the reduction of density ratio below 1.7 is inevitably associated with the appearance of step-like structures in vertical *T–S* profiles (Fig. 8.2). The spatial pattern of staircases is also very sensitive to R_ρ. As the density ratio decreases, staircases become more pronounced and the height of steps sharply increases. This connection is significant. The density ratio is the single most important parameter controlling the intensity of salt fingers and the sensitivity of staircases to variations in R_ρ is one of many signs that staircases are a product of double-diffusion.

More direct evidence for the double-diffusive origin of staircases includes optical observations of salt fingers in interfacial regions using shadowgraph systems, as in Figure 8.3. Images of salt fingers from the Mediterranean outflow staircases bear striking resemblance to the laboratory shadowgraphs (bottom panel of Fig. 8.3). Shadowgraphs taken in the western tropical Atlantic during the C-SALT (Caribbean Sheets and Layers Transect) program also show well-defined salt fingers, frequently tilted by large-scale shear. Microstructure measurements in the interfaces (Magnell, 1976; Gregg and Sanford, 1987; Schmitt *et al.*, 2005) reveal all the expected signatures of salt-finger dynamics. The spectrum of the thermal signal is usually narrow-band, and the spectral peak corresponds to the wavelengths of the fastest growing finger modes. The scaled ratio of thermal variance and energy dissipation is high (recall that fingers are effective mixers of temperature but poor mixers of momentum), which rules out mechanical turbulence as a dominant source of dissipation.

The most recent and, in my opinion, incontrovertible proof of double-diffusive dynamics of staircases comes from the Salt Finger Tracer Release Experiment (SFTRE). During the cruise of 2001, 175 kg of sulfur hexafluoride (SF_6) was released in the interior of the Caribbean staircase. The molecular diffusivity of SF_6 and salt are close and therefore the mixing rate of salinity, which is otherwise difficult to measure, was inferred from the vertical spreading of the tracer. The observed dispersion rates imply an effective vertical diffusivity of salt of $K_S = (0.85 \pm 0.05) \cdot 10^{-4}\,\mathrm{m^2\,s^{-1}}$, which significantly exceeds the temperature diffusivity measured at the same location, $K_T = (0.45 \pm 0.2) \cdot 10^{-4}\,\mathrm{m^2\,s^{-1}}$ – a clear signature of salt fingering. The corresponding flux ratio for the typical density ratio of $R_\rho \approx 1.6$ is

$$\gamma = \frac{F_T}{F_S} = \frac{K_T}{K_S} R_\rho \approx 0.85. \tag{8.1}$$

The flux ratio expected for salt fingers on the basis of DNS, experiments and linear theory (see Fig. 2.7a) is $\gamma_{\mathrm{sf}} \approx 0.6$ – somewhat less than (8.1) but definitely within

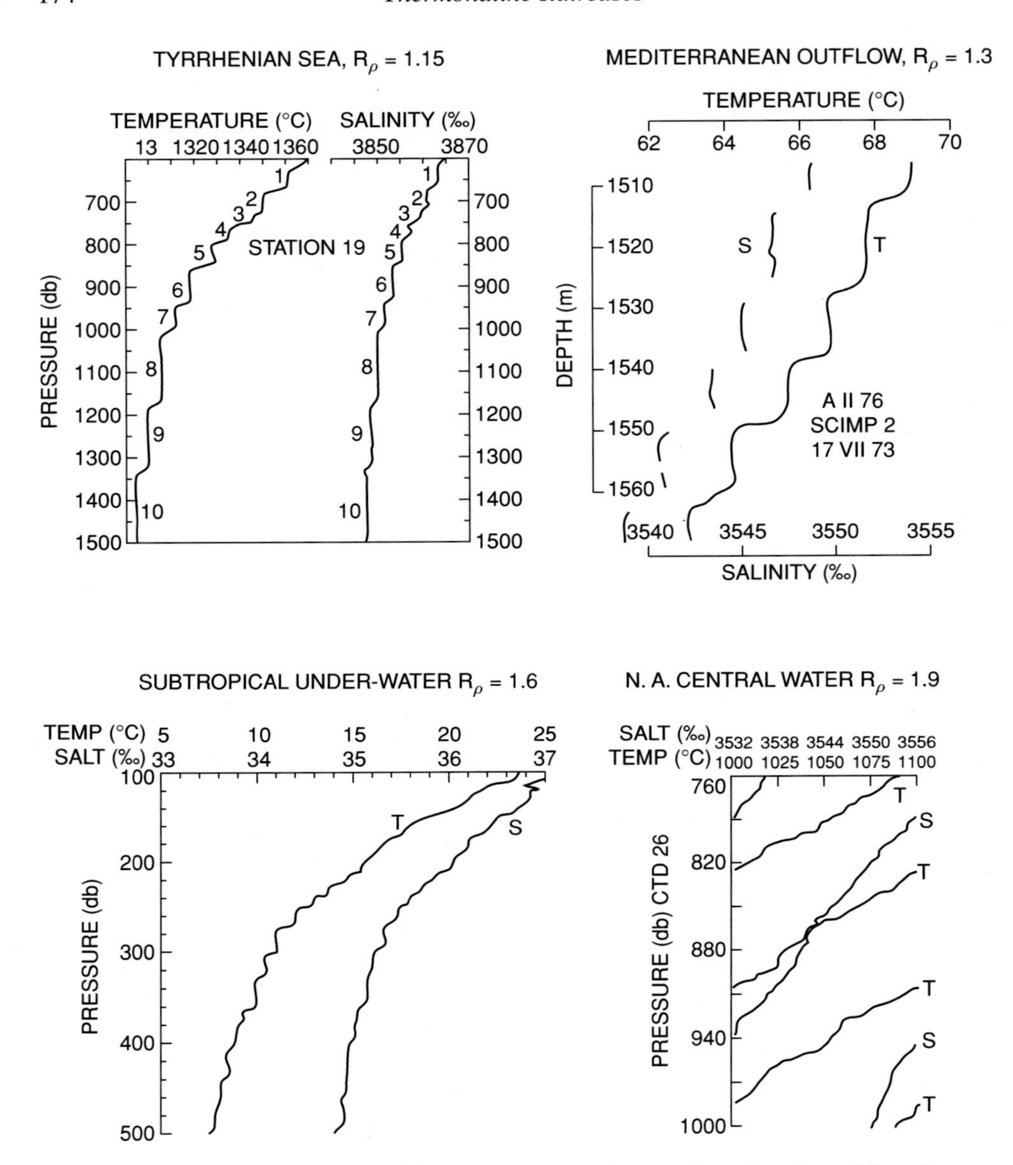

Figure 8.2 Examples of thermohaline staircases from various locations. Note the tendency of staircases to become more pronounced as the density ratio decreases. Redrawn from Schmitt (1981).

a margin of error. The discrepancy could be readily attributed, for instance, to the nonlinearities in the equation of state (McDougall, 1991). The flux ratio in (8.1) cannot be produced by any known mixing process in the ocean except for double-diffusion. For turbulence, which tends to mix heat and salt at equal rates, the

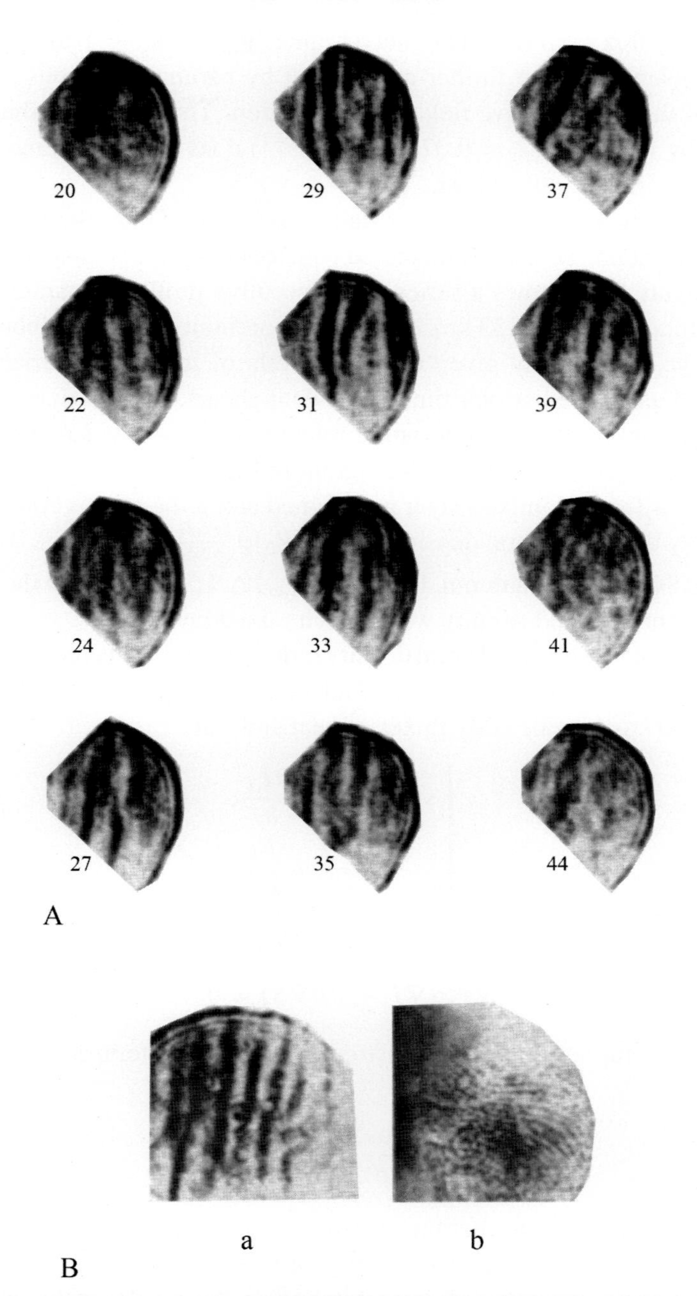

Figure 8.3 (A) Shadowgraph images of salt fingers in the Mediterranean outflow. The dark bands are shadows of descending fingers (high refractive index) and the light bands are shadows of rising fingers (low refractive index). (B) Shadowgraph images in a laboratory tank. Image (a) shows salt fingers. Image (b) is a blank pattern after the tank was stirred. From Williams (1974).

corresponding flux ratio would be much higher ($\gamma_{turb} = R_\rho \approx 1.6$). The secondary role of turbulence can be further ascertained by parameterizations based on the intensity of the gravity wave field at this location. The level of turbulent mixing supported by waves ($K_{turb} \approx 0.02 \cdot 10^{-4}\,\mathrm{m^2\,s^{-1}}$) does not even come close to the actual diffusivities of salt or heat.

The three-dimensional structure of staircases is also of interest. Several layers in the Caribbean staircase were found to be coherent over lateral distances exceeding 100 km. Figure 8.4a shows a series of consecutive profiles separated by 1.3 km within the total distance of 33 km. The spatial continuity of the Caribbean staircase is impressive, particularly given the high levels of mesoscale variability at this location and associated strong time-dependent shears, both horizontal and vertical. The interfaces tilt at a finite angle relative to the isopycnal surfaces, which is revealed most clearly by the *T–S* diagram in Figure 8.4b. The temperature and salinity values for each mixed layer are aligned along the curves corresponding to a remarkably uniform lateral density ratio of $R_l = \frac{\alpha \Delta T_{layer}}{\beta \Delta S_{layer}} = 0.85 \pm 0.02$ in spring and $R_l = 0.84 \pm 0.03$ in autumn. Here, $\left(\Delta T_{layer}, \Delta S_{layer}\right)$ represent the lateral variation in temperature and salinity within each mixed layer.

The observed value of the lateral density ratio in the mixed layers can be rationalized from the balance between the lateral advection in mixed layers and convergence of the vertical small-scale eddy fluxes of heat and salt:

$$\begin{cases} \vec{v} \cdot \nabla T = -\dfrac{\partial F_T}{\partial z}, \\ \vec{v} \cdot \nabla S = -\dfrac{\partial F_S}{\partial z}. \end{cases} \tag{8.2}$$

If the flux ratio is relatively uniform $\alpha F_T = \gamma \beta F_S,\ \gamma \approx const$, then

$$\vec{v} \cdot (\alpha \nabla T - \gamma \beta \nabla S) = 0, \tag{8.3}$$

which, in turn, implies that the downstream variations in temperature and salinity are linked:

$$R_l = \frac{\alpha \Delta T_{layer}}{\beta \Delta S_{layer}} \approx \gamma. \tag{8.4}$$

Equation (8.4) affords an alternative estimate of the flux ratio: $\gamma \approx R_l \approx 0.85$ in agreement with (8.1). Such consistency of two independent observation-based estimates lends credence to our interpretation of the results in Figure 8.4b. The cross-flow variation in properties is more difficult to predict. However, it is not unreasonable to speculate that lateral mixing in layers tends to homogenize water masses in the direction normal to the advecting velocity and therefore all measurements of temperature and salinity in individual layers collapse on the distinct

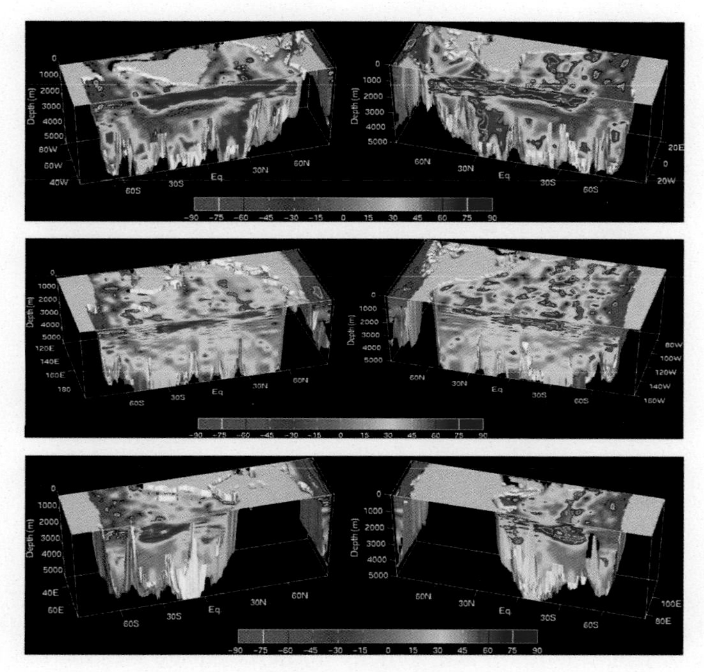

Figure 1.8 Distribution of the Turner angle in the Atlantic (top), Pacific (center) and Indian (bottom) Oceans. From You (2002).

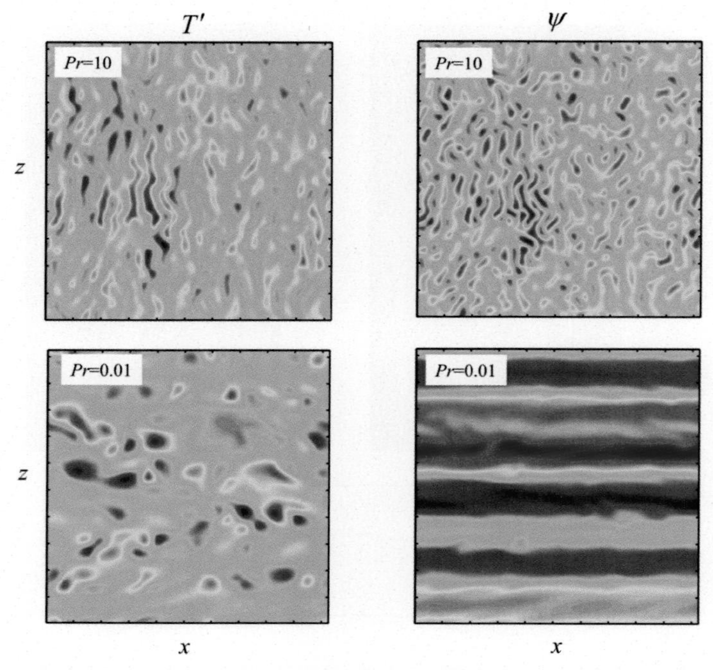

Figure 3.2 Variability of the salt-finger patterns. Instantaneous temperature (left) and streamfunction (right) fields from direct numerical simulations (DNS) are presented for $Pr = 10$ (top) and for $Pr = 0.01$ (bottom). In all cases $R_\rho = 2.8$ and $t = 1/3$. Red color corresponds to high values and low values are shown in blue. From Radko (2010).

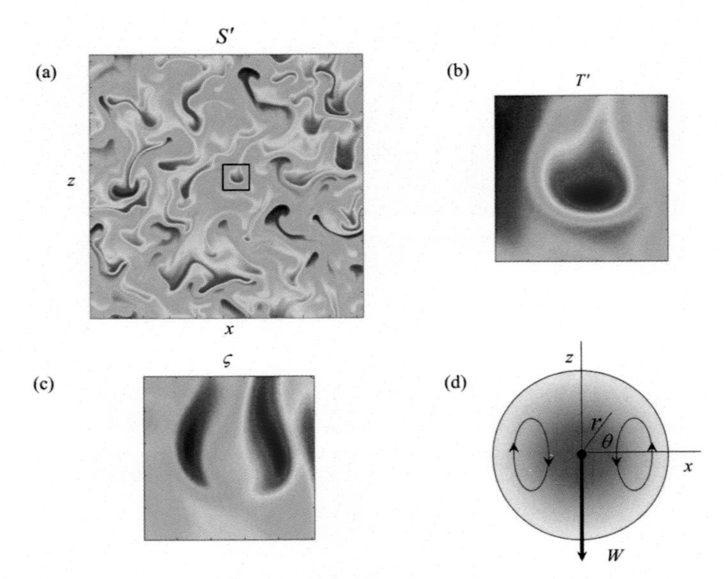

Figure 3.5 (a) Instantaneous salinity field in the numerical experiment with $R_\rho = 1.2$, $Pr = 7$ and $\tau = 1/3$. Red color corresponds to high values of S' and low values are shown in blue. Distribution of temperature (b) and vorticity (c) in the square area marked in (a), which contains a well-defined double-diffusive modon. (d) Schematic diagram of the analytical similarity solution for the downward propagating modon. From Radko (2008).

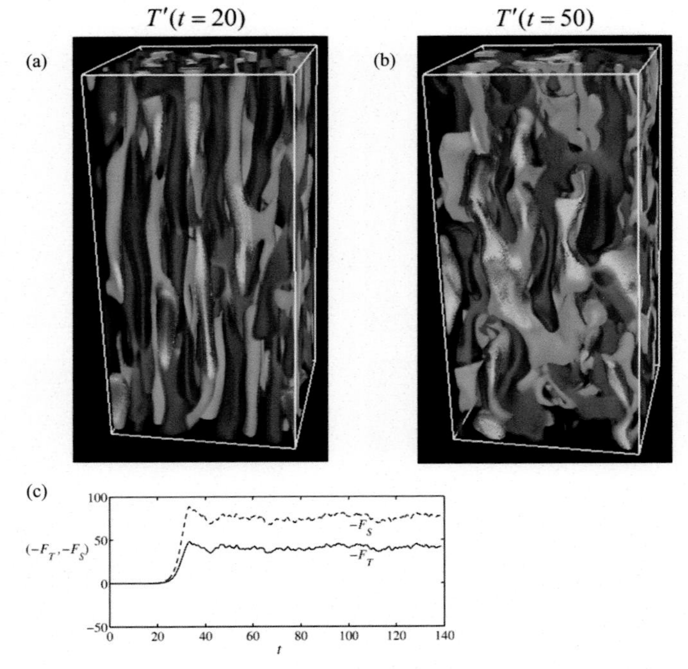

Figure 3.7 Equilibration of salt fingers in the numerical experiment with $Pr = 7$, $\tau = 0.01$ and $R_\rho = 1.9$. The temperature fields are shown for (a) the early stage of linear growth at $t = 20$ and (b) the fully equilibrated state at $t = 50$. Red/green corresponds to high values of T and low values are shown in blue. (c) Time record of the temperature (solid curve) and salinity (dashed curve) fluxes. From Radko and Smith (2012).

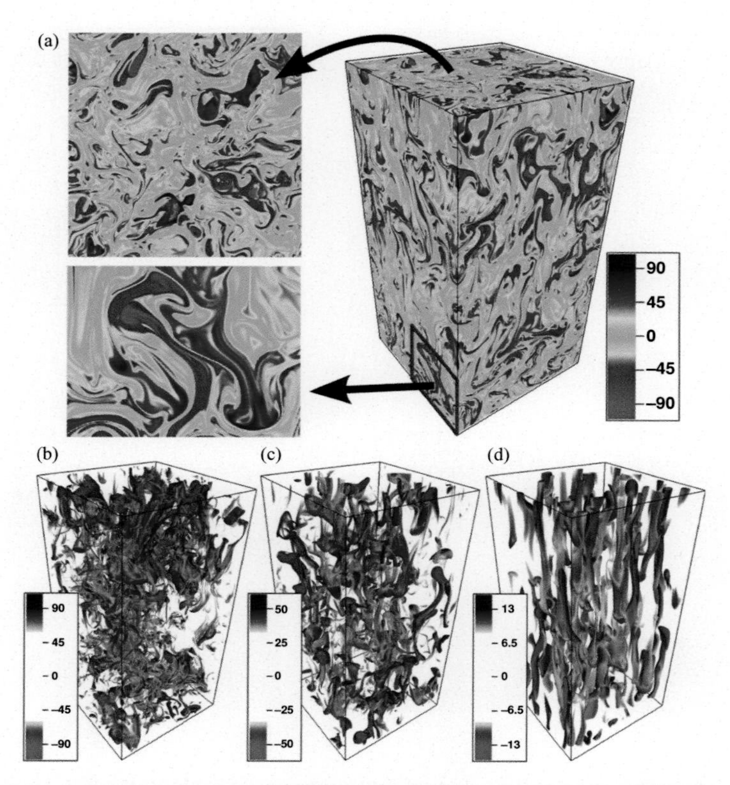

Figure 3.9 Snapshots of the salinity field in simulations of fingering convection for $Pr = 7$, $\tau = 0.01$. (a) Salinity field at $R_\rho = 1.2$, plotted on the three planes of the computational domain. (b)–(d) Volume rendering of the salinity field for $R_\rho = 1.2$, $R_\rho = 2$ and $R_\rho = 10$ (from left to right). From Traxler *et al.* (2011a).

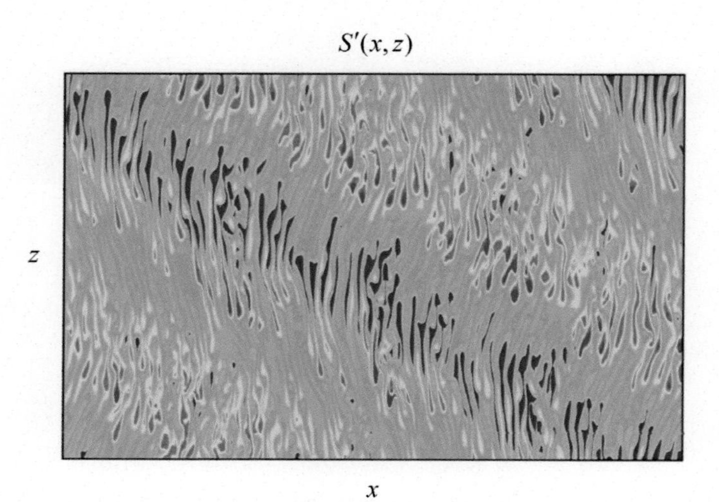

Figure 6.6 Two-dimensional direct numerical simulation showing salt fingers growing in the internal gravity wave. Presented is the departure of salinity from the uniform background gradient. Red color corresponds to high values and low values are shown in blue. From Stern *et al.* (2001).

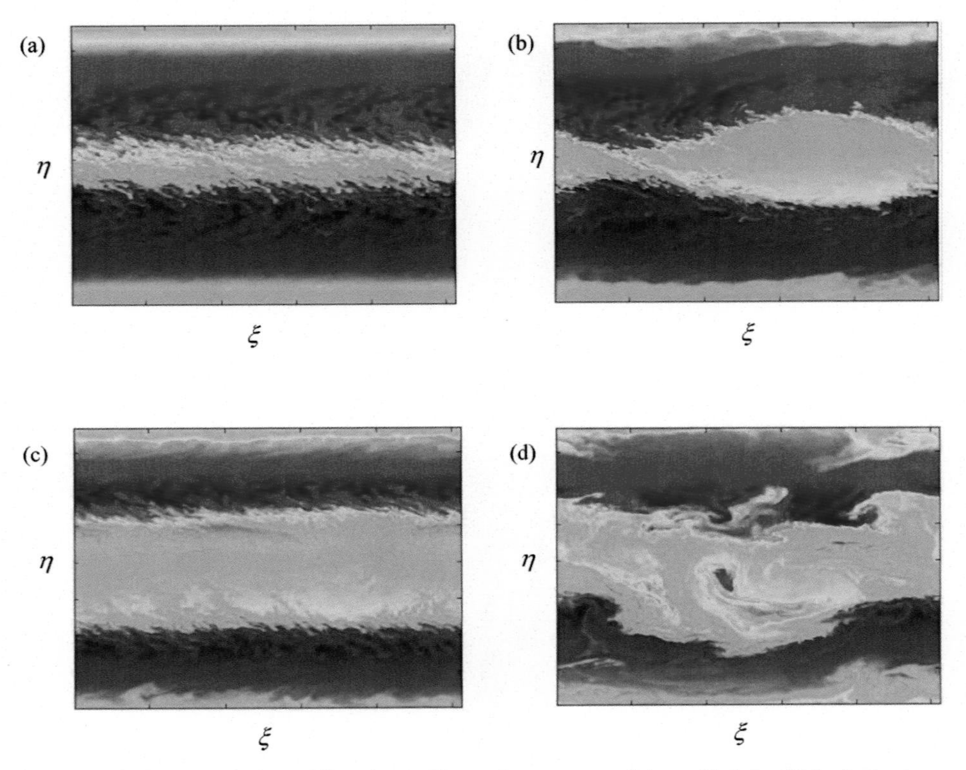

Figure 7.19 Growth and equilibration of intrusions susceptible to Kelvin–Helmholtz instabilities. The temperature perturbation is shown at various stages of intrusion development in (a)–(d). Red color corresponds to high values and low values are shown in blue. From Hebert (2011).

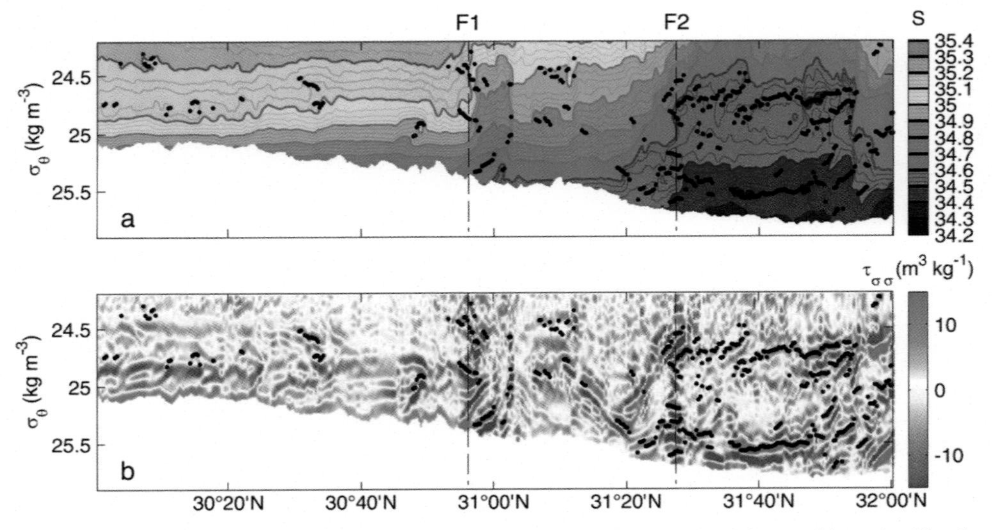

Figure 7.33 The vertical sections of salinity (a) and curvature of spiciness (b) at the North Pacific Subtropical Frontal Zone. The spiciness diagnostics reveal the presence of numerous intrusions that have very limited expression in terms of salinity. From Shcherbina *et al.* (2009).

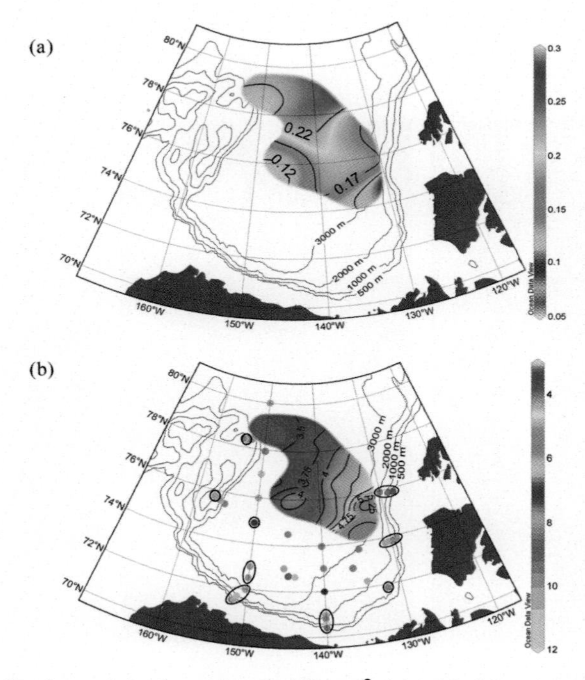

Figure 8.10 Distribution of the heat flux in W m^{-2} (a) and the average density ratio (b) in the diffusive layering region of the Beaufort Gyre estimated from the ITP data. From Timmermans *et al.* (2008).

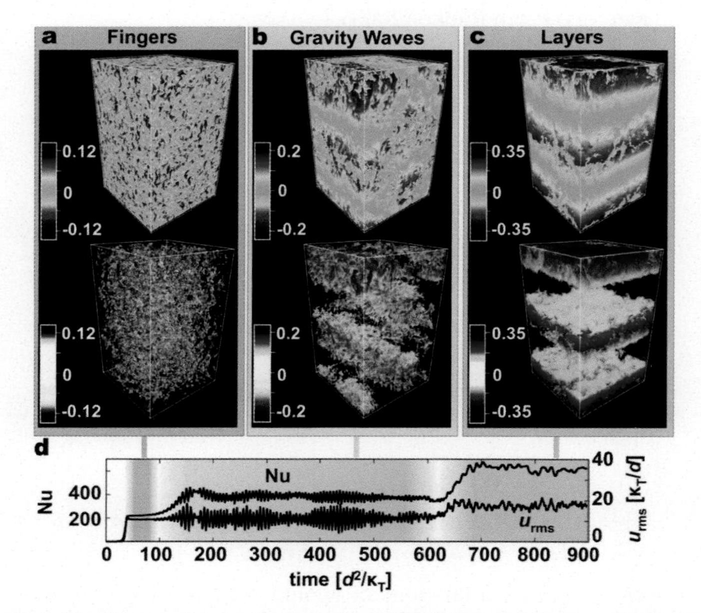

Figure 8.18 Simulated layer formation in a large triply periodic domain. Temperature perturbations are shown in color, with their respective color scales in each panel. The top panels are visualizations on the data-cube faces, while the lower ones are volume-rendered images. Snapshots are shown for three characteristic dynamical phases: (a) homogeneous fingering convection, (b) flow pattern dominated by gravity waves, (c) formation of vigorously convecting layers separated by thin fingering interfaces. (d) Time series of the Nusselt number and the flux ratio. From Stellmach *et al.* (2011).

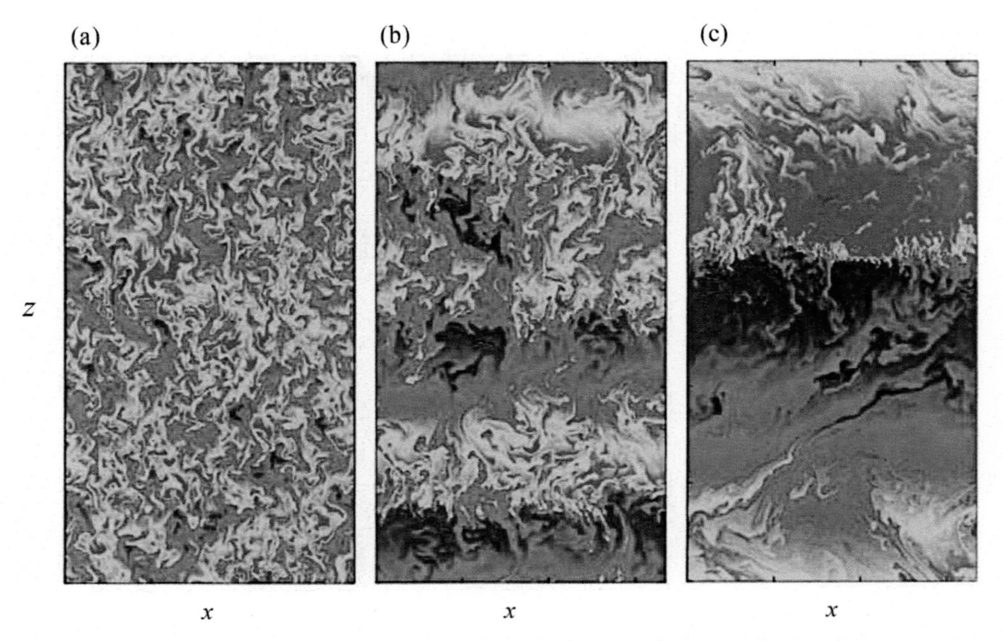

Figure 8.20 Formation and evolution of layers in a numerical experiment. Instantaneous perturbation temperature fields are shown for (a) $t = 50$, (b) $t = 400$ and (c) $t = 800$ (standard non-dimensional units). Note the spontaneous appearance of layers, followed by their systematic mergers. From Radko (2003).

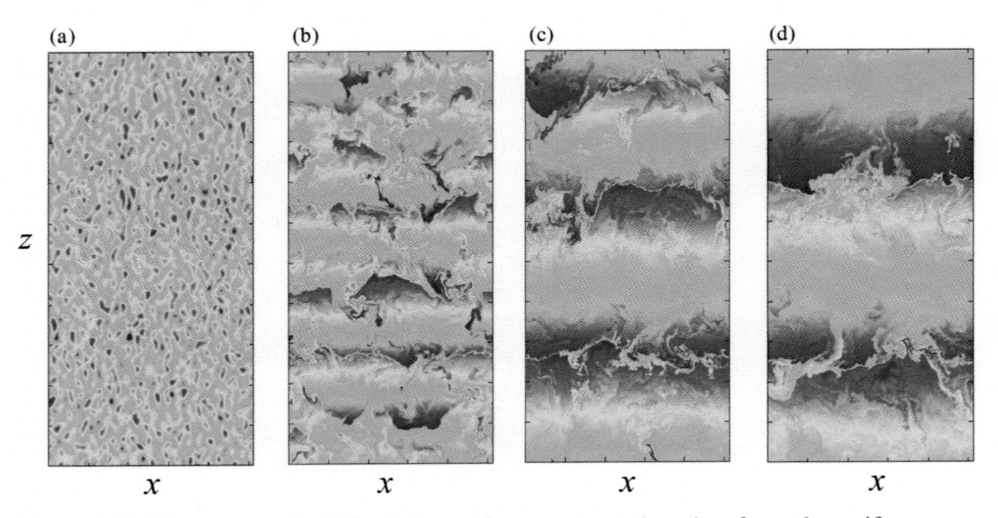

Figure 8.22 Direct numerical simulations of spontaneous layering from the uniform gradient in a diffusively stratified fluid. The small-scale oscillatory perturbations give way to horizontally uniform layers, which merge sequentially until producing the final quasi-equilibrium state. From Prikasky (2007).

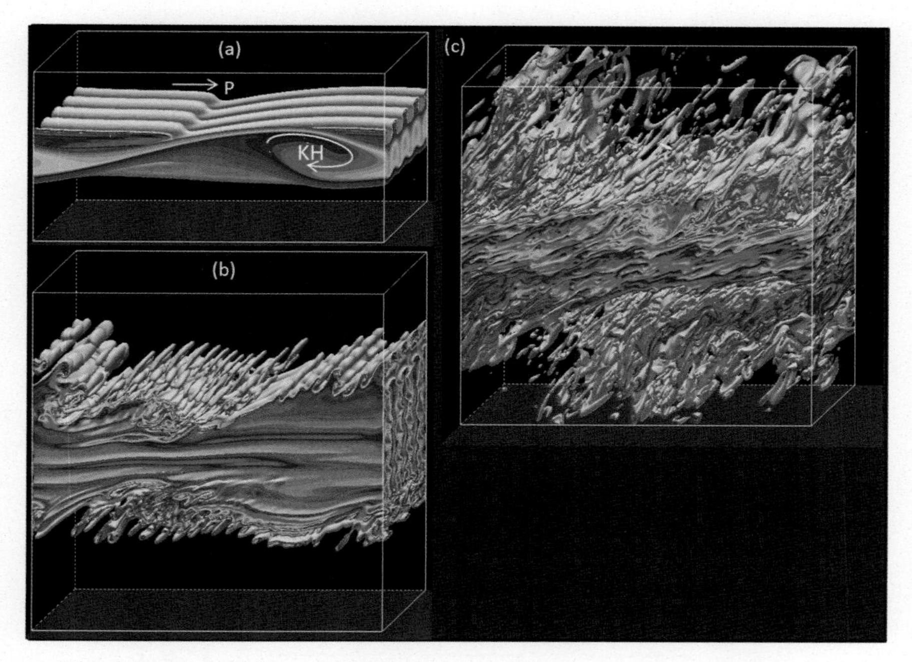

Figure 10.6 Salinity field from DNS of salt fingers in an unstable shear shown at various stages of evolution. From Smyth and Kimura (2011).

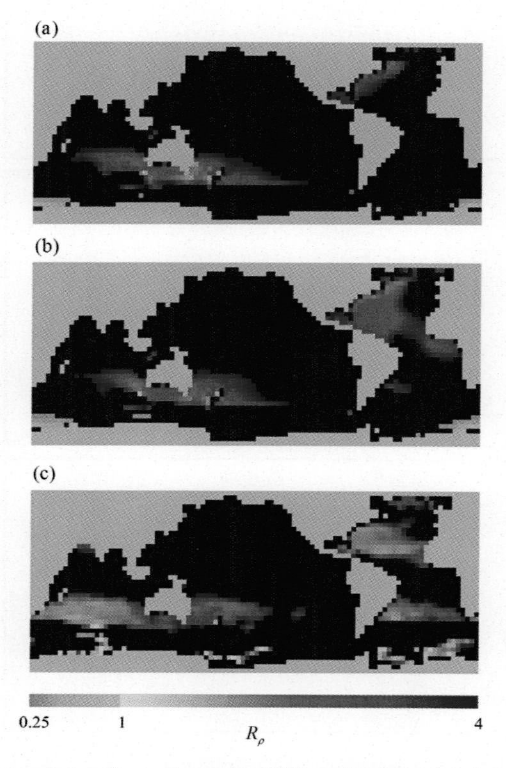

Figure 11.2 The pattern of density ratio at the 700 m depth in (a) the non-double-diffusive experiment, (b) double-diffusive experiment and (c) the annual-mean climatology. From Merryfield *et al.* (1999).

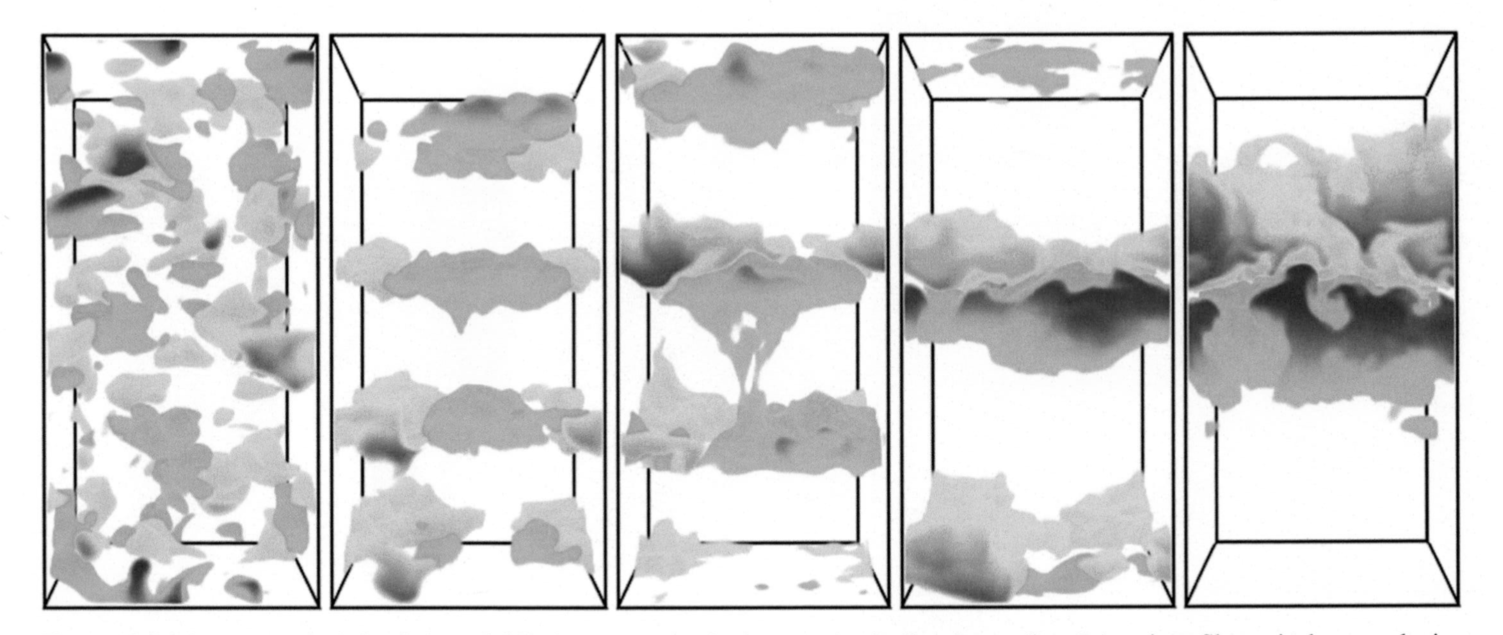

Figure 12.3 Direct numerical simulation of diffusive convection in the astrophysically relevant (low *Pr*) regime. Shown is the perturbation of the composition at various times. Note the formation of the well-defined layers and their sequential mergers. From Rosenblum *et al.* (2011).

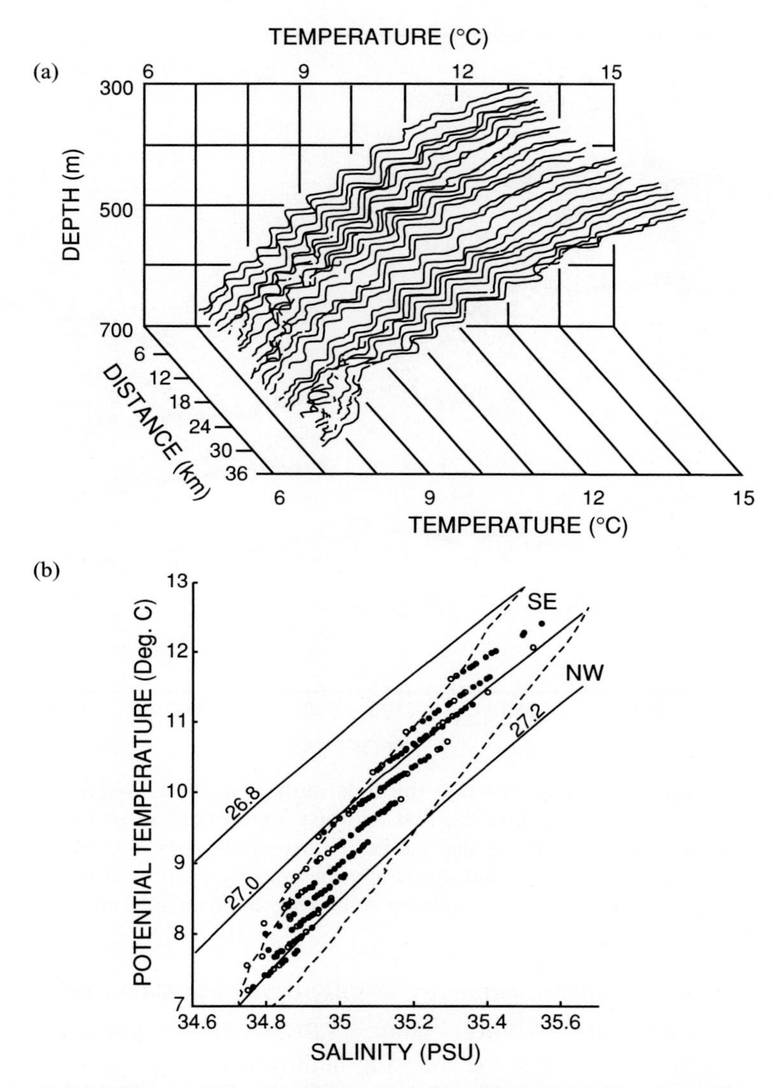

Figure 8.4 (a) Three-dimensional view of the offset temperature profiles taken during the C-SALT experiment. (b) Potential temperature–salinity values of the C-SALT layers. The solid circles are from mixed layers more than 10 m thick, the open circles are from layers 5–10 m thick. Also shown are the *T–S* relations at the northwest and southeast corners of the survey area (dashed curves) and isopycnal surfaces (solid curves). From Schmitt *et al.* (1987).

curves clearly visible in the *T–S* diagram (Fig. 8.4b). What is most striking in this story is the extent to which the data in Figure 8.4b conform to (8.4). The lateral density ratio in layers is uniform to within a fraction of a percent – something as firm and consistent is rarely found in oceanographic field measurements.

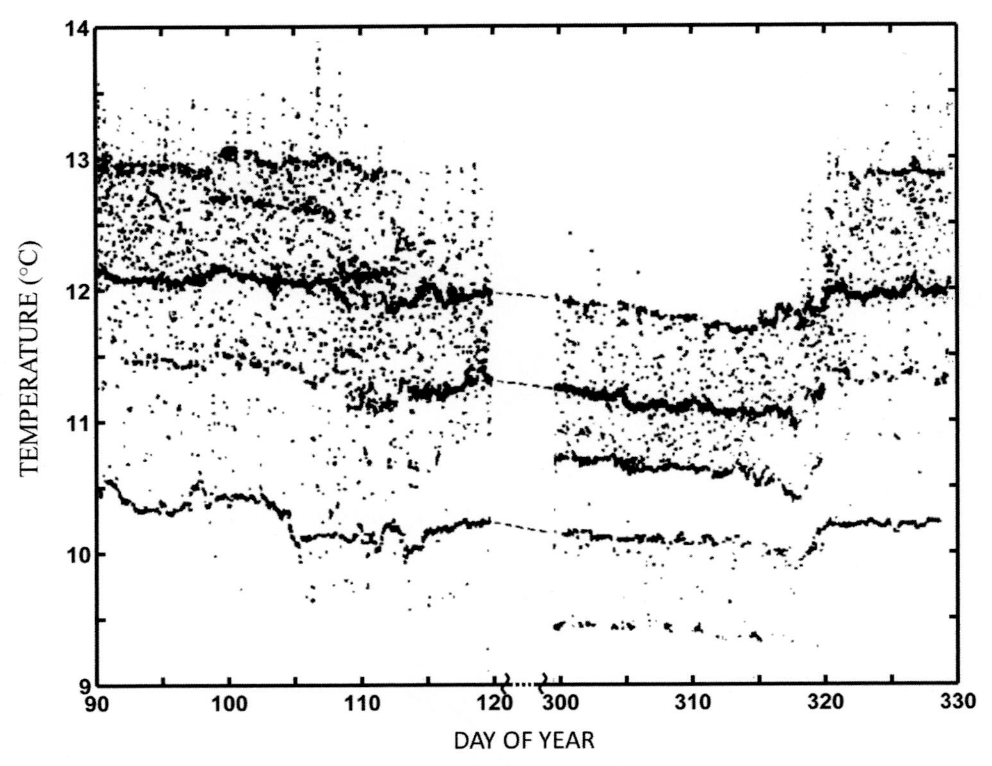

Figure 8.5 Temperature recorded at the mooring site in the C-SALT area in the depth range 325–415 m. Only the first and last 30 days are shown from a continuous eight-month record, as the character of the data does not change during the intervening time. Well-mixed layers appear as heavy concentrations of points, three of which may be traced through the entire record. From Schmitt *et al.* (1987).

The analysis of temporal variability in fully developed staircases reveals their persistence and structural stability. Figure 8.5 presents a continuous record of the vertical temperature profile at the mooring deployed for eight months during C-SALT (Schmitt *et al.*, 1987). Several well-defined layers maintained their identity throughout the entire period of observation, despite the significant short-term variability associated with gravity waves and mesoscale eddies. Figure 8.6 offers a glimpse into the evolutionary dynamics on even longer time scales. It shows a series of temperature and salinity profiles taken in the Tyrrhenian staircase during the period from 1973 to 1992. Two features should be emphasized. Several interfaces in Figure 8.6 can be identified, by their location and *T–S* values, in profiles taken several years apart – a testament to the remarkable resilience of thermohaline staircases. However, occasionally interfaces disappear, resulting in the systematic merging of the adjacent mixed layers. Figure 8.6 shows ten layers in the 1973

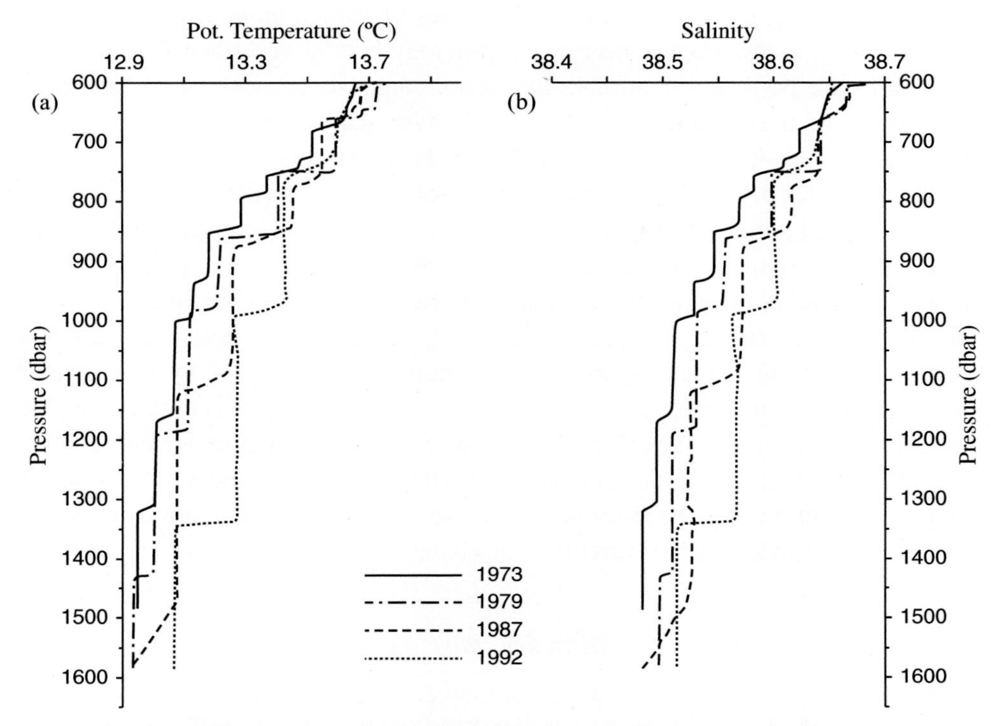

Figure 8.6 (a) Potential temperature and (b) salinity profiles in the central Tyrrhenian basin over a 19-year period. From Zodiatis and Gasparini (1996).

profiles but only four spectacular layers several hundred meters thick are present in 1992. The dynamical mechanisms and ramifications of layer-merging events will be discussed in Section 8.4.

Generally, staircases are not expected to occur in the coastal environment. Strong shears and elevated levels of turbulence near topographic boundaries are usually quite effective in preventing spontaneous layering. However, a curious counterexample has been reported recently. Spear and Thomson (2012) present evidence of a well-defined staircase in Belize Inlet, a British Columbia fjord. Hydrographic measurements suggest an intrusion of cold and fresh oceanic water into the fjord, which created finger-favorable conditions above its level (~150 m) and diffusive conditions below. Salt-finger staircases were observed in the depth range of 70–150 m with step heights of approximately 10 m. Diffusive layers formed in the depth range 150–210 m and were much smaller (~1 m). This example illustrates the ability of thermohaline staircases to withstand substantial levels of adverse ambient forcing, associated in Belize Inlet with strong tidally induced shears and turbulence.

An interesting perspective on the dynamics of staircases is afforded by seismic imaging, a new ocean-observing technology briefly described in Chapter 7. Staircases are particularly suitable for seismic analysis because of (i) the strong reflectivity of high-gradient interfaces, (ii) layer scales that are easily resolved by seismic inversion and (iii) limited horizontal variability. Biescas *et al.* (2010) presents a series of spectacular images of the salt-finger staircase formed below the Mediterranean outflow. In Figure 8.7a, the staircase is visibly perturbed by internal waves and yet remains well-defined and spatially coherent. The interaction with a meddy (Fig. 8.7b) offers another example of the resilience of the staircase. Layers are disorganized in the immediate vicinity of the meddy but remain remarkably regular on the outside. In Figure 8.7c, the staircase is disrupted by the elevated turbulent mixing near a topographic boundary, but the damage is contained within the local area of interaction. A similar seismic image of a region occupied by the Caribbean (C-SALT) staircase (Fer *et al.*, 2010) is shown in Figure 8.8. Once again, acoustic detection tells the story of a staircase struggling to maintain its structure and coherence in the face of external disturbances.

Diffusive staircases

Whilst salt-finger staircases have been the subject of continuous scrutiny since their discovery, diffusive staircases, on the other hand, have historically received less attention. One of the reasons is related to their prevalence in high-latitude locations, where field programs are more demanding. The first documented observations of diffusive staircases were made by Neal *et al.* (1969), whose vertical temperature profiles from the central Arctic revealed regular step-like patterns (Fig. 8.9) in the region overlying relatively warm and salty waters of Atlantic origin. The layering was promptly attributed to double-diffusion and dynamic parallels with salt-finger staircases (Tait and Howe, 1968) were drawn. Aside from the opposite sign of the background temperature and salinity gradients, diffusive staircases in the Arctic are characterized by much smaller steps. Both the thickness and particularly the variation in properties across the steps are much less than in typical salt-finger staircases (see Table 8.1). Subsequent field programs (Neshyba *et al.*, 1971; Neal and Neshyba, 1973; Perkin and Lewis, 1984) revealed that staircases are ubiquitous in the Arctic and are likely to affect the high-latitude water-mass distribution (e.g., McDougall, 1983).

Arctic staircases were explored more systematically during the Arctic Internal Wave Experiment (AIWEX). Padman and Dillon (1987, 1988, 1989) found that staircases in the central Arctic (Canada Basin) were located between 320 and 430 m depth, with steps 1–2 m thick and temperature jumps across interface in the range $0.004\ ^{\circ}\mathrm{C} < \Delta T < 0.012\ ^{\circ}\mathrm{C}$. The corresponding vertical heat fluxes

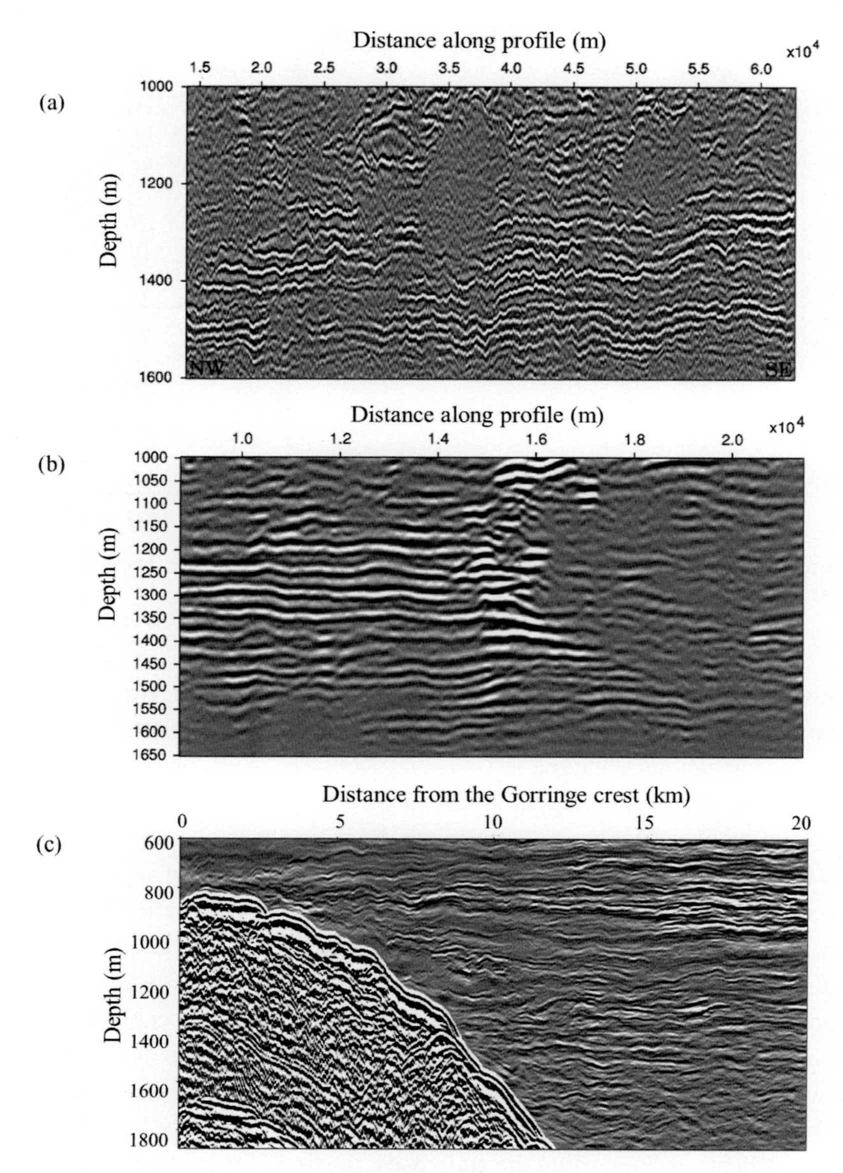

Figure 8.7 Seismic images of the Mediterranean outflow staircase. The staircase in (a) is perturbed by internal waves, in (b) interacts with a meddy and in (c) interacts with the topographic boundary. From Biescas *et al.* (2010).

evaluated using the interfacial laboratory-derived flux laws (Chapter 4) are limited to 0.02–0.1 W m^{-2}. Of course, the accuracy of the 4/3 flux laws is debatable in the context of the observed staircases. It is unclear to what extent the extrapolation of the laboratory tank to the ocean is warranted – it is not unreasonable to expect an error

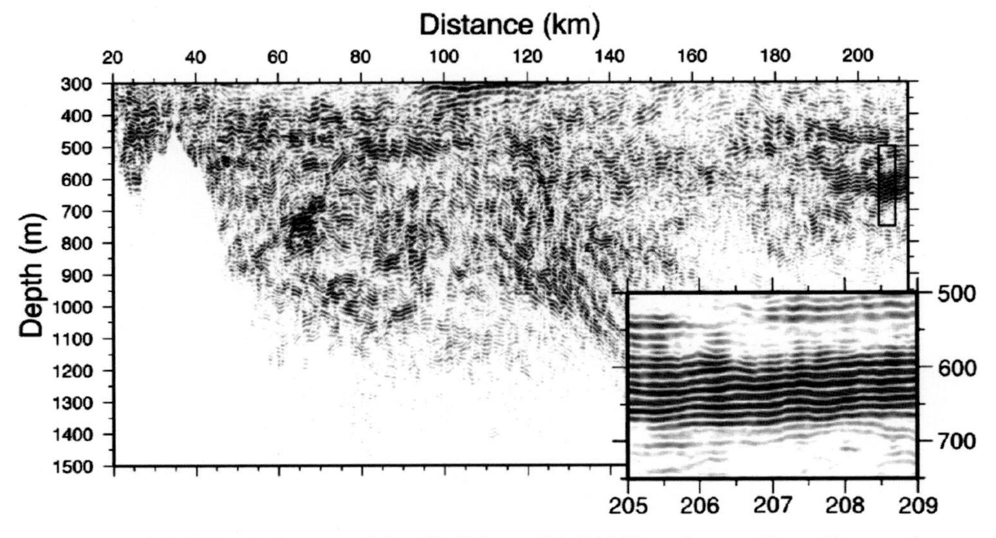

Figure 8.8 Seismic image of the Caribbean (C-SALT) staircase. From Fer *et al.*, (2010).

by a factor of two or more. Nevertheless, such estimates gave the oceanographic community a starting point for discussion of the potential contribution of staircases to the upper-ocean heat budget.

The data coverage of the upper Arctic in general, and our knowledge of staircase characteristics in particular, has improved dramatically since the commencement in 2004 of the Ice-Tethered Profiler (ITP) program, providing repeated sampling of the ice-covered upper-ocean Arctic (Toole *et al.*, 2011). The ITP program is currently active and more than fifty ITPs, distributed over much of the Arctic basin, have already been deployed. The number of observations provided by this program is impressive. The typical lifetime of an ITP is 500 days and each instrument produces on average 741 vertical profiles. Add to the picture their frequent sampling rate (two or more profiles per day) and vertical resolution of 25 cm, and it becomes clear that the ITPs are rapidly and dramatically increasing our knowledge of the Arctic stratification. Regarding the staircases, the first ITP-based results (e.g., Timmermans *et al.*, 2008) have already led to several firm and significant conclusions, of which we emphasize the following:

(i) Staircases are widespread throughout the central Arctic. Well-defined steps were absent in less than 4% of the ITP profiles. The smooth-gradient profiles were confined to selected regions, either located in close proximity to the basin boundaries or within strong mesoscale eddies.

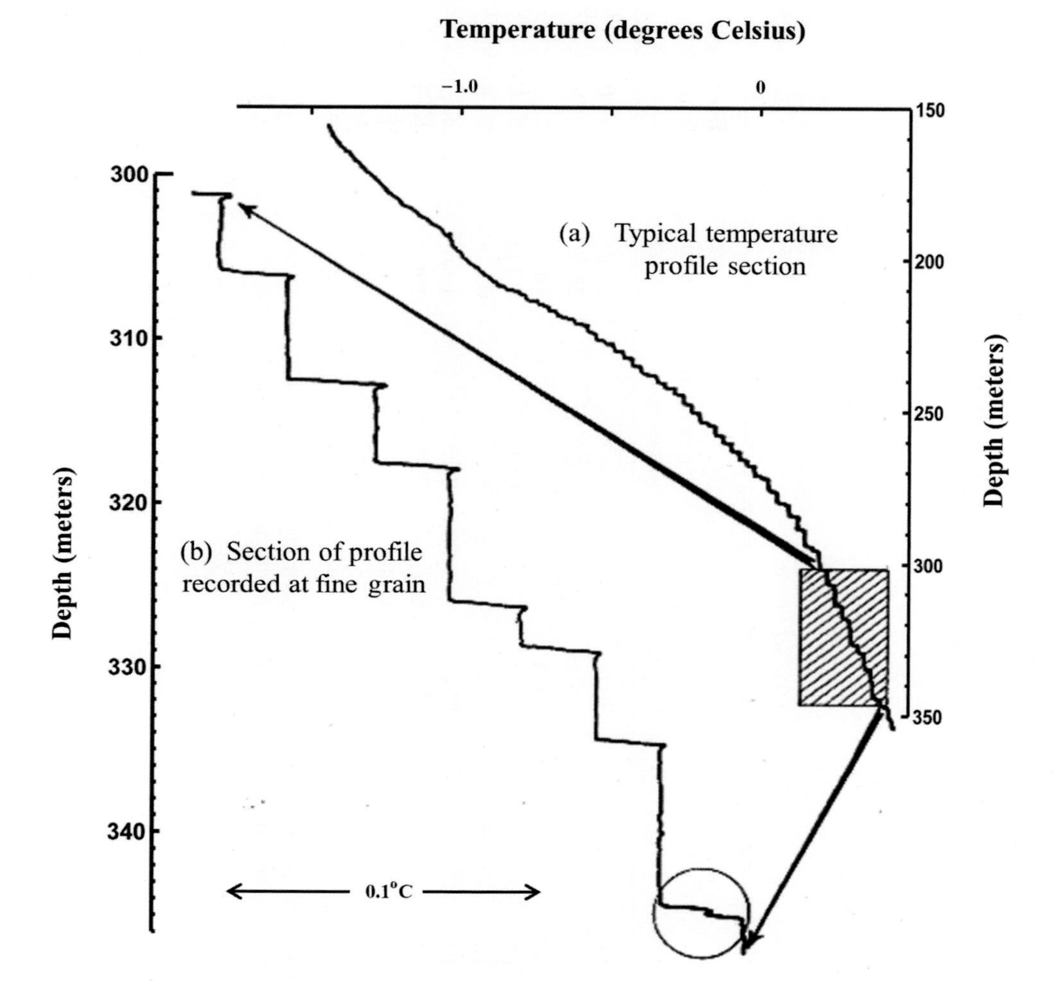

Figure 8.9 An example of temperature profile in the Arctic. From Neal *et al.* (1969).

(ii) Staircases are characterized by a remarkable spatial (hundreds of kilometers) and temporal (years) coherence. Most individual layers can be easily traced by their T–S values across the entire Beaufort Gyre and throughout the lifetime of each ITP.

(iii) Staircases actively respond to the climatic changes associated with warming and shoaling of the Atlantic Water. Relative to the AIWEX 1985 measurements, the present-day staircase region is approximately 100 m shallower and the thickness of steps is twice as large (~3 m average, with most steps in the 1–5 m range). The variation in temperature and salinities across typical

Table 8.1 *Characteristics of major staircases in the world ocean*

Location	Type	Density ratio	Step height	Temperature variation	Interfacial thickness	Buoyancy frequency (overall)
Western tropical Atlantic	SF	1.6	20 m	0.6 °C	2 m	$5 \cdot 10^{-3}$ s^{-1}
Tyrrhenian Sea	SF	1.2	400 m	0.2 °C	20 m	$4 \cdot 10^{-4}$ s^{-1}
Mediterranean outflow	SF	1.3	10 m	0.1 °C	1m	$2 \cdot 10^{-3}$ s^{-1}
Central Arctic	DC	4	3 m	0.05 °C	0.2 m	$4 \cdot 10^{-3}$ s^{-1}
Weddell Sea (upper)	DC	1.52	5 m	0.06 °C	0.1 m	$3 \cdot 10^{-3}$ s^{-1}
Weddell Sea (lower)	DC	1.36	80 m	0.13 °C	1 m	10^{-3} s^{-1}
Black Sea	DC	2	10 m (interior) 400 m (bottom)	0.05 °C	1 m	10^{-3} s^{-1}
Lake Kivu	DC	4	0.5 m	0.003 °C	0.2 m	$4 \cdot 10^{-3}$ s^{-1}

Salt finger and diffusive conditions are denoted as SF and DC respectively.
Thickness of the diffusive interfaces is based on the temperature stratification.

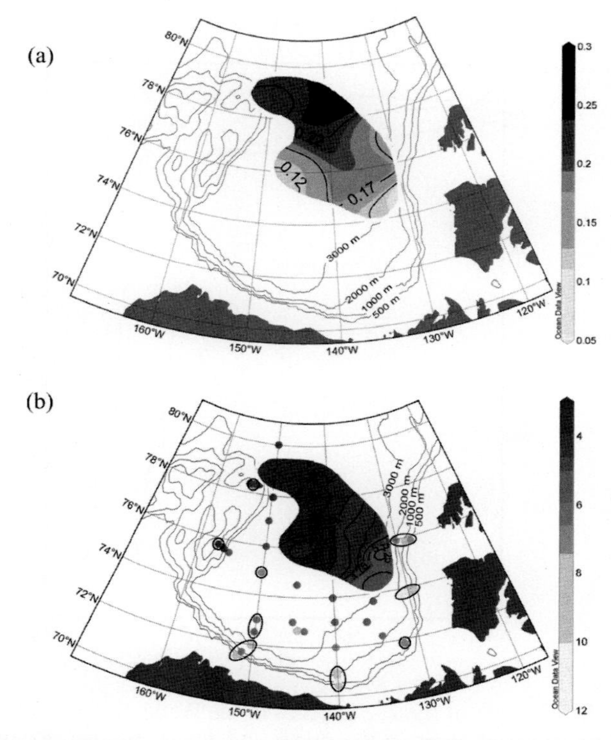

Figure 8.10 Distribution of the heat flux in W m^{-2} (a) and the average density ratio (b) in the diffusive layering region of the Beaufort Gyre estimated from the ITP data. From Timmermans *et al.* (2008). See color plates section.

interfaces are $\Delta T \approx 0.04$ °C and $\Delta S \approx 0.014$ – considerably higher than in the observations of Padman and Dillon (1987).

The distribution of the heat flux through staircases deduced from the 4/3 flux laws (Timmermans *et al.*, 2008) is shown in Figure 8.10a. The average value is $F_H \approx 0.22\,\mathrm{W\,m^{-2}}$, which exceeds earlier estimates (Padman and Dillon, 1987) by at least a factor of two. Figure 8.10b shows the distribution of (across-thermocline) density ratio in the Beaufort Gyre; values are limited to the range $2 < R_\rho^* < 7$ with the spatial average of $R_\rho^* \approx 4$. Layer-merging events have been observed but are rather rare, which suggests that the Arctic staircase is in a mature quasi-equilibrium state.

Figure 8.11 presents an example of a staircase from the Antarctic (Weddell Sea). Generally, Weddell staircases are characterized by much higher spatial and temporal variability. Muench *et al.* (1990) grouped the observed staircases into two distinct classes. Type A staircases are relatively shallow (limited to the upper 180 m) and small-scale (with step sizes of 1–5 m). Type B staircases are located in the

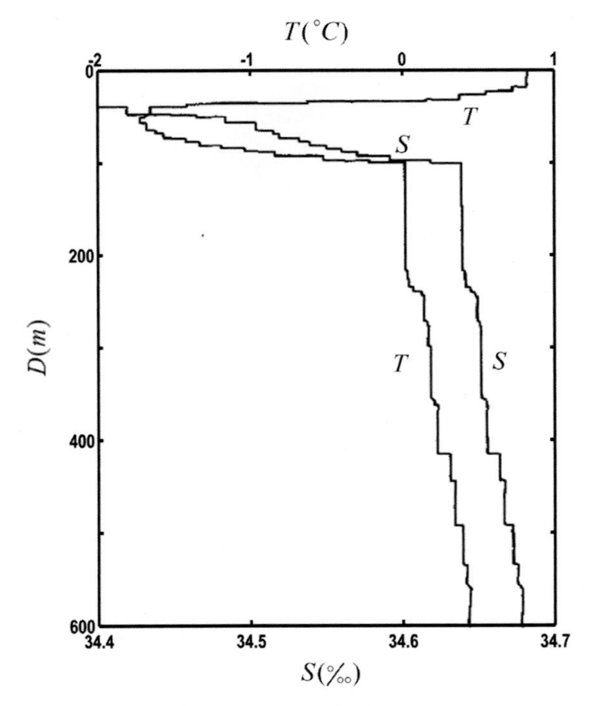

Figure 8.11 An example of the *T–S* profiles in the Weddell Sea (Antarctica). Note the distinct structure of steps in the upper and lower regions. From Foster and Carmack (1976).

weakly stratified portion of the thermocline (200–500 m) and vary considerably in step thickness (10–100 m). Robertson *et al.* (1995) examined data collected from a drifting ice station in the continental slope region of the western Weddell Sea. Their observations revealed the presence of small-scale (Type A) steps, statistics of which varied considerably over the observational period of three months in response to changes in the background stratification. The limited size of Type A steps could be associated with the temporal variability in the staircase, since significant periods of uninterrupted evolution are needed for the staircase to coarsen through a series of merging events. Deep (Type B) staircases, located in a more pristine environment, are likely to be closer to equilibration, which means thicker steps. Density ratio in the Weddell Sea is low ($R_\rho^* \sim 1.5$), which generally implies more active convection, thick steps and high *T–S* transport. Diffusively driven heat fluxes vary from $\sim$2 W m^{-2} in the slope region to $\sim$20 W m^{-2} in the interior, exceeding the Arctic fluxes by one to two orders of magnitude.

In addition to the Arctic and Antarctic examples, where diffusive layering is expected based on surface forcing patterns, diffusive staircases have also been

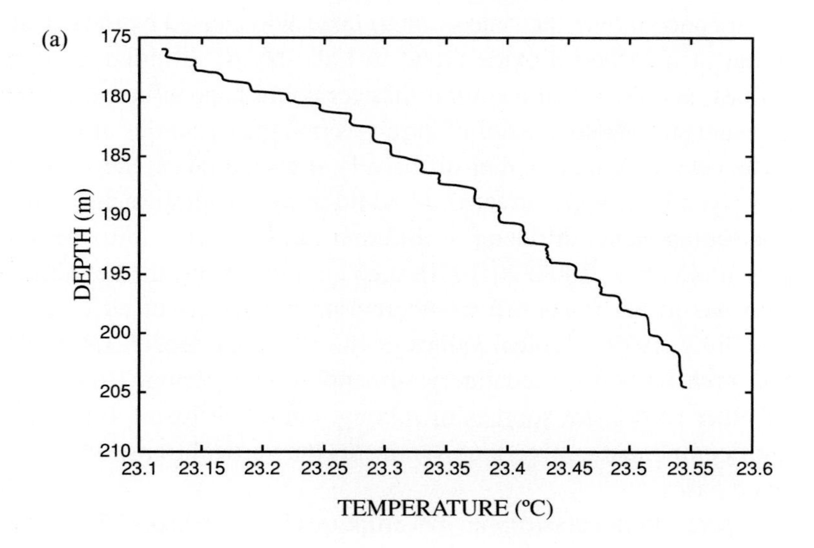

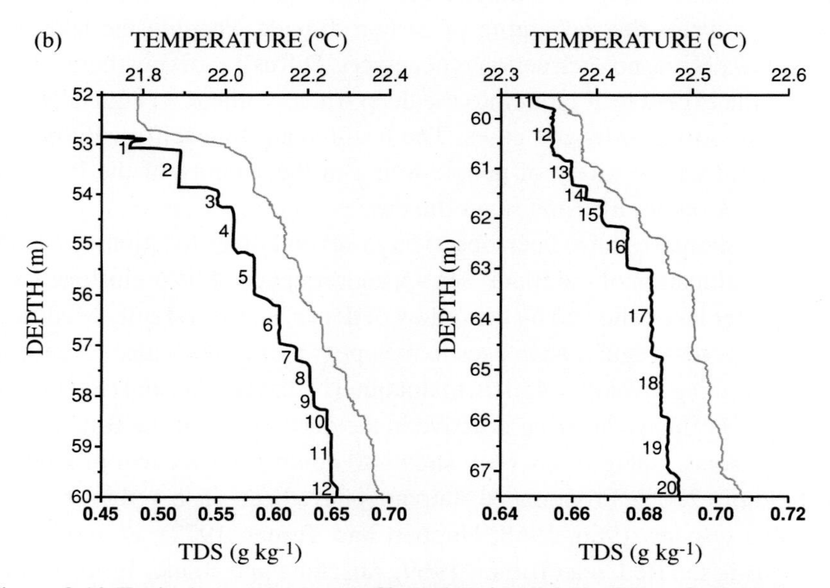

Figure 8.12 Typical temperature profiles taken in (a) Lake Kivu (from Newman, 1976) and (b) Lake Nyos (from Schmid *et al.*, 2004).

observed in low- and mid-latitude regions, albeit on a much smaller scale. Interesting case studies include salt-water lakes. Figure 8.12 shows the temperature profiles taken in Lake Kivu (a) and Lake Nyos (b), which reveal characteristic staircase patterns with numerous small-scale but well-defined steps. The background stratification is predominantly diffusive, with heat at the bottom supplied by geothermal springs. Great concern over the state of these lakes was caused by the catastrophic 1986 eruption of a carbon dioxide cloud in Lake Nyos, resulting in more than 1700 casualties, and fears that a similar disaster could happen in Lake Kivu. The Lake Nyos event stimulated a series of inquiries into the dynamics of vertical mixing. As is the case with many other diffusively stratified lakes, the water density in Kivu and Nyos is strongly affected, in addition to temperature and salinity, by several other components diffusing at different rates – carbon dioxide, methane and silica (Schmid *et al.*, 2004, 2010). In such circumstances, the definition of the density ratio has to be broadened to incorporate the effects of all density components (Griffiths, 1979). Typical values of the modified density ratio ($R_\rho^* \sim 5$) suggest that both lakes are moderately susceptible to layering. However, in the absence of other significant sources of mixing, double-diffusion (or, to be more precise, multicomponent convection) dominates the vertical transport of heat, salt and dissolved gases.

Can we expect future catastrophic gas eruptions in these lakes? The reader can rest assured that it is highly unlikely. The degassing pipe installed in Lake Nyos effectively controls the recharging of carbon dioxide, keeping the lake in a safe state. For Lake Kivu, no such action is necessary. Diffusive mixing there is sufficient to remove the excess heat input into the deep water, without an equivalent upward transfer of dissolved salts and gases. The result is a stable self-regulating system, which guarantees the safety of people living in the vicinity of the lake (Schmid *et al.*, 2010). Double-diffusion saves the day.

Diffusive staircases have been observed in several other locations. For instance, diffusive stratification of the Black Sea – a consequence of its prehistoric existence as a freshwater lake followed by the inflow of dense, warm and salty Mediterranean waters – supports irregular staircases in the upper part of the water column and the bottom convecting layer of ~450 m, maintained by the geothermal heat flux (Ozsoy *et al.*, 1993). Diffusive layering is active in the deep layers of the Brazil–Malvinas Confluence zone, where it has been shown to control the vertical property fluxes (Bianchi *et al.*, 2002). Pronounced staircases have been observed in the Antarctic Lake Vanda (Hoare, 1966, 1968; Huppert and Turner, 1972), in geothermal hot brine basins in the Red Sea (Turner, 1969; Anschutz and Blanc, 1996; Swift *et al.*, 2012) and in the eastern Mediterranean (Boldrin and Rabitti, 1990).

Oceanographic measurements indicate that staircases are likely to form in diffusive stratification as long as the density ratio is in the range $1 < R_\rho^* < 10$. Occasionally, staircases have been observed for even larger values of R_ρ^*. Such a wide layering-favorable range is in contrast with the relatively restrictive conditions for salt-finger staircases, which require significant density compensation of temperature and salinity. In the diffusive case, the gradient of the (unstable) thermal density component needs to be only one-tenth of the corresponding (stable) haline

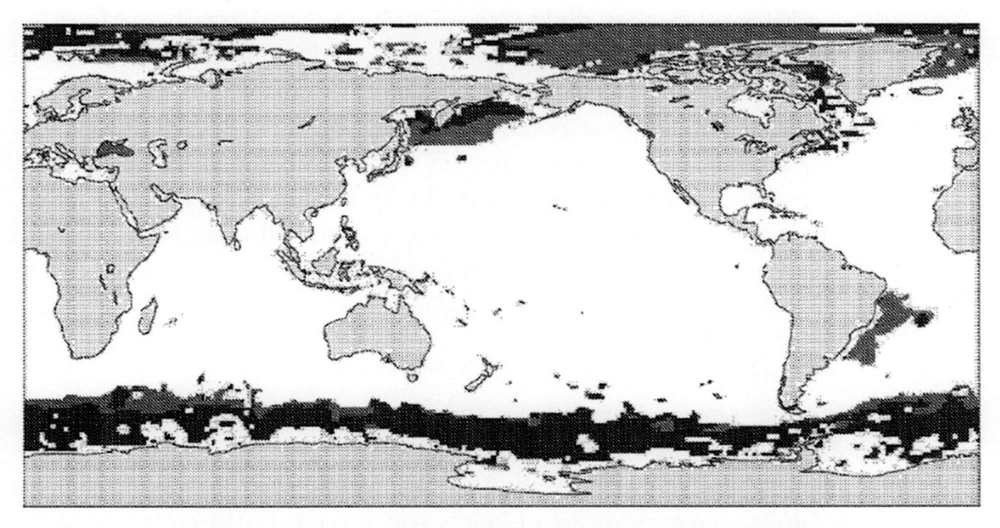

Figure 8.13 Ocean regions that are susceptible to diffusive layering. The light grey indicates areas with density ratios in the range $3 < R_\rho^* < 10$ somewhere in the water column and the dark grey indicates $1 < R_\rho^* < 3$. From Kelley *et al.* (2003).

gradient to produce well-defined staircases. Based on this criterion, it is possible to identify regions of the world ocean susceptible to diffusive layering, which are shown in Figure 8.13. As expected from the statistics of their incidence, diffusive staircase regions are mostly limited to the Arctic and Southern Oceans where, due to low temperatures, density stratification is often controlled by its haline component. In other susceptible areas, layering-favorable conditions are produced by region-specific interactions of distinct water masses.

8.2 Staircase origins

The list of usual explanations for the spontaneous formation of thermohaline staircases is a lengthy one. Staircase origin has been attributed to

(i) collective instability (Stern, 1969);
(ii) thermohaline intrusions that develop into a staircase (Merryfield, 2000);
(iii) metastable equilibria, initially forced by external disturbances (Stern and Turner, 1969);
(iv) applied flux mechanism (Turner and Stommel, 1964);
(v) negative density diffusion (Schmitt, 1994b);
(vi) instabilities of the flux-gradient laws followed by a series of mergers (Radko, 2003).

The overabundance of answers could be as disturbing as their absence. Much too often it is a warning sign that the proposed explanations are deficient in some respect and that the correct one has not been found. Fortunately, the situation with regard to the origin of staircases is not that bleak; we may be sufficiently close to establishing the leading contenders. Recent DNS of spontaneous layering have been particularly illuminating in this regard, although caution is certainly advised. To be objective, let us start by briefly reviewing the strengths and weaknesses of each hypothesis.

Collective instability mechanism

Perhaps the most influential hypothesis for the formation of staircases involves collective instability (Chapter 6). This idea was motivated by laboratory experiments in which staircases formed from the initially uniform *T–S* gradients (Stern and Turner, 1969). The appearance of layers was preceded by a period of active internal wave motion. Stern (1969) suggested that the growing wave might overturn and generate a stepped structure. Some support for this hypothesis comes from observations of oceanic staircases. Waves grow when the Stern number A exceeds unity (Chapter 6) and the measured values of A are indeed of order one in interfaces. However, there are several reasons to question the collective instability mechanism. For instance, in many laboratory experiments, salt and sugar replace heat and salt as buoyancy components and Stern numbers there are extremely low. Lambert and Demenkow (1972) report values as low as $A = 2 \cdot 10^{-3}$, casting doubt on the generality of Stern's (1969) criterion. Another dubious aspect is the assumed link between wave overturning and the formation of permanent well-defined staircases. Wave levels in regions of pronounced staircases (e.g., C-SALT area) are generally rather modest; numerous locations lacking staircases exist where waves are more active. At best, overturning internal waves produce transient irregular steppiness (Lazier, 1973; Lazier and Sandstrom, 1978) that is dynamically and structurally distinct from thermohaline staircases. Despite its undeniable historical interest, we cannot place the collective instability hypothesis too high on our list in terms of plausibility.

Thermohaline intrusion mechanism

The idea that staircases represent the final stage in the evolution of thermohaline intrusions is a relatively new one (Merryfield, 2000). As discussed in Chapter 7, intrusions can evolve either to a state consisting of alternating salt-finger and diffusive interfaces separated by convecting layers, which is common at high density ratio, or to a series of salt-finger interfaces when the density ratio is low

($R_\rho < 1.6$). The latter, argues Merryfield, form the observed oceanic staircases. This proposition is plausible, although it necessarily relies on the presence of lateral property gradients to drive interleaving. Intrusions-transformed-into-staircases are likely to exist in strong *T–S* fronts. For such regions, attempts have already been made to validate Merryfield's hypothesis using field measurements. Morell *et al.* (2006) analyzed a thermohaline staircase in a mesoscale eddy observed in the eastern Caribbean and found that the intrusion hypothesis is consistent with several aspects of the data. Intrusion-like structures, gradually morphing into regular steps, have been observed along the margins of the C-SALT staircase (Zhurbas and Ozmidov, 1983), suggesting that interleaving could play some role in the staircase dynamics.

The biggest question regarding the intrusion hypothesis concerns its generality. Can the intrusion mechanism explain the formation of major staircases, such as observed in the interior of the C-SALT area or Tyrrhenian Sea? Laboratory experiments (Stern and Turner, 1969; Krishnamurti, 2003) and numerical simulations (Radko, 2003; Stellmach *et al.*, 2011) indicate that staircases form spontaneously even in the absence of lateral gradients. Hence, intrusions are not essential for layering. For the intrusion hypothesis to be viable, we have to accept that the origins of the oceanic and laboratory/numerical staircases are fundamentally different. It is not impossible that two explanations are needed, one for the lab and another for the ocean, although most aspects of staircase experiments have proven to be highly suggestive of oceanographic observations.

Metastable equilibria mechanism

It has been speculated that staircases and smooth-gradient configurations represent distinct metastable equilibria. Stern and Turner (1969) suggested that finite amplitude perturbations to the gradient state force the system into a layered regime where it can remain for long periods of time (even indefinitely if the overall variation in temperature and salinity is maintained). In retrospect, it is clear that this mechanism does play some role in the evolution of staircases. Large initial perturbations to the gradient state undoubtedly make the transition to the staircase more likely and expedite the process. Once the staircase is created, the system becomes resilient to further structural changes. In this regard, Stern and Turner's insight is truly impressive, particularly considering that it was based on very early and qualitative laboratory experiments. The ability of sufficiently strong perturbations to transform the smooth gradient into a dramatically distinct state is linked to the subcritical nature of layering instabilities. In subcritical instabilities, nonlinearity tends to destabilize the system, leading to a harder transition to a new regime and also making this transition dependent on the initial perturbation amplitude.

The big question with regard to the metastable equilibria hypothesis is whether the finite amplitude perturbations are necessary for the spontaneous formation of layered states. For the diffusive case, the answer is yes. Calculations by Veronis (1965, 1968) – albeit performed in the context of the bounded model (Chapter 5) – have shown that when the uniform-gradient solution is linearly stable, finite amplitude perturbations can drive the system into a steady nonlinear convecting regime and often do so. For salt-finger staircases, however, the emerging evidence suggests that under typical conditions ($R_\rho < 2$) the uniform-gradient solution is *linearly* unstable. Numerical simulations (Radko, 2003; Stellmach *et al.*, 2011), laboratory experiments (Krishnamurti, 2003) and analytical arguments (Radko, 2003) indicate that layering can be initiated by small-amplitude perturbations. The layering instability takes the form of horizontally homogeneous modes that ultimately transform smooth stratification into well-defined steps. Thus, the metastable equilibria mechanism is not essential for finger-induced layering and plays only an auxiliary role in the transition to staircases.

Applied flux mechanism

A somewhat special mechanism of layering is realized in laboratory experiments in which a stable salinity gradient is heated from below (Turner and Stommel, 1964; Huppert and Linden, 1979). The direct response to the applied heat flux is top-heavy convection in the lower part of the water column. The convecting layer is well mixed and bounded from above by a thin high-gradient interface, which is clearly visible in vertical temperature profiles taken during the experiment (Fig. 8.14). Heat is transferred upward from the convecting layer by a combination of molecular diffusion and entrainment across the interface. Since the molecular transfer of heat exceeds that of salt, the net result is the supply of buoyancy to the region immediately above the interface, which leads to the formation of a second convecting layer. The process then repeats over and over, resulting in a sequence of mixed layers separated by sharp interfaces – the thermohaline staircase.

The basic dynamics of layering in the bottom-heated case are robust and well understood. Similar processes are expected to occur when the diffusive stratification is cooled from above (Molemaker and Dijkstra, 1997) and when the salt flux is applied at the top of the salt-finger favorable stratification. It is, however, unclear whether the prescribed flux boundary condition is appropriate for oceanic layering. The applied flux mechanism is most likely at work in cases when layering is caused by geothermal heating, examples of which were given in the previous section. The dynamics of the more prominent staircases – such as observed in the

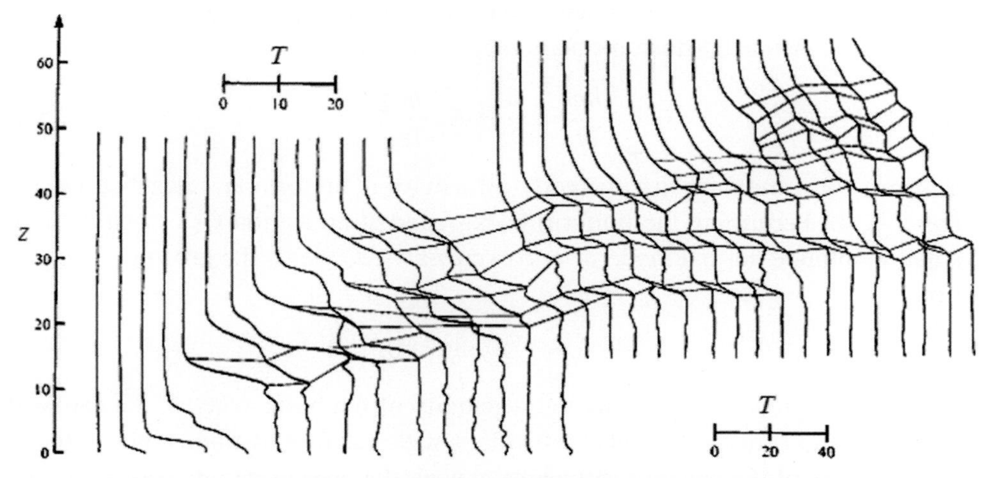

Figure 8.14 A set of temperature–depth profiles taken during the course of a laboratory experiment in which a stable salinity gradient is heated from below. Each successive profile is offset to the right. From Huppert and Linden (1979).

Arctic or in the C-SALT area – are substantially different. The layering-favorable conditions there result from the juxtaposition of distinct water masses rather than from the surface/bottom buoyancy forcing. In such cases, causality is reversed: layering is no longer controlled by the fluxes but, rather, fluxes are determined by the pattern of stratification. Numerical simulations of layering in the effectively unbounded gradients tend to confirm the subservient role of fluxes. As layers form and sequentially merge, the vertical transport of heat and salt dramatically increases.

Negative density diffusion

To explain the mechanism of layering, oceanographers often point out that if the flux of density is counter-gradient, it should have a destabilizing effect on smoothly stratified regions of the ocean (Schmitt, 1994b). The origin of this idea can be traced to the heuristic arguments put forward by Phillips (1972) and Posmentier (1977) for one-component turbulent fluids. At first glance, this explanation comes across as an even better fit to the thermohaline layering problem. Double-diffusion is driven by the release of potential energy of the background stratification; the light upper part of the water column becomes lighter, the heavy lower part becomes even heavier. Hence, the eddy diffusivity of density is negative, which by itself implies the instability of a uniform gradient and a likely transition to a different stratification pattern.

The negative diffusion argument can be made more quantitative by focusing on the one-dimensional diffusion equation of density:

$$\frac{\partial \rho}{\partial t} = \frac{\partial}{\partial z}\left(K_\rho \frac{\partial \rho}{\partial z}\right). \tag{8.5}$$

Suppose for a moment that the eddy diffusivity is uniform and negative ($K_\rho = const < 0$). Perturbing the uniform background stratification ($\bar{\rho}_z = const$) by a small-amplitude normal mode $\rho' = \hat{\rho}\exp(imz + \lambda t)$ yields the growth rate equation:

$$\lambda = -K_\rho m^2 > 0. \tag{8.6}$$

The positive growth rates imply the instability of the basic state, which can lead to layering. One of the technical difficulties with this formulation is related to the ill-posedness of the negative diffusivity model: the growth rate increases without bound with the wavenumber (m). This problem, however, is not fundamental. The concept of eddy diffusivity assumes some scale separation between eddies and the mean field, and the ultraviolet catastrophe in the negative diffusivity model can be prevented by applying (8.5) only to scales substantially exceeding the salt-finger width.

Unfortunately, the negative diffusivity model suffers from at least two other inconsistencies that severely limit its ability to explain the mechanism of layering. First, (8.6) implies that the uniform gradient is always unstable and therefore transition to staircases is expected regardless of the specific conditions. This conclusion is obviously false. The oceanic heat–salt staircases form in a relatively narrow range of density ratios $1 < R_\rho < 2$ – only 1% of the net salt-finger favorable range $1 < R_\rho < \tau^{-1} \approx 100$. Call me a pessimist, but if a model gives a wrong answer in 99% of cases, it must be missing some essential physics. Doubts as to its relevance linger even when the model does make a correct prediction for low density ratios.

The second problem with regard to the negative diffusivity mechanism is related to the strong dependence of diffusivity on the density ratio. In double-diffusive convection, the mixing intensity rapidly diminishes with increasing R_ρ, the significance of which becomes apparent when (8.5) is rewritten as

$$\frac{\partial \rho}{\partial t} = K_\rho \frac{\partial^2 \rho}{\partial z^2} + \frac{\partial K_\rho}{\partial R_\rho}\frac{\partial R_\rho}{\partial z}\frac{\partial \rho}{\partial z}. \tag{8.7}$$

While the first term on the right-hand side of (8.7) is always destabilizing, it is usually less than the second (stabilizing) term. The layering conditions are ultimately controlled by the competition between the "negative diffusivity" and "variable diffusivity" mechanisms. Thus, the negative diffusivity hypothesis in its original form ignores a very significant aspect of the layering problem. To summarize, the negative diffusivity view may have some merit, but only in a very general and qualitative sense. Any meaningful quantitative model of double-diffusive layering should necessarily take into account the dependence of mixing characteristics on the density ratio.

My money is on the sixth and final hypothesis considered here – the instability of flux-gradient laws – which builds on the negative diffusivity model but puts it on a more solid and physical footing. Instead of combining temperature and salinity into a single density term, it follows a more natural, for double-diffusive problems, approach and treats both density components individually. The flux law instability mechanism does not seem to suffer from inconsistencies that plague other theories of double-diffusive layering and it has been successfully tested by DNS. These features warrant a more detailed discussion of the flux law hypothesis, which follows next.

8.3 Instability of the flux-gradient laws

To be specific, we discuss the flux law instability in the context of salt-finger favorable stratification, which has been favored for several reasons (including traditional preference) by most theoretical studies of layering. It should be understood, however, that the following model can be readily adapted for the diffusive regime as well, provided the background stratification is susceptible to the primary oscillatory instability. Consider the one-dimensional (z) large-scale temperature and salinity equations

$$\begin{cases} \dfrac{\partial T}{\partial t} = -\dfrac{\partial}{\partial z} F_T, \\ \dfrac{\partial S}{\partial t} = -\dfrac{\partial}{\partial z} F_S, \end{cases} \tag{8.8}$$

which, in essence, represent the two-component counterpart of (8.5). F_T and F_S are the vertical temperature and salinity fluxes that we attribute to salt fingering. All variables are non-dimensionalized using the standard system (1.11), which makes it possible to express the fluxes in terms of the Nusselt number and flux ratio as in (6.3).

The next step involves the development of a mixing model. Following the treatments of collective instability in Chapter 6 and thermohaline interleaving in Chapter 7, we assume that both the Nusselt number (Nu) and the flux ratio (γ) are uniquely determined by the local density ratio (R_ρ) and (8.8) is linearized with respect to the uniform background gradients. The linear perturbations of temperature and salinity fluxes are given in (6.5)–(6.7), which reduce the governing system (8.8) to

$$\begin{cases} \dfrac{\partial T'}{\partial t} = A_{Nu}\dfrac{\partial^2}{\partial z^2}\left(T' - \bar{R}_\rho S'\right) + Nu(\bar{R}_\rho)\dfrac{\partial^2 T'}{\partial z^2}, \\ \dfrac{\partial S'}{\partial t} = A_\gamma Nu(\bar{R}_\rho)\dfrac{\partial^2}{\partial z^2}\left(T' - \bar{R}_\rho S'\right) + \gamma^{-1}(\bar{R}_\rho)\dfrac{\partial T'}{\partial t}, \end{cases} \tag{8.9}$$

where $\bar{R}_\rho$ is the basic density ratio and $A_{Nu} = \bar{R}_\rho \left.\frac{\partial Nu}{\partial R_\rho}\right|_{R_\rho=\bar{R}_\rho}$ and $A_\gamma = \bar{R}_\rho \left.\frac{\partial \gamma^{-1}}{\partial R_\rho}\right|_{R_\rho=\bar{R}_\rho}$. Stability properties of the resulting linear system are analyzed using the normal modes

$$\begin{pmatrix} T' \\ S' \end{pmatrix} = \begin{pmatrix} \hat{T} \\ \hat{S} \end{pmatrix} \exp(imz + \lambda t). \tag{8.10}$$

When (8.10) is substituted in the linear system (8.9) and $\left(\hat{T}, \hat{S}\right)$ are eliminated, we arrive at the quadratic eigenvalue equation for the growth rate λ:

$$\lambda^2 + \lambda\left(A_{Nu} + Nu(\bar{R}_\rho) - A_\gamma Nu(\bar{R}_\rho)\bar{R}_\rho - \frac{\bar{R}_\rho A_{Nu}}{\gamma(\bar{R}_\rho)}\right)m^2 - A_\gamma Nu^2(\bar{R}_\rho)\bar{R}_\rho m^4 = 0. \tag{8.11}$$

The stability/instability of the system is determined by the coefficients of Eq. (8.11) and, ultimately, by the background density ratio $\bar{R}_\rho$. Of particular significance is the sign of A_γ. Theoretical arguments and numerical simulations (Chapter 2) indicate that the $\gamma(R_\rho)$ dependence is non-monotonic. As the density ratio increases from unity, the flux ratio first decreases, as shown in Figure 8.15, reaches a minimum value ($R_{\min}$) and then starts to increase. For density ratios in the range $1 < \bar{R}_\rho < R_{\min}$, the free coefficient of the quadratic equation (8.11) is negative ($A_\gamma > 0$), which implies that there are two real roots of opposite sign. The existence of a positive root means that the basic uniform gradient is unstable. It can also be shown that, under certain unrestrictive assumptions, the flux-gradient laws are stable for $\bar{R}_\rho > R_{\min}$. Thus, the decrease of γ with R_ρ is both a necessary and sufficient instability condition. This instability can be thought of as a special one-dimensional case of the intrusive γ-instability modes discussed in Chapter 7.

Analysis of the amplitude/phase relationships for T', S', R_ρ and γ in a growing normal mode suggests the following physical explanation of layering. If the amplitude of the temperature perturbation ($\hat{T}$) exceeds the amplitude of the salt perturbation ($\hat{S}$), as shown in a schematic in Figure 8.16, then the density ratio R_ρ reaches its maximum at the location of the largest temperature gradient (that is, $z = 0, 2\pi/m$ in Fig. 8.16). If γ were constant, then the growth rate of the first normal mode in (8.11) would be zero – T and S would not change in time, which means

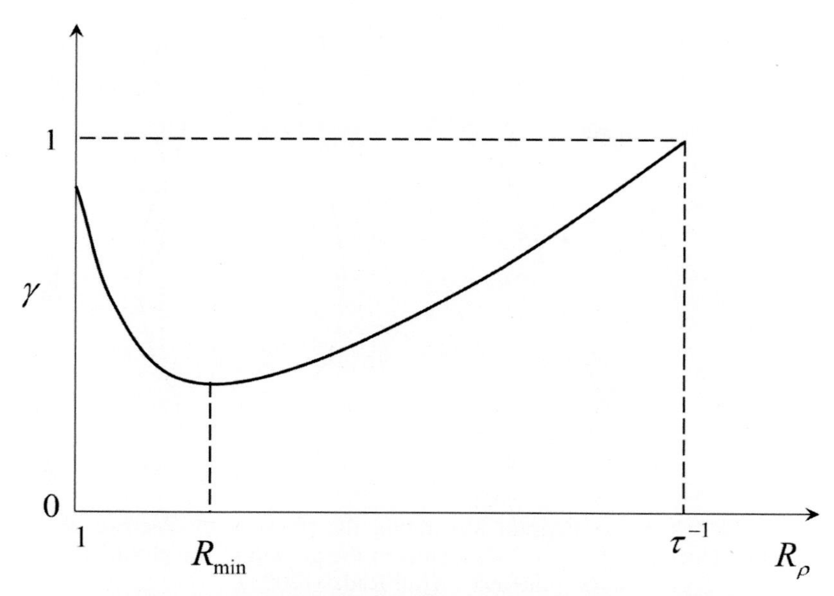

Figure 8.15 Dependence of the flux ratio on the density ratio.

that F_T and F_S are uniform in z. However, when γ is a decreasing function of R_ρ, there is an increase of the heat/salt flux ratio at $z = \pi/m$ and, correspondingly, a decrease at $z = 0, 2\pi/m$. As a result, the temperature flux convergence at $0 < z < \pi/m$ (and divergence at $\pi/m < z < 2\pi/m$) exceeds that for salt. This convergence pattern leads to an enhanced accumulation of heat, relative to the accumulation of salt, in the lower part of the layer in Figure 8.16. This, in turn, is followed by an additional increase in R_ρ at $z = 0, 2\pi/m$ and, correspondingly, further decreases γ there. At $z = \pi/m$, the density ratio further decreases while the flux ratio increases. This self-enhancing mechanism produces a monotonic growth of the perturbation.

It is easy to imagine how the γ-instability can transform a smooth *T–S* gradient into a staircase. Suppose that the growth of the unstable horizontally uniform modes persists long enough for the fluid to develop density reversals, as indicated in the schematic in Figure 8.17. These top-heavy regions overturn and the fingering interfaces become sandwiched between nearly homogeneous convecting layers. Since the finger-driven flux of density across the interfaces is up-gradient (downward), there is a continuous supply of density to the top of each convecting layer; density is removed at the bottom. This forcing pattern maintains convection in mixed layers. Vigorous convection in layers, in turn, prevents salt-finger zones from spreading vertically and the system remains locked in the staircase regime.

Conditions for the γ-instability are generally consistent with the incidences of layering in the ocean. The pattern of $\gamma(R_\rho)$ for the oceanic (heat/salt) parameters

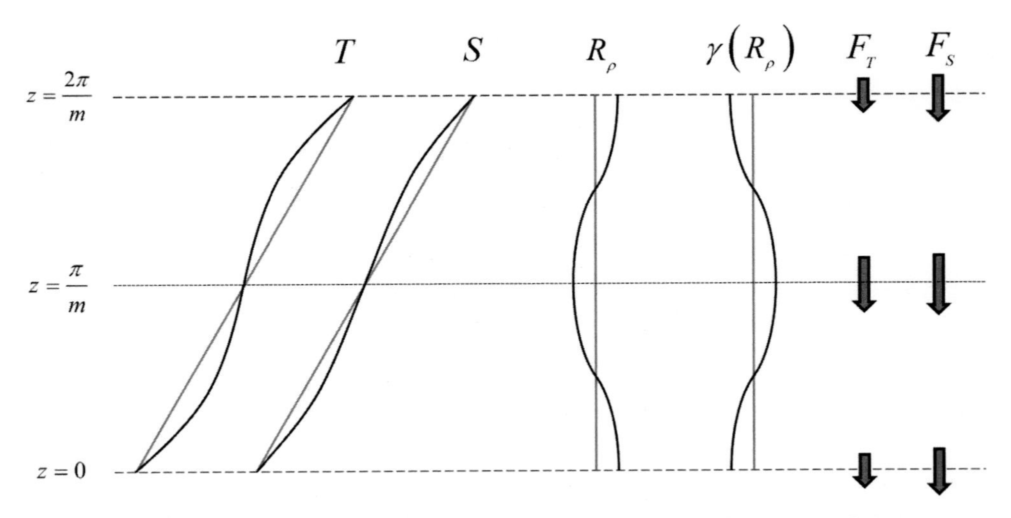

Figure 8.16 Schematic diagram illustrating the physical mechanism of the γ-instability. Decrease in γ with R_ρ results in the growth of the perturbations on a uniform T–S gradient (see the text). After Radko (2003).

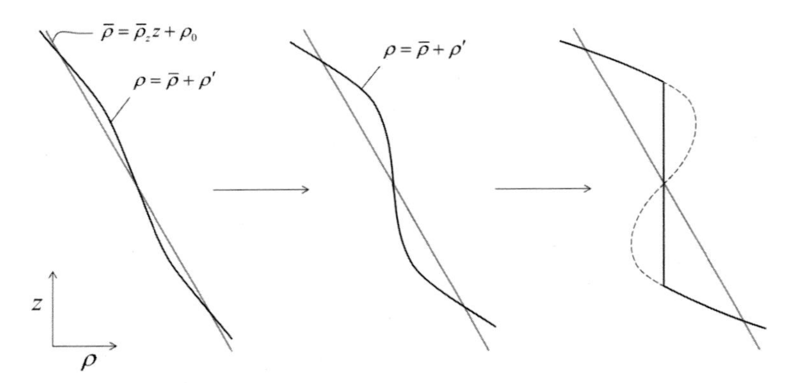

Figure 8.17 Transition from the smooth-gradient to staircase configuration. The growing horizontally uniform perturbation develops top-heavy regions that rapidly overturn, producing mixed layers separated by high-gradient interfaces.

(Fig. 2.7a) is such that the most significant decrease in γ occurs within the interval $1 < R_\rho < 2$, which rationalizes both observations of staircases for low R_ρ and the lack of staircases for $R_\rho > 2$. Another consideration for the ocean involves the contribution of turbulent mixing by overturning gravity waves. Turbulence tends to mix temperature and salinity at equal rates and therefore $\gamma_{\text{turb}} = R_\rho$. Taking into account this contribution shifts the minimum of the net flux ratio towards lower density ratios. The magnitude of this shift is as uncertain as our knowledge of turbulent diffusivity in the thermocline. However, it is not unreasonable to assume

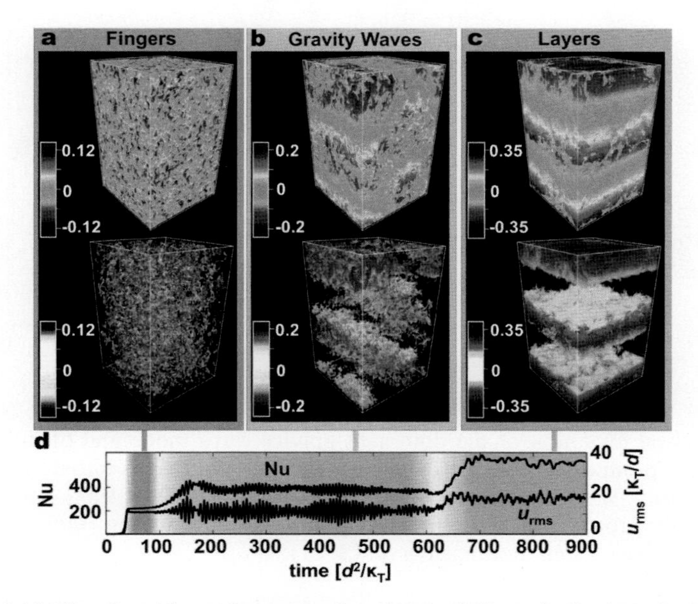

Figure 8.18 Simulated layer formation in a large triply periodic domain. Temperature perturbations are shown in color, with their respective color scales in each panel. The top panels are visualizations on the data-cube faces, while the lower ones are volume-rendered images. Snapshots are shown for three characteristic dynamical phases: (a) homogeneous fingering convection, (b) flow pattern dominated by gravity waves, (c) formation of vigorously convecting layers separated by thin fingering interfaces. (d) Time series of the Nusselt number and the flux ratio. From Stellmach *et al.* (2011). See color plates section.

that the minimum of the $\gamma(R_\rho)$ relation for the combination of double-diffusion and turbulence matches the typical point of staircase/gradient transition in observations ($R_{\text{min}} \approx 1.7$).

Diagnostics of spontaneous layering in DNS – both two-dimensional (Radko, 2003) and three-dimensional (Stellmach *et al.*, 2011) – confirm that layering is indeed driven by the γ-instability. Figure 8.18 shows a sequence of events leading to staircase formation. The experiment was initiated by a uniform salt-finger favorable gradient. The vigorous field of salt fingers emerges very rapidly (Fig. 8.18a) and initially lacks any signs of large-scale self-organization. However, this homogeneous fingering state does not persist for long. The next evolutionary phase is defined by the presence of an internal wave field, which is a direct and expected consequence of the collective instability (Chapter 6). Waves grow until their amplitude is sufficiently large to develop intermittent density overturns. After saturation, the system remains in the wave-dominated state for a substantial period of time ($150 < t < 600$ in standard non-dimensional units). At this stage, the wave

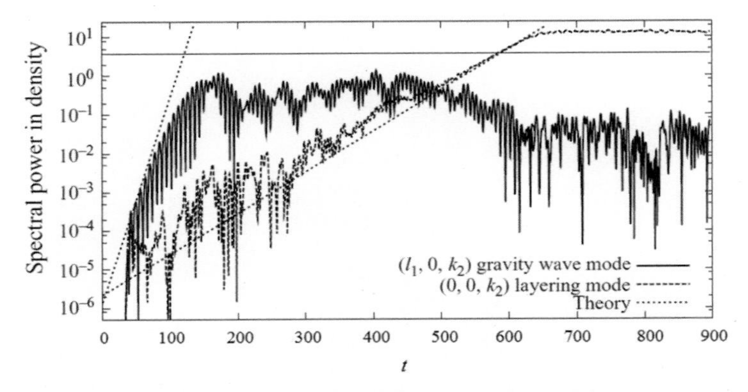

Figure 8.19 Diagnostics of the numerical simulation in Figure 8.18. Shown is the temporal evolution of the dominant gravity-wave mode and of the γ-mode ultimately producing the staircase along with the predictions based on the parametric theory. While the gravity-wave mode rapidly saturates, the γ-mode slowly but steadily grows up to the critical amplitude for overturning (indicated by the horizontal line). From Stellmach *et al.* (2011).

activity is visually most striking (Fig. 8.18b). After observing waves of such magnitude it is easy to appreciate why the early explanations of layering ascribed critical significance to collective instability (Stern and Turner, 1969). In reality, the role of waves is surprisingly limited. The spatially averaged *T–S* fluxes, expressed in terms of Nusselt number and flux ratio in Figure 8.18d, are modulated by waves. However, their time-mean values remain largely unchanged until the system enters the third phase – the emergence of a staircase (Fig. 8.18c).

The two-layer structure in Figure 8.18c is produced by slow but persistent growth of the horizontally uniform Fourier mode with the vertical wavelength of half the domain size. Eventually, this mode generates density inversions, which overturn and produce well-mixed convecting layers, in accord with the scenario depicted in Figure 8.17. A time record of the amplitude of the layering mode (diagnosed from the DNS) is plotted along with the theoretical prediction based on Eq. (8.11) in Figure 8.19; their apparent agreement confirms the relevance and accuracy of the parametric model. It is also instructive to compare the evolutionary patterns of layers and waves in the DNS experiment. The most energetic wave mode, also shown in Figure 8.19, grows more rapidly than the layering mode and reaches saturation much faster. However, even when the wave activity is maximal, it does not significantly affect the layering mode, which continues to grow at the predicted rate until the staircase is formed. Waves are far more fragile. Upon reaching the near-saturation amplitude, the layering mode substantially suppresses the wave motion.

The experiment in Figure 8.18 shows that, overall, waves do little in the layering process – despite spectacular amplitudes, they control neither small-scale mixing by salt fingers nor the growth of the large-scale layering modes. In this regard, it should be acknowledged with sadness that the literature on double-diffusion (or on any exciting topic for that matter) is often polluted with myths and unsubstantiated conclusions. One of the most widespread notions is that the vertical shear induced by internal waves in the ocean has a devastating effect on fingers and on secondary double-diffusive structures. Numerical simulations, such as the DNS reported by Stellmach *et al.* (2011), demonstrate that this notion is without basis.

The final question we wish to discuss in this section concerns the preferred vertical scale of layering modes. Analysis of the growth rate equation (8.11) for the unstable regime ($R_\rho < R_{\min}$) shows that its positive root scales as

$$\lambda_1 \sim -\frac{\partial \gamma}{\partial R_\rho} Nu\, m^2. \tag{8.12}$$

Equation (8.12) implies that the growth rate increases without bound with the perturbation wavenumber (m) and therefore the model suffers from the ultraviolet catastrophe (similar to that in the negative diffusivity model discussed in Section 8.2). This brings in a couple of related complications. The first one is that the model becomes obviously unphysical if the prediction in (8.12) is taken literally – infinitely large growth rates are prohibited in nature.

This criticism is relatively easy to dismiss by simply recalling that the theory assumed from the outset a certain scale separation between large-scale layering modes and salt fingers. The model in general and (8.12) in particular applies only to sufficiently small wavenumbers ($m < m_0$). The range of validity cannot be deduced internally from the theory. However, simulations (Traxler *et al.*, 2011a; Stellmach *et al.*, 2011) indicate that scales exceeding the salt-finger width by an order of magnitude or more are accurately represented by the model; smaller scales are not. Thus, the largest growth rates are expected to occur in the vicinity of the point of failure of the theoretical model $m_0 \sim 0.1 \cdot k_f$, where k_f is the typical horizontal wavenumber of salt fingers. The calculation in Figure 8.18 corroborates this suggestion: the wavelength of the γ-instability mode (L_γ) destined to evolve into the staircase exceeds the fastest growing finger width by a factor of twenty. In dimensional units, evaluated for typical stratification of the mid-latitude thermocline, this translates to

$$L_\gamma \sim 1 - 2\,\text{m}. \tag{8.13}$$

The estimate (8.13) is a cause for concern. The limited scale of layering instability seems to be at odds with oceanographic observations of much thicker steps

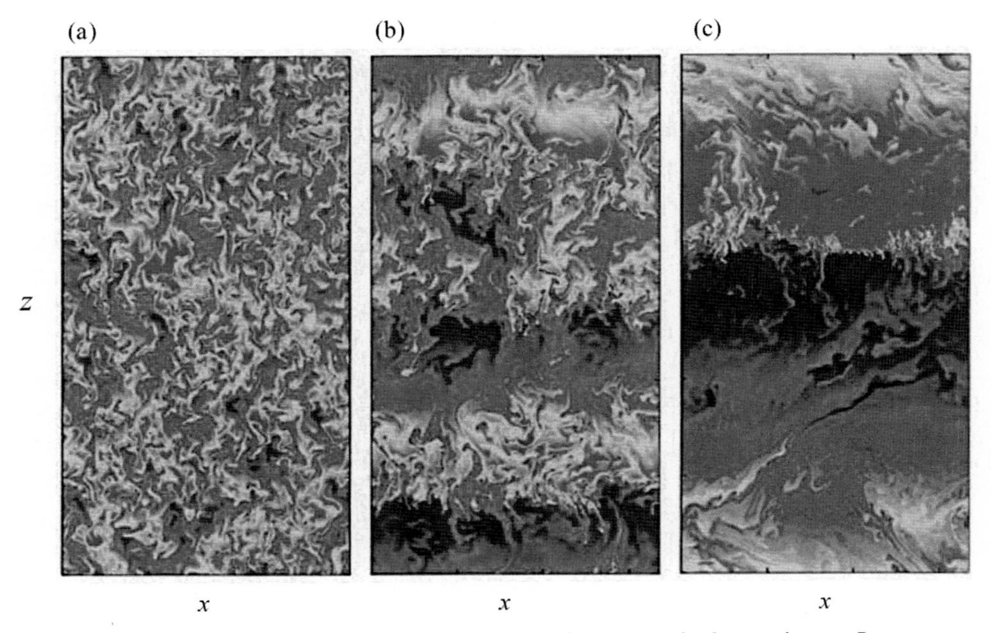

Figure 8.20 Formation and evolution of layers in a numerical experiment. Instantaneous perturbation temperature fields are shown for (a) $t = 50$, (b) $t = 400$ and (c) $t = 800$ (standard non-dimensional units). Note the spontaneous appearance of layers, followed by their systematic mergers. From Radko (2003). See color plates section.

of 10–100 m, which are more common in salt-finger staircases. Fortunately, numerical simulations offer an important hint for the step-size selection puzzle. Figure 8.20 shows the formation and evolution of layers in a two-dimensional DNS (Radko, 2003). As expected, the layers that formed first are relatively thin (~10–20 times the finger wavelength). However, in time they merge sequentially until there is only one interface left within the limits of the computational domain. Profiles of the horizontally averaged density (Fig. 8.21) reveal a general tendency for strong steps characterized by significant temperature and salinity jumps to grow further at the expense of weaker steps, which gradually erode and eventually disappear. This merging pattern – referred to as the "B-merger" mode in Radko's (2007) classification (Section 8.4) – is reminiscent of the evolution of the Tyrrhenian Sea staircase in Figure 8.6. Sequential layer mergers have also been observed in the laboratory experiments on the spontaneous gradient/staircase transition (Krishnamurti, 2003). Figure 8.22 presents the layering simulation in the diffusive regime (Prikasky, 2007). The evolutionary pattern of steps is strikingly similar to that of salt-finger staircases (Fig. 8.20). Layers that develop initially are not steady but merge continuously and the characteristic step size increases substantially.

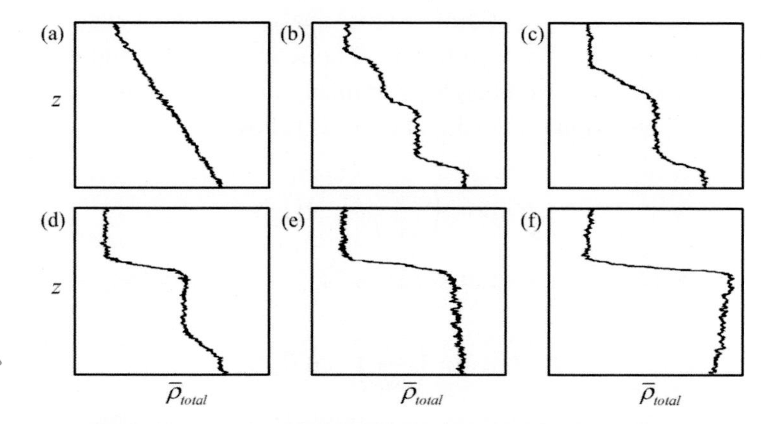

Figure 8.21 Evolution of the horizontally averaged density for the calculation in Figure 8.20. From Radko (2003).

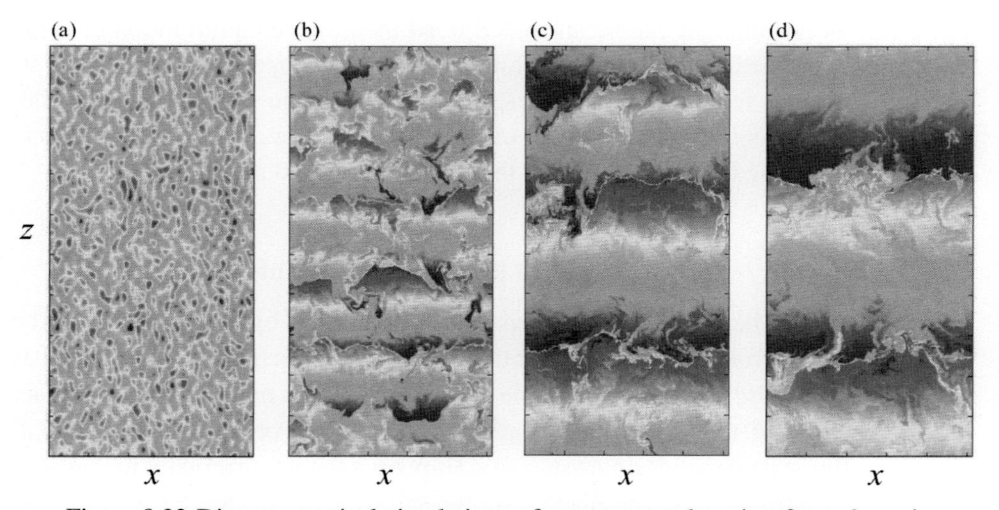

Figure 8.22 Direct numerical simulations of spontaneous layering from the uniform gradient in a diffusively stratified fluid. The small-scale oscillatory perturbations give way to horizontally uniform layers, which merge sequentially until producing the final quasi-equilibrium state. From Prikasky (2007). See color plates section.

If the merging scenario is commonly realized in oceanic staircases, then the initial thickness of layers is of secondary importance. Nature has all the time it needs to complete the transition to the ultimate equilibrium state. In this case, the observed step size is controlled by processes that arrest the coarsening of a staircase and the final thickness of layers can greatly exceed the initial one. Regarding the equilibrium step size, it should be mentioned that some general conclusions could be reached on the basis of dimensional analysis, without even formulating a mechanistic model of staircase evolution. If no length scale is imposed externally, then the non-dimensional step height is uniquely determined by the density ratio, which can be expressed in dimensional units as follows:

$$H_{\text{dim}} = \left(\frac{k_T \nu}{g\alpha \bar{T}_z}\right)^{\frac{1}{4}} H(R_\rho). \tag{8.14}$$

Alternatively, (8.14) can be written in terms of the buoyancy frequency (*N*):

$$H_{\text{dim}} = \left(\frac{k_T}{N}\right)^{\frac{1}{2}} G(R_\rho), \tag{8.15}$$

where $G = H\left[Pr(1 - R_\rho^{-1})\right]^{\frac{1}{4}}$. Kelley (1984) and Kelley *et al.* (2003) used various oceanographic observations of diffusive staircases to demonstrate that (8.15) captures the general trend of step-size distribution. The departures of data from (8.15) are noticeable but can be attributed to an incomplete equilibration of some staircases examined by Kelley *et al.* While it is reassuring that dimensional analysis can predict gross features of staircases, deeper insight into fluid dynamical phenomena usually requires development of explicit physical models. Some results in this direction, based on layer-merging theory, are discussed next.

8.4 Mechanics of layer-merging events

The first ideas on the evolution of staircases were introduced by Huppert (1971). The essential insight from his model is that the dynamics of layers should be treated as a secondary instability problem. A series of identical steps might well be an exact steady solution of the governing equations, but whether or not the system will remain in this state depends on its stability. Huppert's theory, originally formulated for diffusive staircases, is now adapted for the salt-finger case and, for pedagogical reasons, expressed in terms of the simplest system consisting of two layers.

Consider the configuration shown in Figure 8.23. Figure 8.23a represents a basic state consisting of a series of identical thin salt-finger interfaces separated by convecting layers of equal thickness *H*. This steady state is perturbed as indicated in Figure 8.23b: we increase slightly the *T–S* jump at the interface $z = z_1$, but decrease the jump at the adjacent interfaces. The vertical structure is assumed to be periodic with the *z*-wavelength of 2*H*. This state can be thought of as an infinite series of layers in which we simultaneously reduce the magnitude of temperature and salinity jumps (ΔT, ΔS) at all steps with even numbers and correspondingly increase the jumps across the odd steps. Note that such a perturbation does not affect

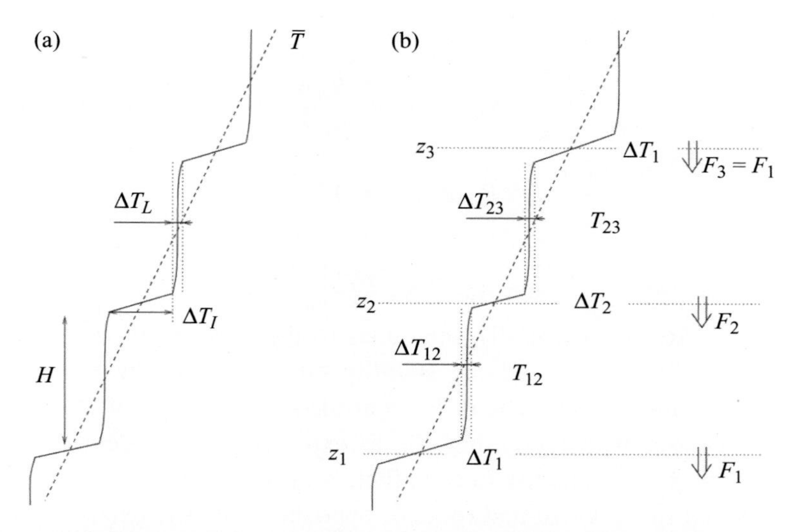

Figure 8.23 Schematic diagram illustrating the stability analysis for an infinite series of interfaces. (a) Basic state consisting of identical steps. (b) Perturbed state in which the *T–S* jumps at the even interfaces are slightly decreased, and the jumps at odd interfaces are increased. From Radko (2005).

the overall *T–S* gradient. Our objective is to determine whether the disturbance will grow in time, implying instability of the basic state in Figure 8.23a, or remain small.

The most basic version of the model (Huppert, 1971; Radko, 2003) invokes the following simplifying assumptions:

(i) the temperature and salinity variations across the convecting layers $(\Delta T_L, \Delta S_L)$ are not taken into account;
(ii) fluxes across the interfaces are parameterized using the 4/3 flux law $F_T = C(R_\rho)\Delta T^{\frac{4}{3}}$, $F_S = \frac{F_T}{\gamma(R_\rho)}$;
(iii) the vertical drift of the interfaces is ignored.

The amplitude of the merging perturbation in Figure 8.23 is conveniently represented by quantities $A = \frac{\Delta T_2 - \Delta T_1}{\Delta T_I}$ and $B = \frac{\Delta S_2 - \Delta S_1}{\Delta S_I}$ – the relative variations in the interfacial jumps of temperature and salinity The heat and salt budgets for each mixed layer are then expressed in terms of A and B, the result is linearized, and the stability of the system is analyzed using the normal modes $(A, B) = (A_0, B_0)\exp(\lambda t)$. The resulting growth rate equation takes the following form:

$$\lambda^2 + \frac{4(\bar{T}_z)^{\frac{1}{3}}}{H^{\frac{2}{3}}}\left[\frac{4}{3}C(\bar{R}_\rho) + A_C - \frac{\bar{R}_\rho A_C}{\gamma(\bar{R}_\rho)} - A_\gamma C(\bar{R}_\rho)\bar{R}_\rho\right]\lambda - \frac{16}{3}\frac{4(\bar{T}_z)^{\frac{2}{3}}}{H^{\frac{4}{3}}}C^2(\bar{R}_\rho)\bar{R}_\rho A_\gamma = 0, \tag{8.16}$$

where $A_C = \bar{R}_\rho \left.\frac{\partial C}{\partial R_\rho}\right|_{R_\rho=\bar{R}_\rho}$ and $A_\gamma = \bar{R}_\rho \left.\frac{\partial \gamma^{-1}}{\partial R_\rho}\right|_{R_\rho=\bar{R}_\rho}$.

Equation (8.16) is structurally analogous to the growth rate equation for the layering instability (8.11) and the stability conditions set by these equations are also very similar – everything is controlled by the sign of A_γ. If the flux ratio (γ) decreases with increasing R_ρ, as expected for low density ratios (see Fig. 8.15) then $A_\gamma > 0$ and the free coefficient of the quadratic equation (8.16) is negative, which implies two real roots of opposite sign. The existence of a positive root means that the basic uniform gradient is unstable. This instability is easy to interpret. In the unstable modes, "strong" interfaces, characterized by larger (ΔT, ΔS) monotonically grow at the expense of weaker interfaces. Numerical solutions of fully nonlinear equations governing the two-layer system indicate that unstable perturbations do not equilibrate until the weaker interface is completely eliminated. In essence, (8.16) represents the mergers of adjacent layers, such as realized in the DNS of staircases (Figs. 8.20–8.22).

While the foregoing merging model explains coarsening of staircases, it still fails to address one critical dynamical element – their eventual equilibration. The growth rate equation (8.16) suggests that as long as the density ratio is sufficiently low ($\bar{R}_\rho < R_{\min}$), the weaker interfaces will be sequentially eliminated. In this scenario, the number of steps continually decreases and the average step thickness increases correspondingly. What then prevents staircases from evolving to the ultimate state consisting of only one high-gradient interface separating two very deep mixed layers? One possibility is that continuous mergers in the equilibrium state could be balanced by splitting of steps – the idea originally advocated by Kelley (1988). However, recent observations (Timmermans *et al.*, 2008) and simulations (Prikasky, 2007; Noguchi and Niino, 2010a,b; Stellmach *et al.*, 2011) indicate that layer-splitting events in staircases are rare and unlikely to play a significant role in the step-size selection. Mergers simply stop when layers become sufficiently thick.

The hypothesis proposed by Radko (2005) invokes the slight inhomogeneity of the convecting layers (an effect ignored by the earlier layer-merging models) as a stabilizing agent controlling the equilibrium height of steps. Taking into account property variations across layers (ΔT_L, ΔS_L) requires parameterization of the convective fluxes. Conventional wisdom (e.g., Turner, 1979) suggests that

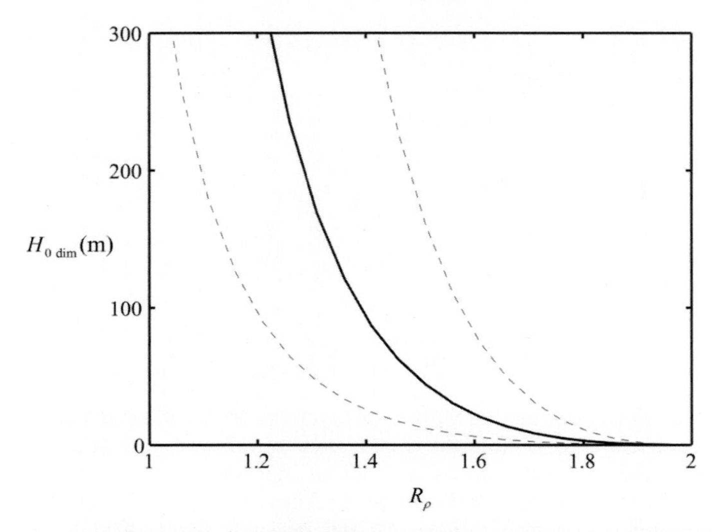

Figure 8.24 The dimensional equilibrium layer thickness as a function of the background density ratio as predicted by layer-merging theory for a fixed background temperature gradient of $\bar{T}_z = 0.03\ ^\circ\mathrm{C\,m^{-1}}$. Dashed lines indicate the plausible range of values due to the uncertainty in the model parameters. From Radko (2005).

the most important parameter controlling the strength of convective mixing is the density-based Rayleigh number:

$$Ra_\rho = \frac{g}{k_T \nu} \frac{\Delta\rho}{\rho} H_{\mathrm{dim}}^3. \tag{8.17}$$

Further assuming that the eddy diffusivities of heat and salt in convecting layers are equal and uniquely determined by (8.17) makes it possible to reproduce the stability calculation for finite $(\Delta T_L, \Delta S_L)$. The new stability model suggests that relatively thin layers successively merge as predicted by the earlier theory (Huppert, 1971; Radko, 2003). However these mergers cease when the thickness of layers exceeds a critical value (H_0), which, in turn, is controlled by background stratification. This critical thickness corresponds to the actual observed step size in fully equilibrated staircases.

Figure 8.24 presents a (dimensional) estimate of the critical thickness based on the layer-merging formulation of Radko (2005) as a function of the density ratio for typical oceanic conditions. Note the dramatic decrease in $H_{0\ \mathrm{dim}}$ with increasing R_ρ. This pattern, as well as the range of layer heights in Figure 8.24 (0–300 m) is generally consistent with observations. For $R_\rho = 1.6$, for example, the model predicts the scale of 20 m, which is close to the thickness of layers observed in the tropical North Atlantic, whereas very thick layers of 300 m and more have been

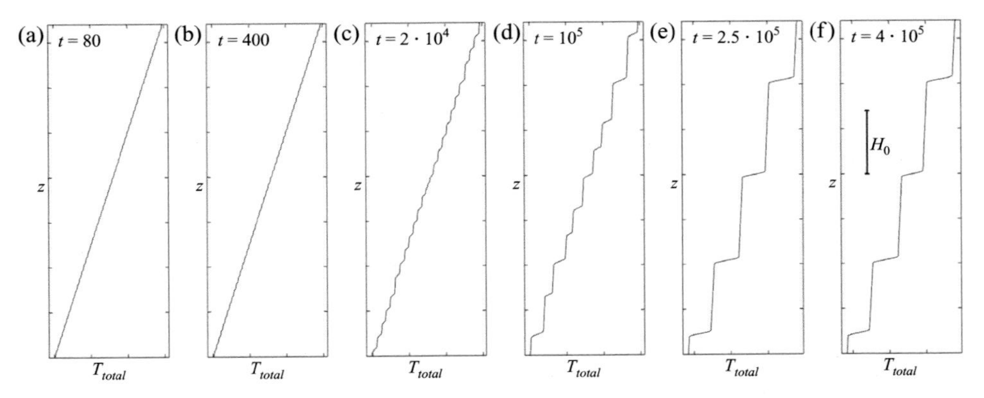

Figure 8.25 Formation and evolution of layers in an experiment with the parameterized vertical fluxes. Note the appearance of mixed layers separated by thin stratified interfaces, followed by a series of merging events, and their eventual equilibration. The dimensional critical thickness of layers evaluated for typical oceanographic parameters is $H_{0\,dim} \sim 20\,\mathrm{m}$. From Radko (2005).

observed in the Tyrrhenian Sea (Zodiatis and Gasparini, 1996) where the density ratio is anomalously low ($R_\rho \approx 1.2$). The uncertainty in theoretical estimates of the equilibrium thickness (indicated by dashed lines in Fig. 8.24) is large. It is primarily associated with the poorly constrained relation between the vertical T–S transport through the convective layers in oceanic staircases and $(\Delta T_L, \Delta S_L)$. However, the general pattern of the equilibrium thickness, characterized by the rapidly decreasing $H_{0\ \mathrm{dim}}(R_\rho)$, appears to be structurally robust.

Support of the merging theory is provided by the fully nonlinear numerical solution of the one-dimensional equations (8.8) in Figure 8.25. In this calculation, the diffusivities of heat and salt are assumed to be controlled by the density ratio in the salt-finger regions ($T_z > 0,\ S_z > 0,\ \rho_z < 0$) and by the Rayleigh number (8.17) in the convecting layers ($\rho_z > 0$). As expected, the layers that emerge first (Fig. 8.23b) are thin and unsteady. Figures 8.25c–e illustrate the subsequent evolution of a staircase: the steps merge continuously. Mergers occur when sufficiently strong interfaces, characterized by large temperature and salinity jumps, grow further, while weaker interfaces decay and eventually disappear, in accord with the layer-merging theory. The number of steps decreases and their characteristic vertical scale increases correspondingly. However, as time progresses, the coarsening of layers becomes less rapid and eventually stops completely. The variation in properties across convecting layers is critical in this regard; failure to take it into account leads to the final state with only one interface within the computational domain. No visible changes in the temperature field occurred between $t = 2.5 \cdot 10^5$ (standard non-dimensional units) in Figure 8.25e and $t = 4 \cdot 10^5$ in Figure 8.25f, suggesting

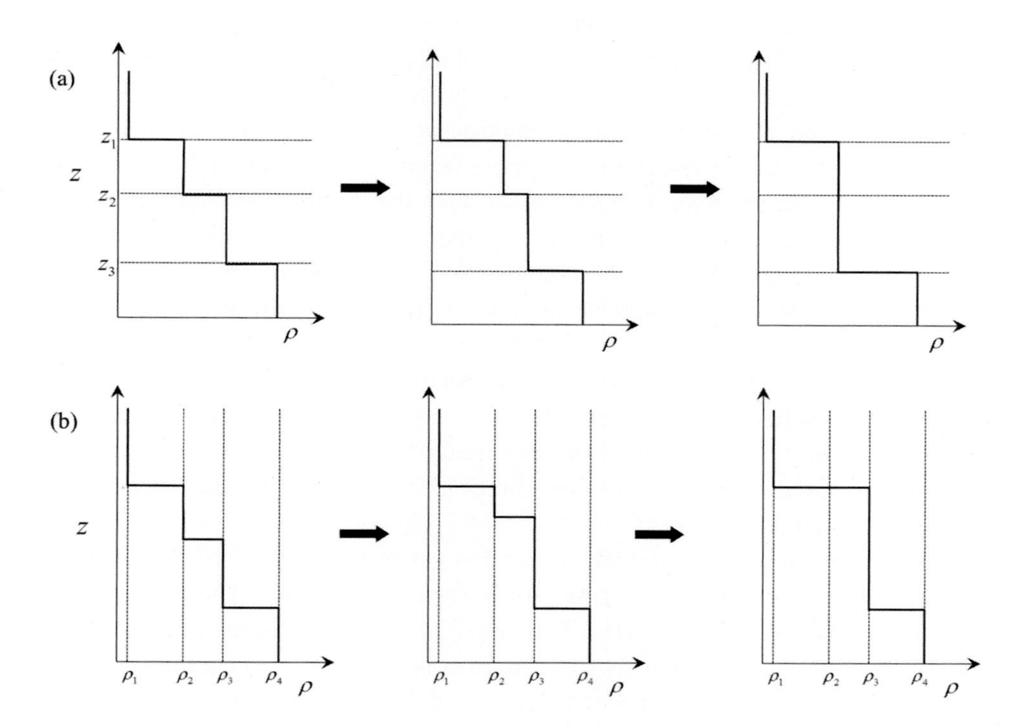

Figure 8.26 Schematic representation of the two possible merging scenarios: (a) the *B*-merger, which occurs when some interfaces gradually erode without moving vertically; and (b) the *H*-merger, which occurs when the interfaces drift vertically and collide. Numerical simulations and field data indicate that *B*-mergers are more common in thermohaline staircases.

that the staircase reached a stable state. The coarsening of the staircase was arrested only when the thicknesses of all layers exceeded the critical value H_0 predicted by the theoretical two-layer model.

The process of staircase equilibration is a lengthy one. In dimensional units, evaluated for typical oceanic scales, the equilibrium state in Figure 8.25e was reached by $t_{\text{dim}} \sim 2.5$ years, exceeding the fingering scale by approximately five orders of magnitude. Such slow predicted evolution of thermohaline staircases is supported by some oceanographic observations – see, for example, Figure 8.6 for the Tyrrhenian staircase.

Two other generalizations of the layer-merging theory should be mentioned. The first one concerns the evolutionary pattern of steps (Radko, 2007; Prikasky, 2007). The schematic in Figure 8.26 illustrates two principal merging mechanisms: (i) the *B*-merger, involving the decay and eventual disappearance of relatively weak interfaces (Fig. 8.26a) and (ii) the *H*-merger, characterized by the vertical drift and collision of adjacent interfaces (Fig. 8.26b). These two merger types are manifestations of two distinct modes of instability, which are, in turn, controlled by the form of the interfacial flux laws. The instability producing B-mergers is caused by the variation in T–S fluxes as a function of property jumps across the steps $(\Delta T, \Delta S)$, whereas the H-instability occurs when fluxes respond to the changes in step height. If the staircase parameters are such that both instabilities are present, the evolutionary pattern is controlled by the instability with the larger growth rate. Whilst mergers through the elimination of weak interfaces (B-mergers) are clearly more common, numerical simulations (e.g., Noguchi and Niino, 2010b) indicate that H-mergers are possible for relatively thin steps in newly formed staircases. As layers grow in size, fluxes become largely insensitive to the step height – the spatially separated interfaces hardly affect each other – and therefore the H-instability is either absent or very weak. Thus, the late evolutionary stages of staircases and their ultimate equilibration are likely to be controlled by the B-merger dynamics.

Another point we wish to make regarding generalizations of the layer-merging theory concerns the parameterization of the interfacial fluxes. The most common versions of the model are based on Turner's 4/3 flux laws (Chapter 4). However, the risk of overreliance on these laws in certain configurations and parameter regimes cannot be ignored. Mergers in the staircase are driven by subtle imbalances of fluxes in the adjacent layers, which can potentially lead to the amplification of the errors associated with the assumed flux laws. This problem motivates a more general formulation of the layer-merging theory (Radko, 2007; Prikasky, 2007). Suppose for a moment that the T–S fluxes across each step are controlled by the property variations $(\Delta T, \Delta S)$ and the step height (H) in some unique but yet unspecified manner:

$$\begin{cases} F_T = F_T(\Delta T, \Delta S, H), \\ F_S = F_S(\Delta T, \Delta S, H). \end{cases} \tag{8.18}$$

In this case, it is possible to reproduce the layer-merging theory without making specific assumptions about the pattern of the one-step flux laws (8.18). The growth rate equation for B-mergers becomes

$$\lambda_B^2 + \frac{4}{H}\left(\frac{\partial F_T}{\partial \Delta T} + \frac{\partial F_S}{\partial \Delta S}\right)\lambda_B + \frac{16}{H^2}\left(\frac{\partial F_T}{\partial \Delta T}\frac{\partial F_S}{\partial \Delta S} - \frac{\partial F_S}{\partial \Delta T}\frac{\partial F_T}{\partial \Delta S}\right) = 0. \tag{8.19}$$

If the appropriate one-step flux laws are known or could be determined (numerically for instance), then the generalized merging theory (8.19) opens an attractive opportunity to theoretically predict the evolution of staircases and their ultimate equilibration.

We conclude the discussion of recent conceptual models of staircases with a cautiously optimistic summary. The instability of the flux-gradient laws mechanism (Section 8.3) and the associated layer-merging models explain several aspects of observations and find support in numerical simulations. The component of theory addressing the step-size selection is still somewhat hypothetical and qualitative. However, the model predicts the right range of scales and could possibly define the way forward by systematically refining its formulation. On the other hand, we should keep in mind that, for half a century, the theory of thermohaline staircases has served as a graveyard of interesting ideas and seemingly plausible explanations. It remains to be seen whether the most recent theory withstands more rigorous tests by field and laboratory experiments.

9

The unified theory of secondary double-diffusive instabilities

In the last three chapters we discussed the major types of large-scale double-diffusive instabilities: collective instability, intrusions and γ-instability. Ironically perhaps, each type has been indentified at some point in history as a precursor of the spontaneous formation of thermohaline staircases. Considerations of clarity forced us to discuss each mode separately, despite the apparent similarities in the formulation of the stability problems and in the analysis. However, a little reflection suggests that at least the linear theory could be united.

The starting point of the unification (Traxler *et al.*, 2011a) is the non-dimensional system (7.1), which has already been used (Chapter 7) to describe interleaving instabilities. This system represents the Boussinesq equations linearized with respect to the motionless basic state uniformly stratified in x and z. The first steps of the unified theory routinely follow the development of all parametric double-diffusive theories (Chapters 6–8). The *T–S* fluxes are expressed in terms of the Nusselt number and the flux ratio (6.3), both of which are assumed to be determined by the density ratio. For simplicity, the model is chosen to be two-dimensional although the extension to three dimensions is straightforward. The stability properties of system (7.1) are analyzed using normal modes (7.2), which yields a cubic equation for the growth rate:

$$\lambda^3 + a_2\lambda^2 + a_1\lambda + a_0 = 0. \tag{9.1}$$

The coefficients of (9.1) are given by algebraic expressions in terms of

$$a_i = a_i\left[k, m, \bar{R}_\rho, Nu(\bar{R}_\rho), \gamma(\bar{R}_\rho), G, A_{Nu}, A_\gamma\right]. \tag{9.2}$$

Unlike previous treatments, Traxler *et al.* (2011a) made no attempt to a-priori isolate various types of secondary instabilities. For each value of the background density ratio ($\bar{R}_\rho$), parameters in (9.2) characterizing the patterns of fluxes $\left(Nu, \gamma, A_{Nu}, A_\gamma\right)$ were diagnosed from a series of salt-finger DNS for the oceanographic heat–salt parameters ($Pr = 7, \tau = 0.01$). Then, the growth rate

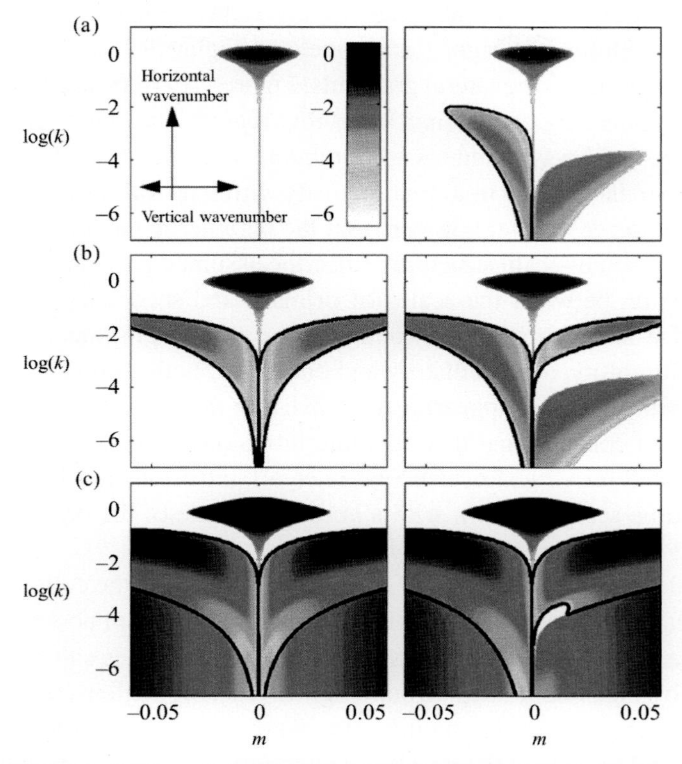

Figure 9.1 The logarithm of the real part of growth rates for the fastest growing perturbations is plotted as a function of wavenumbers. Only positive values of Re(λ) are shown. The horizontal axis shows vertical wavenumber, and the vertical axis shows the logarithm of horizontal wavenumber to capture the broad range of expected scales. The left-hand column shows results in the absence of lateral gradients ($G = 0$), while the right-hand column shows results for $G = 0.01$. Instability regions surrounded by a dark contour show oscillatory behavior. (a) High density ratio ($\bar{R}_\rho = 7$). (b) Mid-range density ratio ($\bar{R}_\rho = 4$). (c) Low density ratio ($\bar{R}_\rho = 1.5$). From Traxler *et al.* (2011a).

equation (9.1) was solved numerically for each wavenumber and the density ratio. The largest real part of its three roots is plotted as a function of the vertical and horizontal wavenumbers in Figure 9.1 for three representative values of $\bar{R}_\rho$. To make the analysis more memorable, Traxler *et al.* (2011a) presented their results in a somewhat peculiar format: the abscissa is the vertical wavenumber and the ordinate is the natural logarithm of horizontal wavenumber. However, the effort was worthwhile – the flower plot in Figure 9.1 is rather amusing.

The interpretation of the unified theory is not as complicated as it might have been since the secondary instabilities generally operate on distinct spatial/temporal scales. As the density ratio decreases, the region of instability systematically grows and its pattern starts to reveal progressively richer dynamics. To help identify intrusive instabilities, the right (left) panels of Figure 9.1 show the growth rate patterns with (without) the lateral gradients. Figure 9.1a (left) for high density ratio ($\bar{R}_\rho = 7$) contains only one distinct instability region. The "bulb" characterized by the high horizontal wavenumbers is a counterpart of the regular fingering mode. This mode persists, albeit in a quantitatively different form, even when the eddy fluxes are set to zero. Note, however, that the parametric theory cannot accurately represent such small-scale structures since the assumed parameterizations require clear separation between the scales of primary and secondary instabilities. The presence of a lateral gradient introduces (Fig. 9.1a, right panel) two additional regions of instability, one oscillatory and one direct, both confined to large vertical and horizontal scales and appearing as "leaves" in the flower plot. The direct mode corresponds to conventional thermohaline intrusions (Chapter 7). In dimensional units (evaluated for typical oceanic parameters), intrusions in Figure 9.1 typically grow on a time scale of a day, with a horizontal scale of the order of a kilometer and vertical scale of a few meters.

For an intermediate density ratio $\bar{R}_\rho = 4$ (Fig. 9.1b, left) the collective instability waves appear at vertical and horizontal dimensional scales of about a meter, with a growth time scale of about 30 hours. Oscillatory instabilities can be recognized in Figure 9.1 by black contours surrounding the unstable regions. The lateral gradient (Fig. 9.1b, right) strongly modifies the waves, increasing both their maximum growth rate and the size of the instability region for positive slopes, while suppressing growth for negative slopes. The inclusion of the lateral gradient also introduces a direct intrusive mode at much larger horizontal scales.

Finally, at low density ratios, exemplified by the $\bar{R}_\rho = 1.5$ calculation in Figure 9.1c, the system becomes strongly unstable to a continuous range of modes on both horizontal and vertical scales. In this regime, the instabilities are dominated by the collective instability (for intermediate range of horizontal scales) and by the γ-instability for much longer horizontal scales. The largest growth rates are attained by the collective instability, which grows on a time scale of a few hours. The regions occupied by the collective and γ-instabilities are largely unaffected by the presence (absence) of lateral gradients in the right (left) panels of Figure 9.1c. The intrusive zone is completely masked by the encroaching collective instability region. The disappearance of intrusions in the growth rate diagram, however, should not be interpreted as a sign of their insignificance. Even though they grow slowly, mixing driven by intrusions may be substantial (owing to their more effective direct mode of operation) possibly exceeding that of collective instability waves.

Overall, the unified analysis did not fundamentally alter our view of secondary large-scale instabilities but, rather, clarified and sharpened our understanding of them. Major insights brought by studies of layering, interleaving and collective instabilities in isolation have been confirmed. New twists, such as the sensitivity of the collective instability to lateral *T–S* gradients and the possibility of the oscillatory intrusive modes, only underscore the richness and complexity of double-diffusive phenomena. It should also be realized that the situation in nature is undoubtedly even more interesting and complicated than in purely double-diffusive models discussed so far. At any given moment, double-diffusive structures in the ocean are forced by ubiquitous large-scale shears, intermittent turbulence and spatial/temporal variability. In the next chapter we attempt to summarize our present, glaringly incomplete, knowledge of the properties of double-diffusion in active environments.

10

Double-diffusion in active environments

In fluid dynamics, the prominent place occupied by double-diffusion is perfectly secure – the indisputable beauty and intellectual challenges of the topic are more than sufficient to maintain the keen interest of dynamicists for many years to come. More applied folks, oceanographers being the prime example, are harder to please. The acceptance of double-diffusion as a core oceanographic subject is conditioned on proving, beyond reasonable doubt, its impact on water-mass composition. The final verdict rests largely on the ability of double-diffusion to remain an effective mixing process in the ocean despite the perpetual forcing by externally driven shears and turbulence. Are salt fingers sufficiently resilient to survive in such an environment? And if they do, how relevant are the good old theories developed for unforced double-diffusion? Concerns of this nature have been expressed continuously throughout the history of the field, and not only by the small and irrational group of double-diffusion haters. An early quote from the founder of the field reflects very serious reservations: "The ease with which organized salt fingers may be disrupted would seem to argue against their playing any significant role in the vertical mixing of a turbulent ocean" (Stern, 1967).

While the precise assessment is yet to be made, the evidence already accumulated suggests that external forcing in the ocean tends to reduce double-diffusive transport, but not everywhere and not dramatically. The very existence of phenomena such as permanent thermohaline staircases indicates that not only is double-diffusion active in the ocean but that it can be the dominant mixing process. Staircases are observed in regions where they are roughly expected on the basis of theoretical considerations (Chapter 8) – considerations that do not take external forcing into account. Thus, if double-diffusion is active enough to generate staircases, then waves, shear and turbulence likely play a secondary role in finescale dynamics. This is not to say that these ocean-specific effects can be dismissed entirely. As we shall see shortly, external forcing can produce detectable changes in stratification and transport even in staircases, but it is certainly no match for double-diffusion. The situation becomes more complicated and uncertain when stratification favors double-diffusion but not necessarily layering, which is the most common situation in the mid-latitude thermocline. In this case, the fate of double-diffusion is determined by a combination of several factors, such as the value of density ratio, the intensity of ambient shear and the frequency of turbulent overturns.

The following discussion is an attempt to quantify the relative impact of various forms of external forcing on double-diffusion in both smooth-gradient and staircase regimes. Our focus is on fingering rather than diffusive convection. There are two (related) reasons for that. First, much more is known about salt fingers in active environments; the diffusive problem has so far generated less excitement and research activity. The second reason is that salt fingers are more likely to be affected by ambient forcing. Observational estimates of mixing in diffusive regions of the ocean are generally consistent with predictions based on unforced laboratory experiments, which is not always the case for salt fingers. The lack of controversy ultimately translates to a lack of interest. So, let us talk about salt fingers and let us start with the effects of shear.

10.1 The interaction of salt fingers with shear flow

Vertical shears are ubiquitous in the ocean. They are usually associated with internal waves generated by tides, topography and atmospheric forcing. Therefore, a very natural development in double-diffusion theory involves the interaction of salt fingers with shear. The first systematic study of this problem was performed by Linden (1974), whose laboratory experiments revealed that active double-diffusive convection is maintained in the presence of a steady shear flow. Using a combination of theoretical arguments and experiments, Linden demonstrated that the preferred pattern of instability is a quasi two-dimensional mode parallel to the mean flow, which was referred to as salt sheets. These remarkably regular structures can be seen very clearly in the top-view shadowgraph images of Linden's experiments (Fig. 10.1).

The interest in the role of shear in salt-finger dynamics was invigorated by the 1985 C-SALT program (Schmitt *et al.*, 1987), which brought two unexpected findings. The first one is the presence of nearly horizontal small-scale laminae (Fig. 10.2) clearly visible in the shadowgraphs taken through high-gradient interfaces (Kunze, 1990). The second surprise is very significant, an order-of-magnitude discrepancy between the observed heat fluxes and predictions based on the laboratory-calibrated Turner's 4/3 flux law (Chapter 4). Both effects are apparently related to some aspect of the external forcing and the most common off-hand

(a)

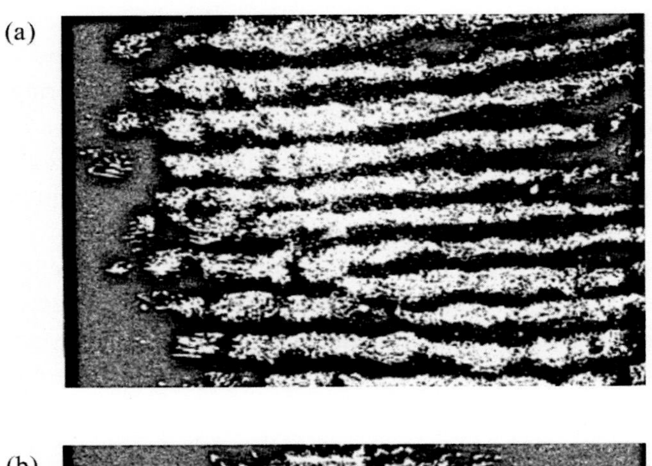

(b)

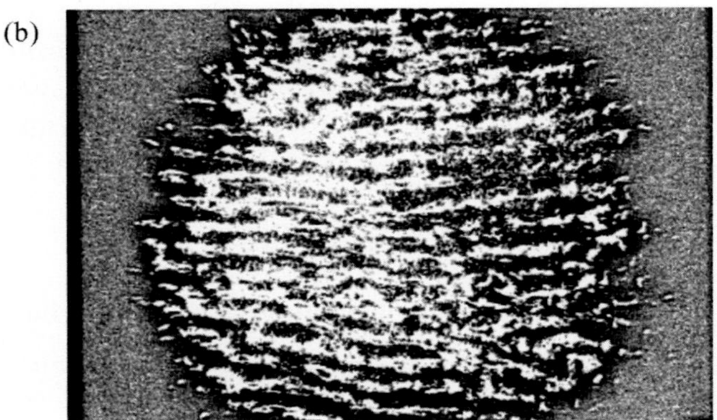

Figure 10.1 Laboratory experiment on salt fingers in steady shear flow. The top-view photographs reveal the horizontal structure of salt sheets aligned in the direction of shear. From Linden (1974).

reaction of an observer is to attribute them to the vertical shear. The strength of shear measured in C-SALT interfaces was moderate, with typical Richardson numbers of $Ri \sim 6$ (Gregg and Sanford, 1987). Shears of such strength are sufficient to tilt salt fingers and there is little doubt that inclined fingers are less effective in mixing temperature and salinity than vertical ones. However, is it reasonable to expect an order-of-magnitude reduction in fluxes, as the failure of the 4/3 flux laws seems to indicate?

Several arguments suggest that the impact of shear on finger transport is relatively mild. For instance, it is clear that while steady unidirectional shear can be effective in suppressing the along-flow variability, it cannot directly affect the cross-flow motion. The appearance of quasi two-dimensional salt sheets in Linden's (1974) experiments (Fig. 10.1) is a sign of highly anisotropic dynamics. The anisotropy

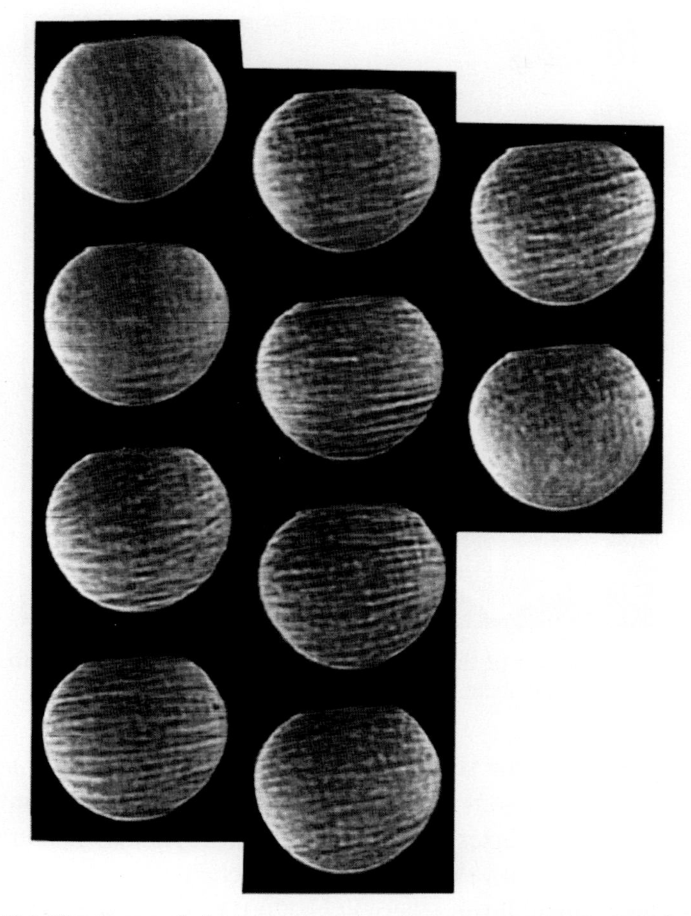

Figure 10.2 Shadowgraph images through a fingering-favorable interface of the Caribbean staircase. From Kunze (1990).

can be readily rationalized from the general structure of the governing (Navier–Stokes) equations. These equations are invariant with respect to the transformation in which patterns that are homogeneous in one spatial direction [$T = T(y, z, t)$] are augmented by a steady shear in the same direction [$u = u_{\text{shear}}(y, z)$]. Hence, shear and two-dimensional cross-shear modes are effectively uncoupled. Even if shear is strong enough to completely eliminate all downstream variability, the cross-shear modes will remain largely intact.

The quasi two-dimensional interpretation of the finger/shear dynamics is supported by numerical experiments. Figure 10.3 presents the direct numerical simulation of salt sheets by Kimura and Smyth (2011). While the cross-flow patterns are remarkably similar to salt fingers in two-dimensional simulations, the downstream

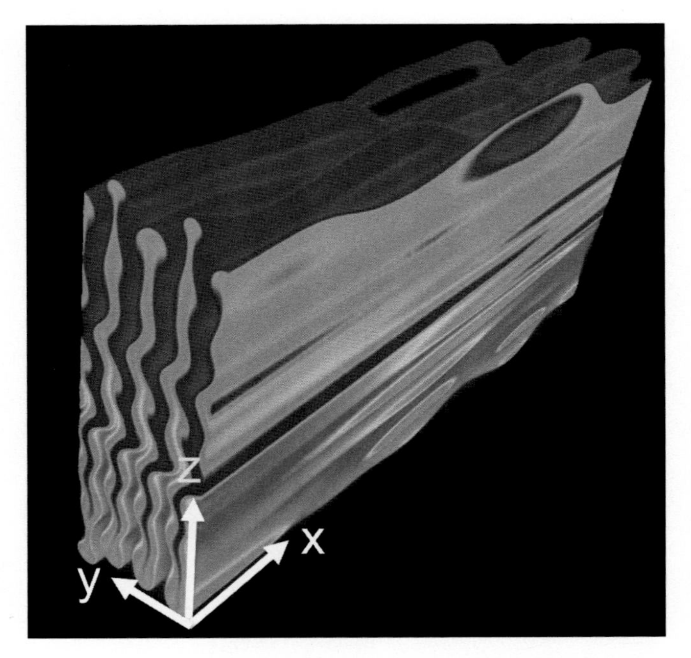

Figure 10.3 A snapshot of the salinity field from DNS initialized with four salt sheets in a background shear. From Kimura and Smyth (2011).

variability is reduced to the small-scale, nearly horizontal laminae suggestive of the shadowgraph observations in Figure 10.2. Three-dimensional simulations also indicate that salt-finger fluxes in laminar shear (Smyth and Kimura, 2007) are approximately halved relative to their unsheared counterparts and are in fact very close to two-dimensional fluxes (Stern *et al.*, 2001).

Another suggestive effect observed in simulations (Kimura *et al.*, 2011) and laboratory experiments (Linden, 1974; Fernandes and Krishnamurti, 2010) involves the very limited response of fingers to variation in the strength of shear. Even the sign of this response could be questioned. Linden (1974) predicted a weak increase of fluxes with increasing shear, whereas the opposite trend was reported by Kimura *et al.* (2011) and Fernandes and Krishnamurti (2010). The lack of significant flux/shear dependence is consistent with our interpretation of the zero-order physics at play. Shear rapidly makes the pattern and dynamics of salt fingers effectively two-dimensional, at which point shear and fingers stop interacting. Thus, while the presence of shear matters, its strength generally does not. The whole story is of course more complicated; the extent of downstream homogenization and secondary three-dimensional instabilities of salt sheets are of interest (Kimura and Smyth, 2011) and should be explored further. However, as far as the magnitude of double-diffusive mixing in shear is concerned, the bottom line is clear. It is unlikely that steady shear can reduce vertical transport by more than a factor of 2–2.5 – the typical difference between finger fluxes in two and three dimensions.

An interesting idea on finger dynamics in shear was put forward by Kunze (1990), who emphasized the time dependence of a background flow. In the C-SALT staircase, as elsewhere in the ocean, shear is predominantly near-inertial. As shear makes a full rotation, it successively suppresses salt-finger variability in all horizontal directions, not just one. In this sense, rotating and steady shears are different, the former being potentially more effective in constraining double-diffusive mixing. What casts some doubt on the relevance of this mechanism is the disparity of temporal scales. While the inertial shear varies on the time scale of a day, the salt-finger growth period for C-SALT parameters is measured in minutes. It therefore appears that the steady-state approximation should be adequate, although more precise quantification of time-dependent effects is highly desirable. A straightforward and convenient way of addressing the problem would be to use direct numerical simulations, since computational capabilities are now sufficient to model salt fingers on the inertial time scales.

10.2 Low fluxes and thick interfaces

What are the other factors that could potentially reduce salt finger transport in the ocean? We have mentioned an order-of-magnitude mismatch between the predictions of the laboratory calibrated flux laws and heat flux measurements in the C-SALT staircase, which still has not been fully explained after three decades of deliberation. A similar mismatch has been noted (Hebert, 1988) for steps observed in a salt lens of Mediterranean origin (meddy Sharon). Taking into account the effects of shear reduces the inconsistency, but does not resove the problem completely. A potentially important hint (Kunze, 1987) for the too-low-flux puzzle is provided by the analysis of the interface thickness (h). The interfacial similarity law (4.8) makes it possible to estimate h based on the extrapolation of laboratory measurements and to compare it with observations. The outcome is intriguing: the typical observed thickness of C-SALT interfaces ($h \sim 2\,\mathrm{m}$) is much higher than the lab-based prediction ($h \sim 0.2\,\mathrm{m}$). This discrepancy is fully consistent with the limited T–S transport in the C-SALT staircase: thick interfaces mean weak gradients; weak gradients mean low fluxes. However, thickness analysis does not reveal the ultimate cause of the mismatch of fluxes, immediately prompting the question: "Then why are interfaces in the ocean so disproportionally thick?"

It has been suggested that, conceptually, it is more appropriate to view interfaces in the ocean as a stack of several thin sub-interfaces and that Turner's similarity laws should be applied to the sub-interfaces, rather than to the net gradient region (e.g., Kunze, 1987). Some support for this interpretation is provided by observations of small-scale vertical non-uniformities of temperature gradient in interfaces (e.g., Molcard and Williams, 1975). The intensity of microstructure is also modulated, with the most active salt fingering restricted to regions with large temperature gradients. Such banded, horizontally coherent structures, of various strengths and patterns, have been observed in all major salt-finger staircases. The most pronounced form of the interfacial substructure is represented by "mini-staircases" with well-defined steps of 1–2 m nested within much thicker parent interfaces (Marmorino, 1989). The origin of such phenomena remains unknown.

A notable attempt to rationalize the discrepancy between the structure of oceanic and laboratory interfaces was made by Linden (1978). A distinguishing feature of laboratory experiments is the initial setup, which usually consists of two homogeneous layers separated by a sharp interface. To examine the ramifications of imposing this somewhat arbitrary initial condition, Linden conducted a series of experiments with thick interfaces. He demonstrated that, in time, such interfaces either shrink significantly or split into several interfaces separated by convecting layers. In either case, the thicknesses of individual interfaces inexorably evolved towards a certain preferred value, typically much less than the initial thickness. Linden's (1978) findings suggest that the role of initial conditions in the selection of interfacial thickness is secondary. This is an interesting result in its own right but, unfortunately, it also implies that the explanation of the difference between oceanic and laboratory interfaces lies elsewhere. One possibility is related to intermittent oceanic turbulence, which is discussed next.

10.3 The interaction with intermittent turbulence

On the qualitative level, the physics of the interaction between salt fingers and turbulence is reasonably well understood. Intermittent density overturns, induced by dynamic instabilities of internal waves, generate energetic turbulent billows. When turbulent events occur in the ocean regions occupied by fingering convection, they disrupt the double-diffusive microstructure. However, the lifespan of turbulent patches is rather short. When turbulence subsides, salt fingers reemerge, gain strength, and then continue to operate at near-equilibrium level until the next turbulent event wipes them out and the cycle repeats. There is a general consensus that the ability of salt fingers to effectively mix temperature and salinity in the ocean can be controlled by the frequency of turbulent disruptions.

Specific estimates of the impact of turbulence on fingers are more controversial and difficult to come by. The frequency of turbulent overturns can be roughly inferred from the so-called intermittency coefficient (I) which measures the fraction

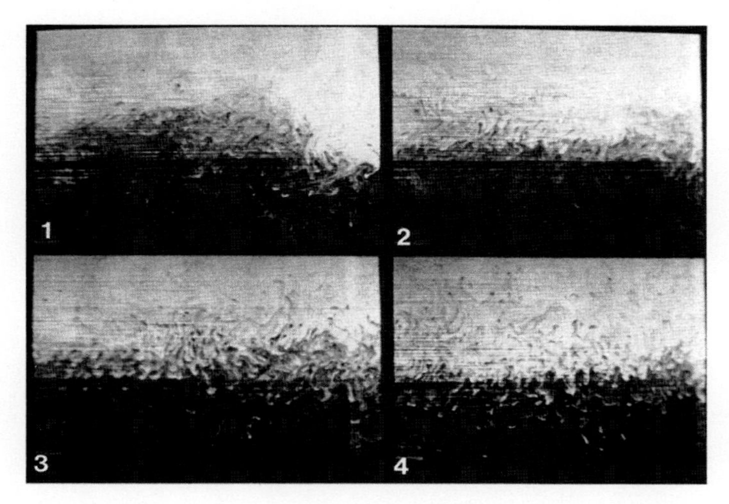

Figure 10.4 Dye fluorescence images of salt fingers establishing after a turbulent event. From Taylor (1991).

of the volume occupied by turbulence at any given time (Monin and Ozmidov, 1985). Observations indicate that the typical value of the intermittency coefficient in the main thermocline is $I \sim 0.05$ and its plausible range extends to $0.01 < I < 0.10$ (e.g., Polzin *et al.*, 2003). Numerical and laboratory experiments (Itsweire *et al.*, 1986, 1993) suggest that turbulence is suppressed on the time scale of one buoyancy period ($2\pi/N$), which leads to the following estimate of the period between subsequent turbulent events at a given location:

$$\Delta t \sim \frac{2\pi}{N}\frac{1-I}{I}. \tag{10.1}$$

It should be noted that this estimate assumes that bursts of turbulence are random in space and occur at comparable time intervals; both assumptions are readily questioned.

The period (10.1) should be compared with the time it takes for salt fingers to fully develop from the rubble left behind by turbulent overturns. Their recovery is by no means instantaneous. Figure 10.4 shows the reestablishment of salt fingers after turbulent disruption modeled in the laboratory by a grid falling through the tank (Taylor, 1991). Grid turbulence decayed rather rapidly ($t \approx 3.8/N$) giving way to wave motions. Salt fingers became visually detectable at $t \approx 10/N$, followed by an even longer period of growth and equilibration. It is intuitively clear that the recovery time for salt fingers (t_{rec}) is directly linked with their maximal growth rate (λ) and laboratory studies (Taylor, 1991; Wells and Griffiths, 2003) suggest

$$3\lambda^{-1} < t_{rec} < 7\lambda^{-1}. \tag{10.2}$$

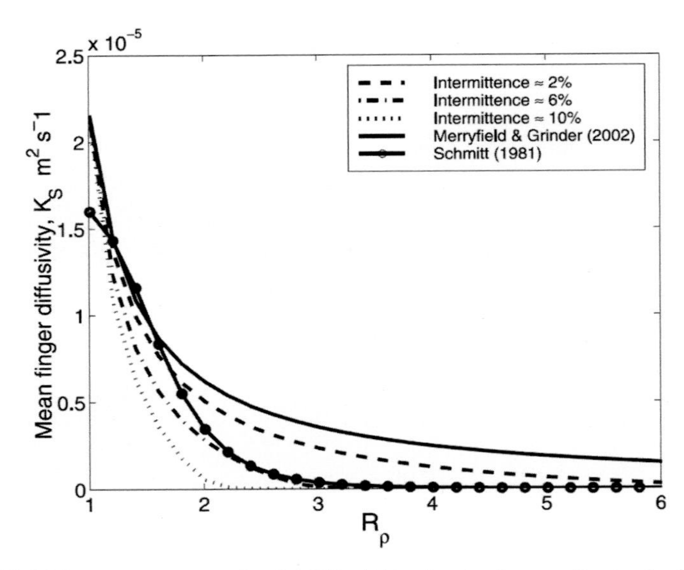

Figure 10.5 The time-averaged salt diffusivity due to fingers for typical oceanographic conditions. The intermittency of turbulent events varies in the range 2% $< I <$ 10%. The solid curve indicates the equilibrium diffusivity in the absence of turbulence and the curve with circles corresponds to the ad hoc parameterization of salt-finger diffusivities in Schmitt (1981). From Wells and Griffiths (2003).

If the recovery time (10.2) is much shorter than the periodicity of turbulence (10.1) then the mean finger-driven transport will be largely unaffected by the turbulence; otherwise it will be considerably reduced. The outcome is largely determined by two parameters: the intermittency coefficient I, controlling Δt, and the background density ratio R_ρ, which determines the growth rate of salt fingers and thereby t_{rec}. In order for salt fingers to remain active, both I and R_ρ have to be sufficiently low. For typical conditions in the Atlantic thermocline, $R_\rho \sim 2$ and $I \sim 0.05$, the ratio of the recovery/turbulence time scales in (10.1) and (10.2) is $\frac{t_{\text{rec}}}{\Delta t} \sim 0.1$ and therefore the effects of turbulence are relatively mild. However, this ratio can increase significantly for more (less) active turbulence (double-diffusion).

A more detailed theoretical analysis was performed by Wells and Griffiths (2003). Figure 10.5 shows time-mean eddy salt diffusivity (K_S) as a function of density ratio (R_ρ) in the absence of turbulent disruptions along with the equivalent relations for the cases of weak ($I = 0.02$), moderate ($I = 0.06$) and strong ($I = 0.1$) turbulence. The results in Figure 10.5 reveal high sensitivity of finger fluxes to the intensity of turbulence. Weak turbulence has only a secondary effect on salt fingers and could be largely ignored for most intents and purposes. The effects of moderate turbulence are, well, moderate. Strong turbulence, on the other hand,

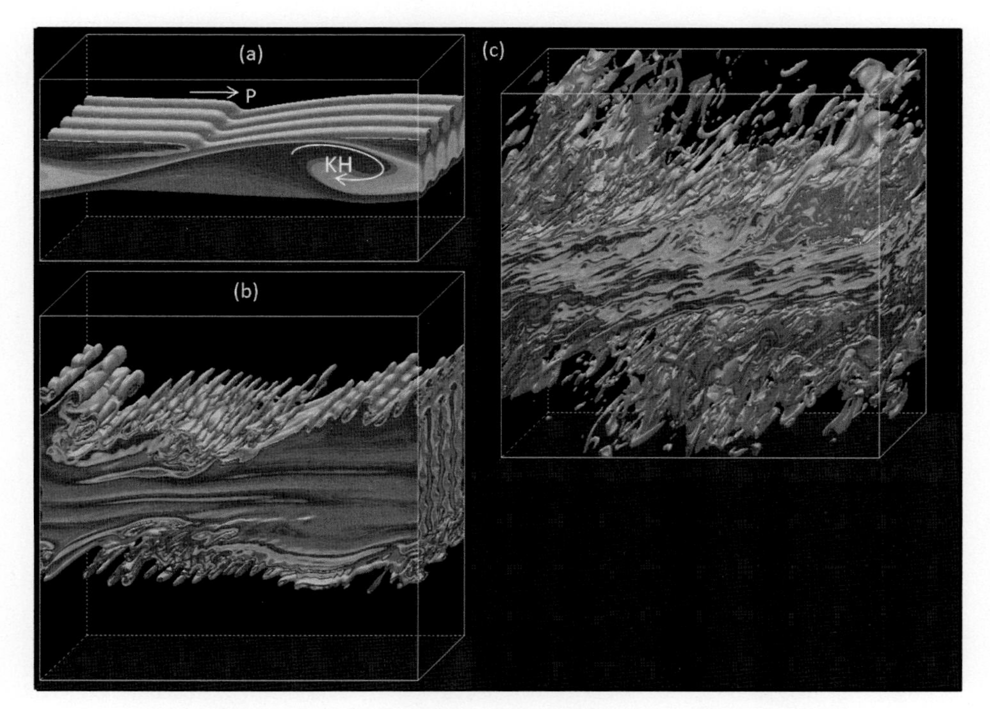

Figure 10.6 Salinity field from DNS of salt fingers in an unstable shear shown at various stages of evolution. From Smyth and Kimura (2011). See color plates section.

substantially diminishes K_S even for very active salt fingering ($R_\rho < 2$) and for $R_\rho > 2$ finger-induced mixing is all but eliminated.

The general picture of the fingers/turbulence interaction and even some specific predictions of the Wells and Griffiths' model are consistent with laboratory and numerical experiments. Laboratory studies have been performed in both heat–salt (Linden, 1971; Taylor, 1991) and salt–sugar (Wells and Griffiths, 2003) configurations; their evolutionary similarities can be taken as evidence of robust physics, independent of experimental media. Due to technical complications, laboratory finger/turbulence experiments have so far have been based exclusively on the grid-stirring method of turbulence production. A key strength of the numerical experiments, relative to their laboratory counterparts, lies in their ability to explicitly represent shear instabilities and the transition to turbulence. Despite substantial differences in setup, the simulated interaction of turbulence and fingers in many ways mirrors the lab experiments. The simulation in Figure 10.6 (Smyth and Kimura, 2011) was initiated by the dynamically unstable shear flow ($Ri = 0.18$) in the salt-finger favorable ($R_\rho = 2$) stratification. The first stage of the experiment is destabilization of the shear flow and formation of a turbulent Kelvin–Helmholtz

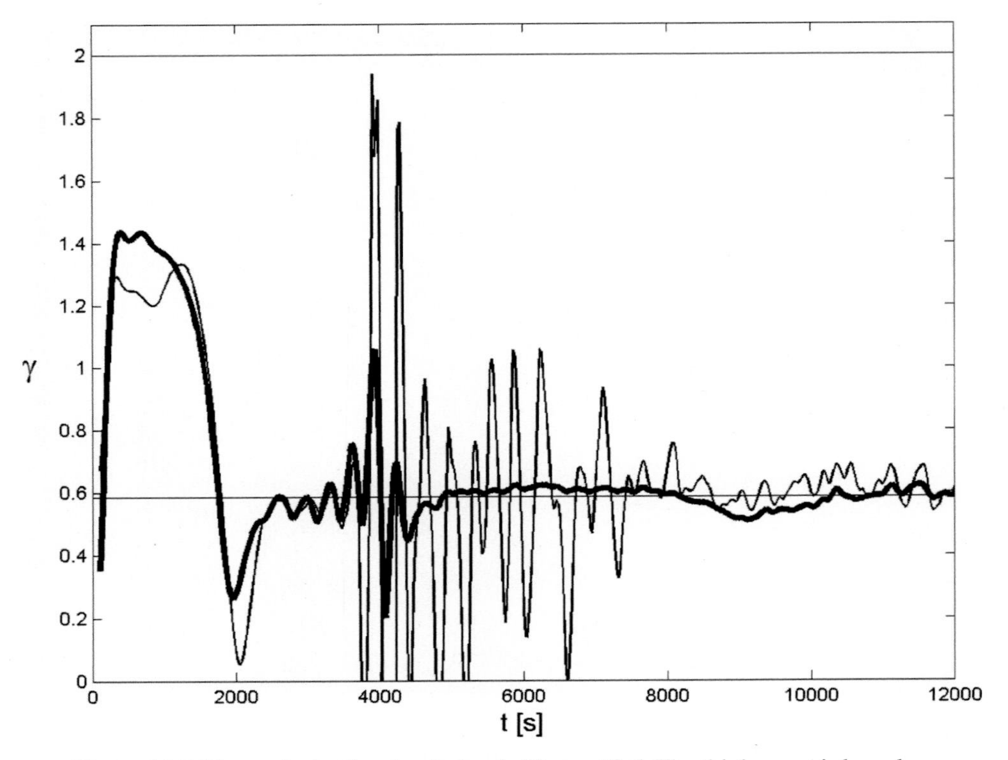

Figure 10.7 Flux ratio for the simulation in Figure 10.6. The thick curve is based on volume-averaged fluxes and the thin curve on fluxes at the centerline. Horizontal lines indicate the value for shear-driven turbulence ($\gamma = 2$) and for the fastest growing finger mode ($\gamma = 0.58$). From Smyth and Kimura (2011).

billow. The subsequent evolution of the system is characterized by the familiar, from lab studies, pattern of fingers/turbulence interaction – a brief period of active turbulence is followed by the more gradual growth and equilibration of fingers.

On the quantitative level, numerical experiments (Smyth and Kimura, 2011) tend to ascribe greater significance to finger transport than is typically seen in laboratory realizations. Despite the visually stunning appearance of turbulent overturns (Fig. 10.6) their direct contribution to the net mixing of temperature and salinity is considerably less than that of salt fingers. A convenient global indicator of the relative significance of fingers and turbulence is given by the heat/salt flux ratio (γ). The value expected for turbulence, which tends to mix temperature and salinity at equal rates, is $\gamma_{\text{turb}} = R_\rho = 2$, whilst for salt fingers it is $\gamma_{\text{sf}} \approx 0.6$. Figure 10.7 presents a time record of the flux ratio for the simulation in Figure 10.6. For the most part, γ hovers about the salt-finger value (γ_{sf}), never coming close to γ_{turb}. Even when turbulence is at its peak, it does not completely suppress fingering – a significant difference relative to the laboratory experiments. Readers are also reminded of the simulations in which salt fingers interact with collective instability waves (Stellmach *et al.*, 2011; Chapter 8, Fig. 8.18). After equilibration, these waves start to overturn, continuously generating random patches of turbulence, yet their impact on mixing is minimal (Fig. 8.18d). Such differences between laboratory and numerical predictions could reflect the significance of the method of turbulence production: grid-stirring experiments introduce a bias towards turbulence and simulations reproducing shear instability favor finger-driven mixing.

So far, our discussion has been focused on finger-driven fluxes and it should be emphasized that the effects of turbulence on the total *T–S* transports, which include both turbulent and finger components, are less pronounced. Turbulence suppresses salt fingers but contributes to vertical mixing on its own and therefore certain compensation is to be expected. Even the sign of the effect is not always clear. Depending on the particular experimental or environmental conditions, adding turbulence can increase, decrease or have little effect on the total transport. Linden's (1971) early laboratory experiments on the interaction of fingers with grid-generated turbulence revealed a weak non-monotonic dependence of the net salt flux on the frequency of turbulent disruptions. Elevated transports were found for predominantly fingering and predominantly turbulent regimes, and the net salt flux was minimal for comparable contributions from fingers and turbulence.

The role of turbulence in the mixing of density is less ambiguous. Double-diffusion is characterized by the counter-gradient (downward) density flux, whereas turbulence acts in the opposite sense, transporting density upward. Thus, taking turbulence into account increases the upward density flux through two distinct mechanisms: (i) by directly contributing to mixing and (ii) indirectly, by reducing the downward finger flux. Laboratory experiments (Wells and Griffiths, 2003) have demonstrated very clearly that, as the frequency of turbulent disruptions increases, the net density flux monotonically increases. The sign of this flux changes from negative in the finger-controlled regime to positive when turbulence dominates. Numerical simulations in which turbulence was generated by moderately unstable shear (Smyth and Kimura, 2011) are characterized by downward time-mean density flux, as expected for predominantly double-diffusive mixing. The instantaneous density flux, however, changes from slightly positive during the brief period of turbulent mixing to negative immediately after the decay of turbulence.

Needless to say, knowledge of the direction – never mind the exact value – of the vertical density transport in the thermocline is essential for many general oceanographic problems. The sign of density flux is controlled by the relative magnitudes of double-diffusive and turbulent mixing. Although both are still poorly constrained by observations, most of the evidence to date indicates that they are comparable. Shear instabilities are violent and intense; they explode into a turbulent convective mess, which rapidly mixes temperature and salinity while the event lasts. In contrast, the instantaneous transport characteristics of salt fingers are relatively moderate. Nevertheless, double-diffusion contributes significantly to diapycnal mixing because the lower intensity of fingers is compensated by their continuous action during the long periods between turbulent events (Taylor, 1991). Time will tell of course, but it would not be surprising, for instance, to find an approximate cancellation of double-diffusive and turbulent density fluxes in the mid-latitude thermocline of the world ocean.

Regarding the large-scale implications of mixing, there is another critical point at stake. A number of modeling studies (e.g., Gargett and Holloway, 1992) reveal high sensitivity of large-scale circulation patterns to the ratio of turbulent diffusivities of heat and salt $r = K_T/K_S$. For many applications, this ratio matters even more than the magnitudes of K_T and K_S. In terms of expected values of r, double-diffusion and turbulence are very different. Turbulence tends to mix temperature and salinity at equal rates ($r_{\mathrm{turb}} \approx 1$), whereas for salt fingers the turbulent diffusivity ratio is significantly lower, typically being restricted to the range $r_{\mathrm{sf}} = \frac{\gamma_{\mathrm{sf}}}{R_\rho} \sim 0.3\text{–}0.6$. Such a difference makes it imperative that the relative strengths of salt fingers and turbulence are quantified – only then will we have a chance of adequately representing small-scale mixing in ocean models. In the next section, we review the techniques used to discriminate between turbulence and salt fingers in oceanographic measurements.

10.4 Microstructure signatures of salt fingers in the ocean

The challenges involved in the detection, let alone the quantification and analysis, of salt fingers using ship-board measurements are obvious. Fingers are small scale and time dependent; they interact with shear and turbulence; they could be embedded in large-scale coherent structures, staircases and intrusions, but more often they are not. Yet, to provide the ultimate proof of importance, field measurements are essential. The ocean is an incredibly complex system and any unsubstantiated ivory-tower-theoretical attempt to assess the magnitude of double-diffusive mixing is likely to be met with suspicion.

Several qualitative methods of identifying double-diffusion in data have already been mentioned. Inferences could be made based on the values of density ratio, shadowgraph observations and the presence of steps in vertical temperature and salinity profiles. Unfortunately, all these signs are rather imprecise indicators of active double-diffusion. For instance, we can claim with some confidence that salt fingers play a substantial role in ocean mixing for $R_\rho < 2$ and that they are generally too weak to compete with turbulence for $R_\rho > 3$. This leaves vast regions of the world ocean in the grey area where double-diffusion may or may not be important. The presence of well-defined staircases is sufficient to predict the dominance of double-diffusion but is far from necessary. Optical observations, which are yet to be included in the standard arsenal of microstructure measurements, can be effective in identifying salt fingers but offer little help in terms of quantifying their magnitude.

As difficult as the problem of identifying salt fingers in field measurements might be, the ocean gradually gives up its secrets. Numerous past successes and future prospects are associated with high-resolution O(cm) microstructure measurements. These methods were introduced more than forty years ago (Grant *et al.*, 1968; Osborn and Cox, 1972) and have become progressively more sophisticated and accurate ever since. The two most common approaches to the detection of salt fingers in microstructure measurements involve the analysis of (i) spectral characteristics inferred from horizontally towed conductivity and temperature sensors and (ii) dissipation rates of kinetic energy and thermal variance.

Spectral characteristics

Salt fingering is a narrow-band phenomenon. This is a direct consequence of their dynamics; fingers are created and maintained by molecular dissipation. While much larger secondary structures (intrusions, collective instabilities and staircases) could be generated occasionally, the primary signal is tightly confined to the dissipative range – the energy enters and exits the system at comparable spatial scales. In this regard, turbulence is fundamentally different. A substantial separation between the forcing and dissipative scales, the presence of a broad inertial subrange, and the associated direct cascades of energy and tracer variances are the essential attributes of three-dimensional turbulence (Kolmogorov, 1941; Batchelor, 1959). The resulting distinction between the spectral imprints of fingers and turbulence has been profitably and extensively exploited to interpret microstructure measurements. Since the preferred width of salt fingers (unlike their vertical scale) can be readily evaluated on the basis of linear theory, spectral methods are particularly effective in the analysis of horizontal variability.

The first attempt of this nature was made by Magnell (1976), who analyzed conductivity fluctuations in a horizontal plane recorded by a towed microstructure instrument in the Mediterranean outflow. The Fourier spectrum of the conductivity record revealed a well-defined peak at wavenumbers closely corresponding to the fastest growing finger modes, which was readily interpreted as clear evidence of salt fingering. Of course, the double-diffusive origin of microstructure in the Mediterranean outflow could have been anticipated, given the low value of density ratio ($R_\rho \sim 1.3$) and the persistence of staircases. In contrast, the unique spectral signatures of salt fingers detected in microstructure data from the North Pacific (Gargett and Schmitt, 1982) must have raised some eyebrows. Gargett and Schmitt analyzed the temperature record from a horizontally towed sensor and concluded that salt fingers operate in patchy regions with density ratios as high as $R_\rho \sim 3$. The limited-amplitude narrowband spectrum of temperature in these regions was markedly different from the broadband turbulent signal (Fig. 10.8a). Furthermore, the spectrum attributed to salt fingers was close to the prediction of the corresponding theoretical finger model (Fig. 10.8b), leaving little doubt in the interpretation of the microstructure signal.

The initial success of the spectral analysis of towed data stimulated numerous extensions. Marmorino (1987) used spectral analysis to identify fingers in the dataset obtained from towed fast-response conductivity sensors deployed in the Sargasso Sea. His conclusions matched those of Gargett and Schmitt (1982) – salt-finger patches are ubiquitous and, normally, could be distinguished from turbulent regions by characteristically narrowband spectra. A more detailed look at clear-cut instances of fingering microstructure (Fig. 10.9) reveals a very specific dependence of the conductivity gradient spectra on the wavenumber (k) on the low-k side: the power spectral density increases as k^2. Such dependence was observed in laboratory measurements (Taylor and Bucens, 1989) and in direct numerical simulations of salt fingers (Shen, 1993). Shen and Schmitt (1995) developed a simple model rationalizing the spectral slope of salt fingers from first principles. For turbulence, on the other hand, spectral slopes $\propto k$ are most common, which is consistent with classical turbulence theory (Batchelor, 1959). Such a definite distinction opened a very attractive opportunity to discriminate between fingers and turbulence on a quantitative, rather than a descriptive, level. A discriminant was badly needed to classify ambiguous cases of microstructure that possess features of both fingers and turbulence. The spectral slope of the temperature or conductivity gradient (S_L) held a promise of removing, or at least greatly reducing, the guesswork inevitable in the visual characterization of indistinct spectra.

An alternative approach was proposed by Holloway and Gargett (1987). These authors capitalized on a difference in the turbulent and fingering patterns of the probability density functions (PDFs) of tracers. Specific examples were presented for temperature fluctuations, although this technique could be applied just as easily to other measured quantities, such as conductivity or its gradient. The large signal variability and intermittence associated with turbulence are reflected in PDFs with relatively long tails. In fingering convection, on the other hand, fluctuations are restrained by secondary instabilities of salt fingers (Chapter 4) resulting in compact, nearly normal PDFs. For any measured quantity φ with zero mean value, the tightness of its PDF is conveniently measured by kurtosis $K = \frac{\langle \varphi^4 \rangle}{\langle \varphi^2 \rangle^2}$, which is

(a)

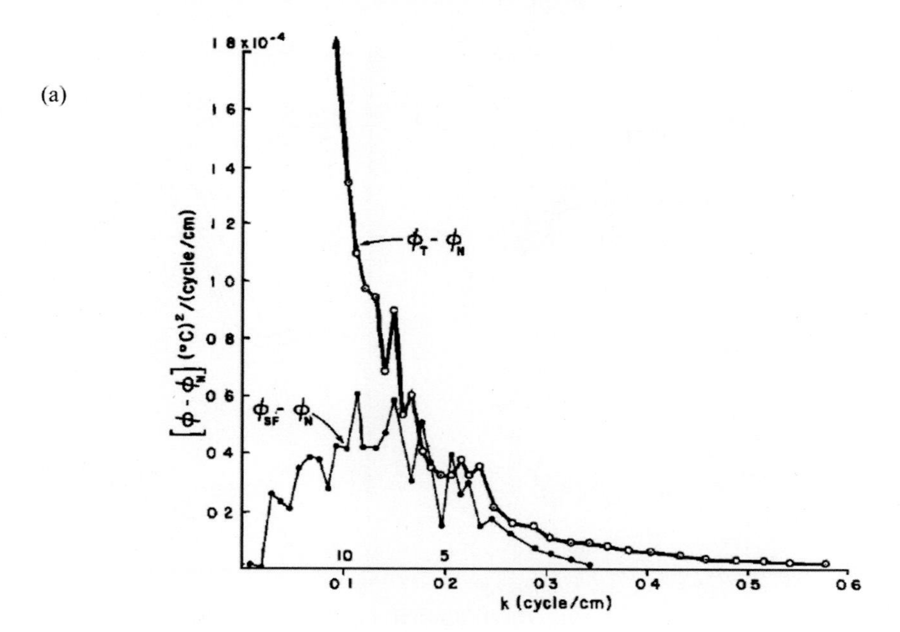

(b)

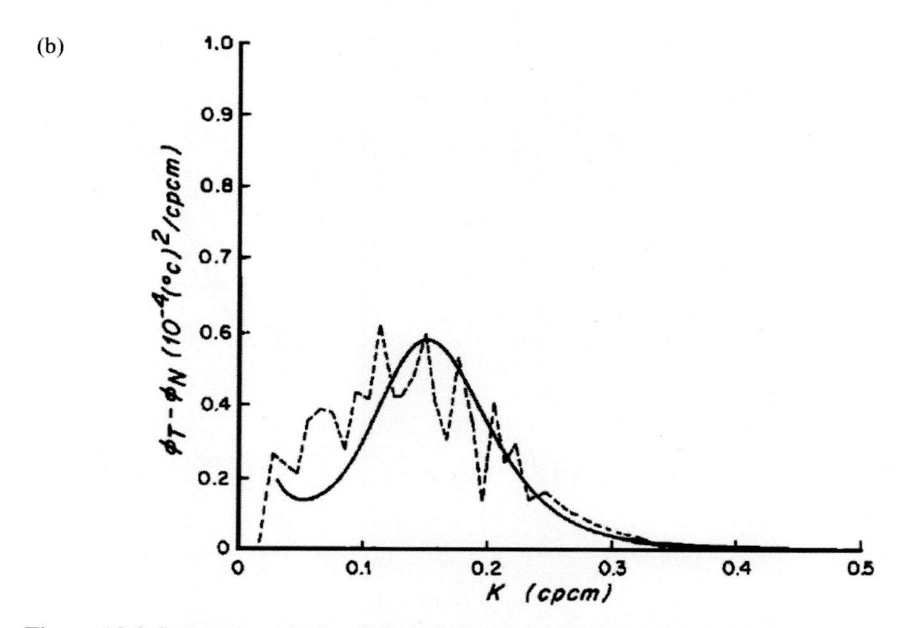

Figure 10.8 Spectral analysis of the temperature signal measured by a horizontally towed temperature sensor. (a) The horizontal temperature spectra show the limited bandwidth of the fingering signature and the characteristic dominance of the turbulent spectrum at low wavenumbers. (b) Comparison of the observed spectrum (dashed curve) with the corresponding theoretical model. From Gargett and Schmitt (1982).

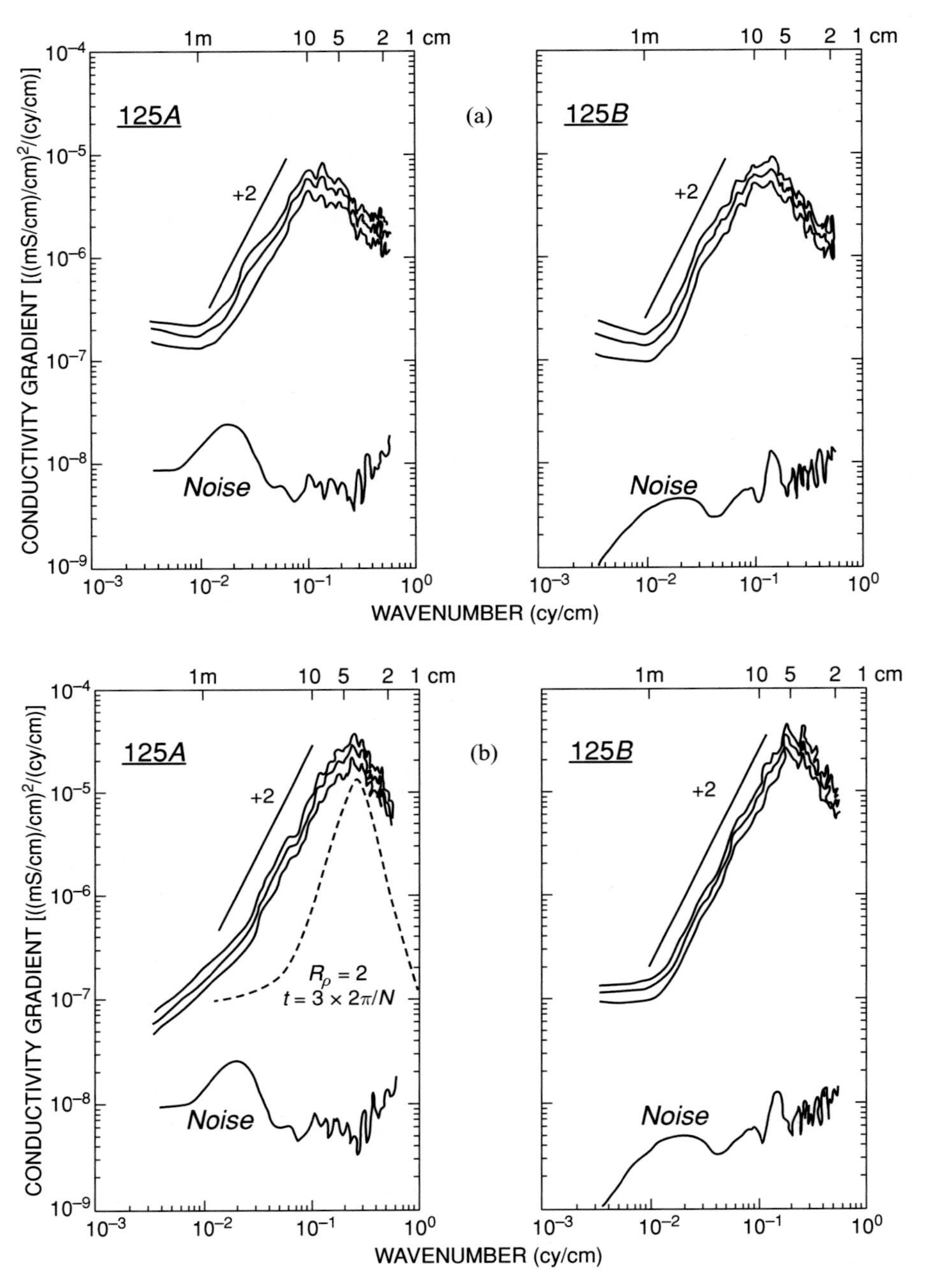

Figure 10.9 Conductivity gradient spectra from horizontally towed microstructure measurements in the regions of active fingering. Two different periods are analyzed in (a) and (b). The power spectral density increases as k^2 on the low-wavenumber side. Redrawn from Marmorino (1987).

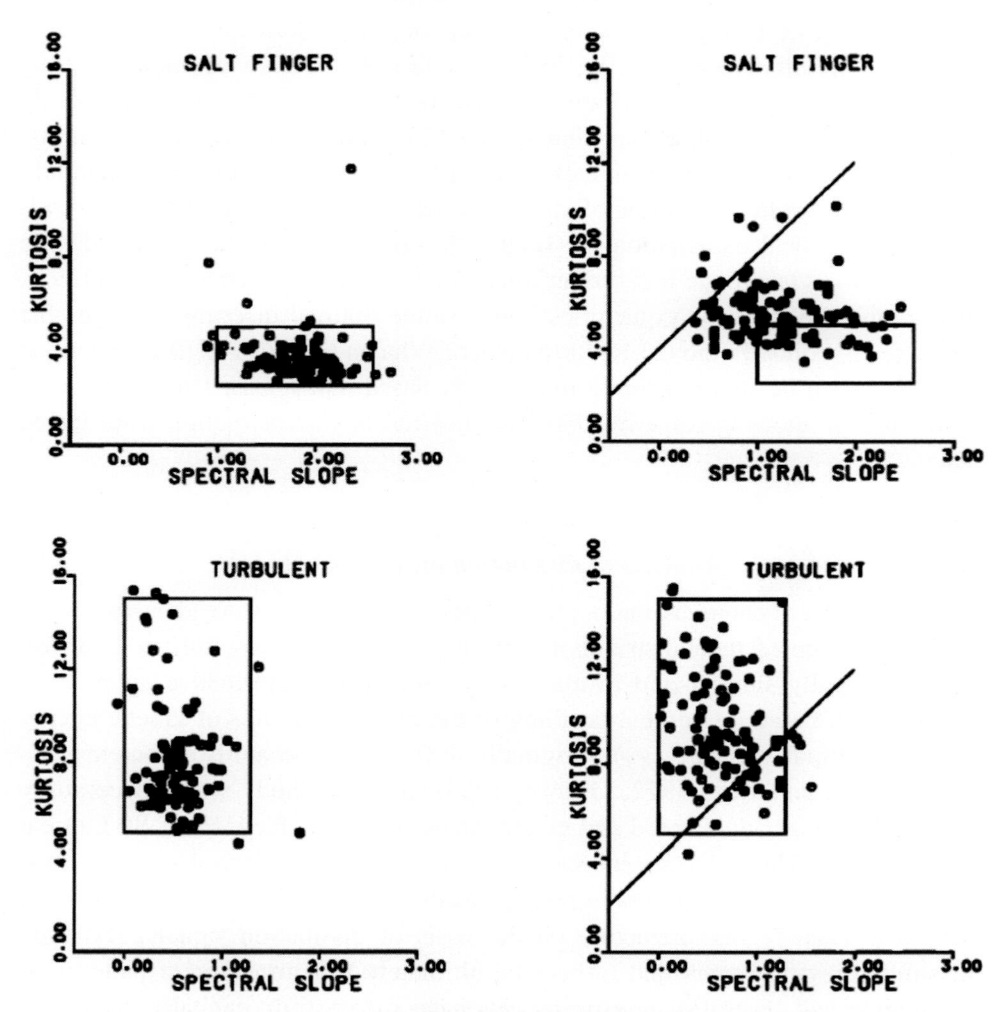

Figure 10.10 Scatter-plots of kurtosis versus spectral slope for data collected in the Sargasso Sea (right panel) and in the interfaces of the C-SALT staircase (left panel). Rectangles outline the Sargasso Sea populations for reference. The sloping line roughly divides the kurtosis–slope plane into fingering and turbulent populations. From Marmorino and Greenewalt (1988).

expected to be large for turbulence but low for fingers. Based on the analysis of unambiguous occurrences of fingers and turbulence, Holloway and Gargett suggested that $K_{\text{finger}} \sim 3$ and $K_{\text{turb}} \sim 7$.

The application of both kurtosis and slope diagnostics produced encouraging results. Figure 10.10 (Marmorino and Greenewalt, 1988) shows the scatter-plots of kurtosis (K) versus spectral slope (S_L) for data collected in the Sargasso Sea (right panel) and in the interfaces of the C-SALT staircase (left panel). Both datasets indicate strong correlation between slope and kurtosis – low S_L, high K signal for turbulence; high S_L, low K for fingers – a sign of mutual consistency and efficacy of S_L-based and K-based methods of classification. Curiously enough, the Sargasso data were characterized by a very pronounced bimodal distribution of turbulent/finger signals, whereas the C-SALT interfaces apparently contained a significant fraction of mixed microstructure. To the best of our knowledge, no physical explanation for such a difference has yet been offered. The slope–kurtosis technique was further improved by Mack and Schoeberlein (1993), who combined these two quantities into a single optimal discriminator variable using the likelihood ratio (LLR) formalism of Whalen (1971). The LLR technique was shown to be more accurate than the S_L-based and K-based methods taken individually, allowing for a reliable classification of microstructure from towed data.

Analysis of dissipation measurements

While spectral techniques undoubtedly improved our ability to identify fingers and knowledge of their distribution, recent years have seen a shift to a different, logistically simpler and in many ways even more informative approach – an approach based on vertical profiling of the dissipation rates of kinetic energy (ε) and thermal variance (χ_T). Originally developed to quantify turbulent mixing (Osborn and Cox, 1972; Osborn, 1980), these methods were successfully adapted for finger-dominated and mixed environments (Oakey, 1988; St. Laurent and Schmitt, 1999). Concurrent measurements of ε and χ_T have become a standard component of field programs focused on diapycnal mixing, further motivating efforts to classify microstructure on the basis of dissipation signals. Dynamic dissimilarities of fingers and turbulence ultimately lead to different patterns of dissipation and several diagnostic models have successfully capitalized on such differences. The challenge lies in selection of the most general and least ambiguous criterion for interpreting the dissipation signals. The basics of the ε–χ_T diagnostic framework are reviewed next, followed by examples of its application to field measurements.

The first set of ideas that we wish to discuss is based on the energetics of fluid motion. The turbulent kinetic energy (TKE) budget, averaged in time and space over large (relative to the size and duration of a turbulent event) scales, is conveniently written (Osborn, 1980) as a balance between energy production (P), buoyancy flux (J_b) and molecular dissipation (ε):

$$P = J_b + \varepsilon. \tag{10.3}$$

The energy production term in (10.3) can be expressed as

$$P = -\overline{u_i'u_j'}\frac{\partial \bar{u}_i}{\partial x_j}, i, j = 1, 2, 3, \tag{10.4}$$

where $(x_1, x_2, x_3) = (x, y, z)$ and $(u_1, u_2, u_3) = (u, v, w)$. Vertical shears in the ocean greatly exceed the horizontal ones and therefore (10.4) reduces to

$$P = -\overline{u'w'}\frac{\partial \bar{u}}{\partial z} - \overline{v'w'}\frac{\partial \bar{v}}{\partial z}. \tag{10.5}$$

The vertical buoyancy flux in (10.3) is represented by

$$J_b = \frac{g}{\rho_0}\overline{\rho'w'}, \tag{10.6}$$

and the rate of loss of the kinetic energy to dissipation can be expressed as

$$\varepsilon = \frac{\nu}{2}\overline{\left(\frac{\partial u_i'}{\partial x_j} + \frac{\partial u_j'}{\partial x_i}\right)^2}. \tag{10.7}$$

Since most oceanographic measurements are one-dimensional, either horizontal or vertical, it is nearly impossible to evaluate (10.7) directly from data. The problem is often bypassed by assuming that microstructure is roughly isotropic at the dissipation scale. The accuracy of this approximation has been questioned repeatedly in both turbulent and double-diffusive contexts and there is a general consensus that the isotropic assumption could easily introduce an error in ε of a factor of two or so. Such an error can be tolerated in some studies, a-posteriori corrected in others, or – if you are particularly unlucky – it may also lead to questionable conclusions. There is more on this issue in the next section, but for now it is sufficient to say that the benefit of the isotropy assumption lies in the reduction of (10.7) to a more usable form:

$$\varepsilon \approx \varepsilon_V = \frac{15}{4}\nu\left(\overline{u_z'^2} + \overline{v_z'^2}\right). \tag{10.8}$$

The thee-way balance of the TKE (10.3) can be described using either the flux Richardson number

$$R_f = \frac{J_b}{P} = \frac{J_b}{J_b + \varepsilon}, \tag{10.9}$$

representing the ratio of potential energy gain to kinetic energy input, or the efficiency factor

$$\Gamma_e = \frac{J_b}{\varepsilon} = \frac{R_f}{1 - R_f}, \tag{10.10}$$

the ratio of potential energy gain to the kinetic energy loss. These quantities are sensitive to the type of microstructure and ultimately make it possible to distinguish between fingers and turbulence. Turbulence is driven by the energy released by the instabilities of the background flow, which is represented by the P-term in the TKE balance. Most of the energy input is dissipated by molecular viscosity and only a small fraction is converted into potential energy. Based on laboratory experiments at the time, Osborn (1980) suggested $R_f \leq 0.15$ and therefore $\Gamma_e \leq 0.2$. It has become common practice in oceanography to use Osborn's upper limit for the efficiency factor in turbulent events:

$$\Gamma_{e\,\mathrm{turb}} = 0.2, \tag{10.11}$$

and subsequent studies (Barry *et al.*, 2001; Shih *et al.*, 2005) confirmed that (10.11) offers a reasonable estimate for much of the relevant parameter range. Fingers, on the other hand, are driven by the release of potential energy stored in the salinity stratification, which renders the production term P inconsequential for their establishment and maintenance. In the limit $P \to 0$, the TKE balance reduces to $J_b = -\varepsilon$ or $\Gamma_{e\,\mathrm{fingers}} = -1$.

The second component of the diagnostic framework, which to some extent mirrors the treatment of the TKE budget, is based on the temperature variance equation (Osborn and Cox, 1972). After averaging in time and space, this equation reduces to the balance between the variance production by the large-scale flow and molecular dissipation:

$$-2\overline{u_i'T}\frac{\partial \bar{T}}{\partial x_i} = \chi_T, \tag{10.12}$$

where the dissipation term is given by

$$\chi_T = 2k_T\overline{\left(\frac{\partial T'}{\partial x_i}\right)^2}. \tag{10.13}$$

The production term in (10.12) represents the effects of both lateral ($i = 1{,}2$) and vertical ($i = 3$) advection. Lateral advection can be essential for variance production in frontal regions with sharp horizontal temperature gradients, which often harbor pronounced intrusive structures (see Chapter 7). However, in much of the ocean, lateral gradients are relatively weak and the production of thermal variance is dominated by its vertical component, which reduces (10.12) to

$$-2\overline{w'T'}\frac{\partial \bar{T}}{\partial z} = \chi_T. \tag{10.14}$$

In order to evaluate (10.14) from one-dimensional microstructure measurements, the isotropic assumption is invoked once again:

$$\chi_T \approx \chi_{TV} = 6k_T \overline{\left(\frac{\partial T'}{\partial z}\right)^2}, \tag{10.15}$$

which reduces the temperature variance equation to

$$-\overline{w'T'}\frac{\partial \bar{T}}{\partial z} \approx 3k_T \overline{\left(\frac{\partial T'}{\partial z}\right)^2}. \tag{10.16}$$

The simplified integral budgets of TKE (10.3) and temperature variance (10.16) form the foundation of the microstructure diagnostic methods. They provide an explicit link between the dissipation characteristics (which are measurable) and the vertical fluxes of heat, density and momentum (which affect the large-scale dynamics of the ocean). While it is straightforward to formulate a similar budget for the salinity variance, it would be of limited utility since technical complications currently preclude direct measurements of salt dissipation (χ_S). It should be mentioned though that there is a significant interest in the technology that would make it possible to infer χ_S from data (e.g., Nash and Moum, 1999) and the prospects of developing sensors capable of measuring salinity microstructure are very real.

The question now arises of how to use the production–dissipation balances to discriminate between turbulence and fingers. Fluid dynamics is a science of non-dimensional numbers and the selection of the most appropriate ones is vital for the success of any diagnostic model. For the background large-scale flow, the two relevant non-dimensional parameters are the density ratio and Richardson number; the former indicating propensity for fingering, the latter for turbulence. Criteria used to characterize the microstructure signals are more complex and the choices are less obvious.

A non-dimensional measure of temperature dissipation is given by the Cox number:

$$Cx_T = \frac{1}{\bar{T}_z^2}\overline{\left(\frac{\partial T'}{\partial x_i}\right)^2} = \frac{\chi_T}{2k_T\bar{T}_z^2}, \tag{10.17}$$

which, in view of (10.14), reduces to the Nusselt number (*Nu*) used throughout this monograph. The physical interpretation of the Cox number is straightforward – it measures the relative magnitudes of the eddy and molecular diffusivities of temperature:

$$K_T = Cx_T k_T. \tag{10.18}$$

Salt fingers and turbulence in the thermocline are typically characterized by comparable Cox numbers ($Cx_T = 10$–100). This rules out Cx_T as an effective discriminator between fingers and turbulence, although it can be conveniently used to quantify the intensity of small-scale mixing.

A non-dimensional number often invoked to characterize the dissipation of kinetic energy is the buoyancy Reynolds number:

$$Re_b = \frac{\varepsilon}{\nu N^2}. \tag{10.19}$$

Unlike the Cox number, Re_b does offer some insight into the origin of microstructure. Turbulence is an ineffective mixer – (10.11) implies that the energy lost to frictional dissipation is much higher than the energy invested directly into the vertical transport of properties. Frictional dissipation due to fingers, on the other hand, is very limited. The typical values of the flux ratio $\gamma \sim 0.6 - 0.7$ mean that the rate at which potential energy is gained by temperature stratification exceeds the frictional loss. Thus, for comparable rates of temperature mixing – as measured by Cx_T, χ_T, or K_T – turbulence has much stronger expression in terms of ε and therefore higher values of Re_b. Inoue *et al.* (2007) capitalized on this property to identify finger-dominated microstructure in the Oyashio/Kuroshio/Tsugaru currents system. The key element of the diagnostic model was the criterion $Re_b < 20$ ($Re_b > 20$) for fingers (turbulence). This criterion was supplemented by a set of necessary conditions: mixing was assumed significant only for $\chi_T > 5 \cdot 10^{-9}\,{}^{\circ}\mathrm{C}^2\,\mathrm{s}^{-1}$ and microstructure was classified as fingering only for $1 < R_\rho < 3$ and $0 < \gamma < 1$. The estimated diffusivities for finger-dominated regions were consistent with the expected salt-finger patterns and in reasonable agreement with extant parameterizations (Zhang *et al.*, 1998). While the Re_b-based criterion is expected to perform well under typical conditions, the possibility exists for misclassification of microstructure in the extreme cases of anomalously low-amplitude turbulence or high-amplitude fingering.

The most effective and widely used approach to classification of microstructure combines measurements of ε and χ_T into a single discriminator variable known as the dissipation ratio:

$$\Gamma = \frac{\chi_T N^2}{2\varepsilon \bar{T}_z^2}. \tag{10.20}$$

This method is based on exactly the same principle as the Re_b-based criterion – frictional loss of energy in turbulent flows is much higher than in fingering regions. Hence, the dissipation ratio should be low for turbulence and high for fingers. This statement can be made more precise as follows. Turbulence mixes temperature, salinity and density at approximately the same rate $K_{T\,\mathrm{turb}} = K_{\rho\,\mathrm{turb}}$ or, equivalently,

$$\frac{\overline{w'T'}}{\bar{T}_z} = \frac{\overline{w'\rho'}}{\bar{\rho}_z} = \frac{J_b}{N^2}. \tag{10.21}$$

Drawing together (10.10), (10.14) and (10.21), we reduce (10.20) to

$$\Gamma_{\text{turb}} = \Gamma_{e\,\text{turb}} \sim 0.2. \tag{10.22}$$

Thus, the dissipation ratio in purely turbulent flows is identical to the efficiency factor. This property often leads to the indiscriminate usage of the terms "dissipation ratio" and "efficiency factor" in microstructure literature – the same variable (Γ) is often assigned for both quantities. All this may be acceptable if the topic is limited to turbulence. However, the lack of firm conventions could lead to confusion in discussions of double-diffusive or mixed microstructure. To eliminate the risk of misinterpretation, it is important to emphasize the principal differences between the efficiency factor and the dissipation ratio. The primary definition of the dissipation ratio (10.20) is based on the dissipation signals. The efficiency factor quantifies the TKE balance and it is unambiguously defined by (10.10). Purely turbulent events aside, the dissipation ratio (Γ) and efficiency factor (Γ_e) are different, numerically and conceptually, and should be treated as such.

For the case of pure fingering, the dissipation ratio can be explicitly evaluated (e.g., Hamilton *et al.*, 1989) by assuming that the production term (P) in the TKE budget is negligible and

$$J_b \equiv g\frac{\overline{w'\rho'}}{\rho_0} = -\varepsilon. \tag{10.23}$$

Breaking down the density flux into its components $F_T = \overline{w'T'}$ and $F_S = \gamma^{-1}F_T = \gamma^{-1}\overline{w'T'}$ simplifies (10.23) to

$$g\alpha\left(\gamma^{-1} - 1\right)\overline{w'T'} = -\varepsilon. \tag{10.24}$$

Combining (10.24) with (10.16), we express the dissipation ratio (10.20) in terms of the flux ratio and the density ratio, whose values are relatively well constrained:

$$\Gamma_{\text{finger}} = \frac{N^2}{\bar{T}_z g\alpha\left(\gamma^{-1} - 1\right)} = \frac{1 - R_\rho^{-1}}{\gamma^{-1} - 1}. \tag{10.25}$$

Thus, for typical values of $R_\rho \sim 2$ and $\gamma \sim 0.6 - 0.7$, we expect $\Gamma_{\text{finger}} \sim 1$, which is considerably higher than the turbulent value (10.22).

The credit for the first application of the criterion $\Gamma_{\text{finger}} > \Gamma_{\text{turb}}$ to data goes to Oakey (1988), who analyzed profiler measurements of ε and χ_T through meddy Sharon. Oakey reported several incidences of unusually high dissipation ratio ($\Gamma > 1$), which invariably occurred in regions with double-diffusively favorable stratification – a clear sign of the efficacy of the Γ-based classification method. McDougall and Ruddick (1992) further developed this approach by proposing an algorithm to infer the salinity flux from microstructure data. The first step involves

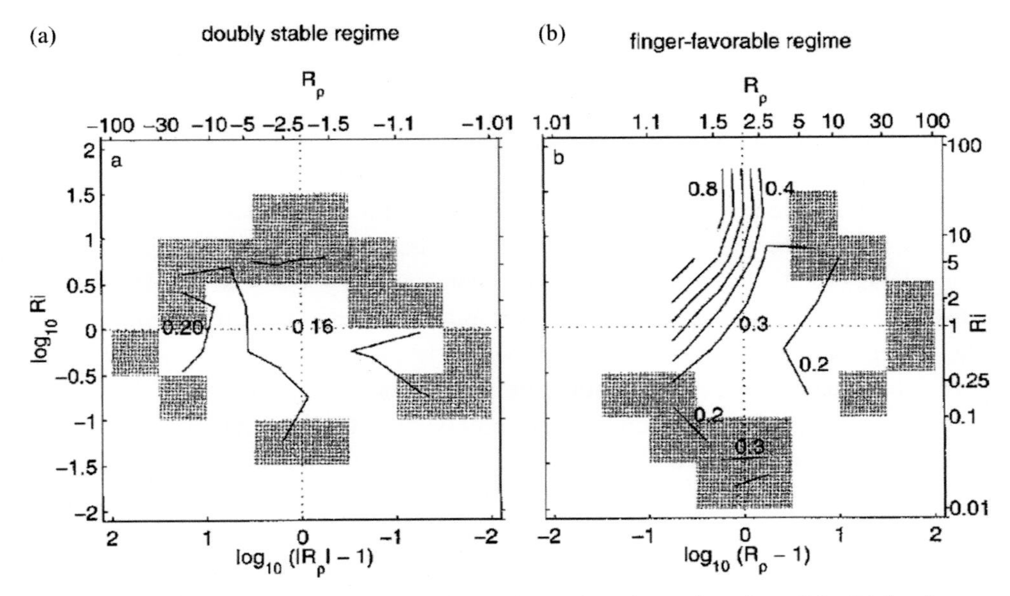

Figure 10.11 The variation of the dissipation ratio (Γ) as a function of the Richardson number (Ri) and the density ratio (R_ρ) in (a) double-diffusively stable and (b) salt-finger favorable regimes. Regions of darker shading denote bins where the uncertainty in Γ exceeds 25% of the mean. The finger-favorable regime is characterized by the elevated dissipation ratios, particularly in the $1 < R_\rho < 2$, $Ri > 1$ region of parameter space. From St. Laurent and Schmitt (1999).

estimation of the relative contributions of fingers and turbulence to mixing from the position of the observed dissipation ratio Γ_{obs} within the interval $\Gamma_{turb} < \Gamma < \Gamma_{finger}$. Knowledge of the composition of microstructure, in turn, makes it possible to determine both the temperature and salinity fluxes from ε and χ_T. The technique was shown to be reasonably accurate; the errors in the inferred salt and heat fluxes are primarily associated with the uncertainty in the assumed finger flux ratio. Since direct measurement of salinity microstructure is currently beyond our technological capabilities, the diagnostics of the salt flux, singularly afforded by the McDougall and Ruddick method, proved to be most valuable for the analysis of mixed microstructure.

A definitive test of the Γ-based approach was performed by St. Laurent and Schmitt (1999). These authors examined the extensive microstructure dataset from the site of the North Atlantic Tracer Release Experiment (NATRE). The NATRE area is only moderately favorable to fingering and is actively turbulent, which presents a major challenge for reliable detection and quantification of finger-induced mixing. Nevertheless, St. Laurent and Schmitt were able to discern the salt-finger signal by focusing on the statistics of the dissipation ratio. The exploration of the two-dimensional parameter space of density ratio and Richardson number (R_ρ, Ri) was particularly informative in this regard (Fig. 10.11). The range

$1 < R_\rho < 2, Ri > 1$ (weak turbulence, strong fingers) was characterized by elevated values of Γ, which are consistent with the finger model (10.25) and cannot be attributed to turbulence (10.22). Outside of this parameter region, the dissipation ratio was low and better described by the turbulence model.

St. Laurent and Schmitt (1999) also inferred the diffusivities of heat and salt (K_T and K_S) from the NATRE microstructure data using methodology analogous to that of McDougall and Ruddick (1992). The results revealed a significant difference between the average values of K_T and K_S in the region – a difference that is caused entirely by fingers:

$$K_T = 0.8 \cdot 10^{-4}\,\mathrm{m^2\,s^{-1}},\ K_S = 1.3 \cdot 10^{-4}\,\mathrm{m^2\,s^{-1}}. \tag{10.26}$$

The microstructure-based estimate of K_S in (10.26) is remarkably close to the tracer diffusivity

$$K_{\mathrm{tracer}} = 1.2 \cdot 10^{-4}\,\mathrm{m^2\,s^{-1}} \tag{10.27}$$

evaluated from the vertical spreading of a passive chemical (sulfur hexafluoride) released during NATRE. The actual eddy diffusivity of the tracer is expected to be equal to that of salt because their molecular diffusivities are close. In this regard, the agreement between (10.26) and (10.27) is encouraging. It supports inferences based on microstructure analysis and emphasizes the significance of double-diffusive mixing in the presence of active turbulence.

In summary, the microstructure diagnostics discussed in this section provide irrefutable evidence of vigorous fingering in much of the upper ocean. Particularly promising are the methods based on the concurrent measurements of ε and χ_T, which carry the twin benefits of (i) classifying the microstructure as finger-dominated or primarily turbulent and (ii) making it possible to evaluate the vertical diffusivities of temperature, salinity and momentum. A questionable aspect of the dissipation-based methods is the assumed isotropy of microstructure, commonly invoked to infer three-dimensional dissipation from one-dimensional measurements. The ramifications of the anisotropic effects and the prospects for reduction of the associated error are discussed next.

Anisotropy of salt fingers

At first, the idea of resorting to an isotropic model of microstructure, particularly in the context of salt fingers, appears to be overly simplistic and unwarranted. Most double-diffusive theories picture salt fingers as anisotropic and vertically oriented (e.g., Holyer, 1984; Kunze, 1987). Even the name "salt finger" immediately conveys the sense of an elongated structure, something that perhaps should not be modeled as a round eddy. In reality, the effects of anisotropy are rather moderate. The microstructure estimates of vertical transport – the estimates based on the isotropic model – are generally consistent with the tracer release measurements. This has been demonstrated for the staircase (Schmitt *et al.*, 2005) and smooth-gradient (St. Laurent and Schmitt, 1999) regimes, which suggests that anisotropic effects are limited. Of course, since tracer- and microstructure-based estimates are indirect, one cannot dismiss the possibility that both methods may be biased and the agreement is, to some extent, fortuitous. Furthermore, even if the isotropic model does not lead to major inconsistencies, the anisotropy should be quantified and an effort made to reduce the associated error.

While the conceptual theory of finger anisotropy is noticeably missing, several attempts have already been made to quantify the effect. Taylor (1993) conducted a series of laboratory experiments aimed at determining the aspect ratio of salt fingers and examining its variation with the density ratio. The operational definition of the aspect ratio was based on the dominant vertical and horizontal wavelengths of temperature spectra. The average aspect ratio for $R_\rho < 5$ was found to be 0.58, which means that fingers are about twice as long as they are wide. Taylor noted the systematic trend of salt fingers to become more isotropic at low values of the density ratio, with the aspect ratio approaching unity for $R_\rho \to 1$. These results are consistent with the *in situ* oceanic measurements of salt fingers by a horizontally towed device (Lueck, 1987). Lueck found very little coherence between the temperature records of two sensors vertically separated by 6 cm – a distance comparable to the dominant scale of primary salt-finger instability at the observational site. The lack of vertical coherence supports the isotropic model of salt fingers.

Numerical studies (e.g., Shen, 1989; Stern *et al.*, 2001) repeatedly challenged the notion of long vertical fingers, particularly at low density ratios ($R_\rho < 2$). An important step was made by Caplan (2008), who used DNS to evaluate and correct the systematic bias in the profiler approximation of dissipation quantities. Salt-finger experiments with various density ratios were diagnosed to evaluate the actual dissipation ε and χ_T, given in (10.7) and (10.13) respectively, as well as the dissipation inferred from the vertical profiles (ε_V and χ_{TV}) in (10.8) and (10.15) using the isotropic approximation. The difference between the actual and inferred dissipation was expressed in terms of anisotropic coefficients A_{VV} and A_{TV} such that

$$A_{VV} = \frac{\varepsilon}{\varepsilon_V} = \frac{\overline{\sum_{i,j} \frac{1}{2}\left(\frac{\partial u'_i}{\partial x_j} + \frac{\partial u'_j}{\partial x_i}\right)^2}}{\frac{15}{4}\nu\left(\overline{u'^2_z} + \overline{v'^2_z}\right)}, \quad A_{TV} = \frac{\chi_T}{\chi_{TV}} = \frac{\overline{\sum_i \left(\frac{\partial T'}{\partial x_i}\right)^2}}{3\overline{\left(\frac{\partial T'}{\partial z}\right)^2}}. \tag{10.28}$$

The results of the anisotropy analysis were suggestive. The anisotropic coefficient for temperature was sufficiently close to unity ($1 < A_{TV} < 1.3$), which implies that the profiler approximation offers a fairly accurate estimate of χ_T. The profiler approximation of ε is more problematic. The anisotropic coefficient for TKE monotonically increased from $A_{VV} = 1.3$ at $R_\rho = 1.1$ to $A_{VV} = 2.0$ at $R_\rho = 1.9$. Such high values of A_{VV} imply that the profiler approximation could quite significantly, by as much as a factor of two, underestimate the rates of TKE dissipation. The anisotropy characteristics of salt fingers in two- and three-dimensional simulations were similar – a sign of the robust dynamics of the processes controlling anisotropy. To estimate the error introduced by the isotropic approximation in the analysis of towed microstructure data, Caplan (2008) also diagnosed the horizontal anisotropy coefficients $A_{VH} = \frac{\varepsilon}{\varepsilon_H}$ and $A_{TH} = \frac{\chi_T}{\chi_{TH}}$. The horizontal coefficients conveyed the same message: salt fingers can be considered roughly isotropic for the purpose of evaluating χ_T but significant error is to be expected for ε. A suggestion was made to reduce the anisotropy-related errors of the data-based dissipation estimates by incorporating the anisotropy coefficients, recorded as a function of R_ρ, in post-processing of microstructure measurements. Caplan's anisotropy analysis was extended to the DNS of fingers in a sheared background flow by Kimura *et al.* (2011), who found that laminar shear has a relatively minor effect on the anisotropy coefficients.

10.5 Inverse modeling of thermohaline staircases

This chapter would be deplorably incomplete without mentioning one more method of evaluating double-diffusive transport in the ocean – the inverse modeling of thermohaline staircases (Lee and Veronis, 1991; Veronis, 2007). This method is yet to receive the level of attention given to microstructure-based and tracer-release studies or, for that matter, the level of attention it deserves. Nevertheless, the technique is promising. In view of the success of inverse modeling in other branches of oceanography and beyond (e.g., Bennett, 2002), we can safely assume that its future applications will constrain estimates of mixing in staircases and generate new insights into their dynamics.

Inverse models attempt to calculate the unknown model parameters by fitting measurements to the governing equations in a manner that minimizes the suitably defined integral error. Lee and Veronis (1991) used the 1985 C-SALT dataset for the Caribbean staircase, which was particularly appropriate for inversion. The steps in the staircase were steady and well-defined. The observations were regularly spaced, making this dataset ideal for discretization. The governing equations were taken to be the temperature and salinity advection–diffusion equations in divergence form:

$$\nabla_H (\vec{v}_H C) + \frac{\partial (wC)}{\partial z} = \frac{\partial}{\partial z}\left(K_C \frac{\partial C}{\partial z}\right), \tag{10.29}$$

where subscript H refers to horizontal components and C may be T or S, and the continuity equation

$$\nabla_H \vec{v}_H + \frac{\partial w}{\partial z} = 0. \qquad (10.30)$$

The governing equations were vertically integrated over each layer and discretized by staggering all variables in space in the form of an Arakawa C grid (Fig. 10.12). The result is finite-difference forms of (10.29) and (10.30), which were applied to each of the six layers and sixteen horizontal locations in the central part of the staircase indicated in Figure 10.12. The temperature, salinity and depth of each layer at observational locations were treated as known variables, velocities and diffusivities as unknown. After discretization, the linear homogeneous governing system reduced to

$$\mathbf{A} \cdot \vec{\mathrm{X}} = 0, \qquad (10.31)$$

where $\vec{\mathrm{X}}$ is the vector of unknowns and $\mathbf{A}$ is the matrix that contains the coefficients of these unknown variables in the finite-difference tracer and continuity equations. In order to complete the inversion, the system was cast in inhomogeneous form by dividing all the equations by an arbitrarily chosen reference unknown X_r, and rewriting (10.31) as

$$\mathbf{A}^* \cdot \vec{\mathrm{X}}^* = \vec{b}, \qquad (10.32)$$

where $\vec{\mathrm{X}}^*$ is a vector consisting of relative unknowns. In Lee and Veronis' model, (10.32) represented an over-determined system of 288 equations with 260 unknowns and its best-fit solution was determined using the method of total least squares (Golub and van Loan, 1983). The solution, of course, was only for the relative unknowns and therefore all original unknowns ($\vec{X}$) were determined up to a constant factor – the complication that stems from the homogeneous nature of the governing equations (10.29) and (10.30). At this point, the inversion yielded the ratios of unknowns but not their absolute values. For instance, the inversion was remarkably successful in predicting the flux ratio:

$$\gamma_{\mathrm{inv}} = \frac{K_T}{K_S} R_\rho = 0.89 \pm 0.07, \qquad (10.33)$$

which closely matched the estimate based on the lateral density ratio in C-SALT layers, $\gamma \approx 0.85$ (Schmitt *et al.*, 1987).

However, the inversion remained incomplete without invoking some additional consideration that would make it possible to normalize the unknowns. The original study (Lee and Veronis, 1991) made use of the horizontal velocity measurements by a current meter at one of the stations during C-SALT. While this seemed a reasonable solution at the time, the resulting diffusivities were unrealistically high,

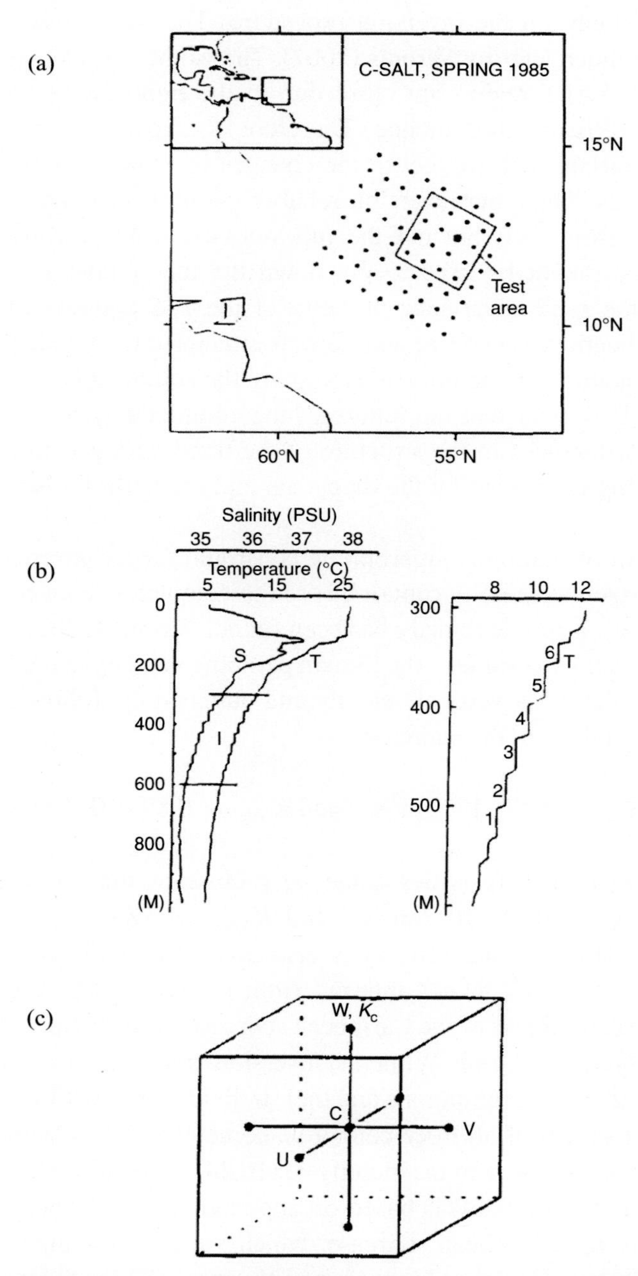

Figure 10.12 Inverse model of the Caribbean staircase. (a) The map of the survey region in C-SALT; the inverse model is applied to the 4 × 4 array in the boxed area. (b) The left panel presents the vertical *T*–*S* profiles at the location marked by a solid square in (a). The section of the temperature profile that contains the six layers used for the inversion is shown in the right panel. (c) The seven-point grid used for discretization of the governing equations. From Lee and Veronis (1991).

casting some doubt on the inversion procedure. The root cause of the problem was identified much later by Veronis (2007). The isotherms, isohalines and isopycnals in the C-SALT region are close due to the tightness of the *T–S* relation in the double-diffusive thermocline. Therefore, the governing equations (10.29) are largely invariant with respect to the changes in velocity component directed along isopycnals. Thus, the inversion reliably predicts only the flow component normal to the isopycnals but not the full velocity, which means that the horizontal velocity cannot be used to pin down the magnitudes of the unknowns. Incidentally, the same invariance property of the *T–S* equations resulted in yet another complication. When Lee and Veronis attempted to include the geostrophic momentum equations in the governing system, the solutions became inconsistent. In retrospect, it is clear that the failure of the combined system is related to the fundamental differences in the structure of the tracer and geostrophic equations, the former being controlled by the diapycnal and the latter by isopycnal velocity components.

The problem of finding an appropriate constraint for the inversion is a serious one. If the horizontal velocity cannot be used to complete the inversion, then what can? Fortunately, a simple remedy has been found. Veronis (2007) recalibrated the inversion data using the value of the Ekman pumping velocity in the C-SALT region as a proxy for the mean vertical velocity and obtained the following estimates of the mean diffusivities in the staircase:

$$K_{T\text{inv}} = 0.47 \cdot 10^{-4}\,\text{m}^2\,\text{s}^{-1} \text{ and } K_{S\text{inv}} = 0.89 \cdot 10^{-4}\,\text{m}^2\,\text{s}^{-1}. \qquad (10.34)$$

The inversion-based diffusivities came very close to the microstructure-based prediction $K_{T\text{micr}} = 0.45 \cdot 10^{-4}\,\text{m}^2\,\text{s}^{-1}$ and $K_{S\text{micr}} = 0.85 \cdot 10^{-4}\,\text{m}^2\,\text{s}^{-1}$ (Schmitt *et al.*, 2005). The estimate (10.34) is also consistent with the salt diffusivity $K_{S\text{tr}} = (0.8 - 0.9) \cdot 10^{-4}\,\text{m}^2\,\text{s}^{-1}$ inferred from the spreading of a passive tracer released in the central part of the Caribbean staircase, which brings the whole story to a very satisfying resolution. While the inversion, tracer and microstructure based methods have their own limitations and their individual predictions could be questioned, the agreement of all three cannot be coincidental. The actual diffusivities are likely to be somewhere in the vicinity of (10.34). The only estimate that significantly deviates from (10.34) is based on application of the laboratory-calibrated 4/3 flux law to the Caribbean staircase, which yields much higher diffusivities: $K_{S\text{lab}} \sim 6 \cdot 10^{-4}\,\text{m}^2\,\text{s}^{-1}$ (e.g., Lambert and Sturges, 1977). As discussed earlier, the elevated values of $K_{S\text{ lab}}$ are readily attributed to very different conditions in the ocean and laboratory. Thus, on the basis of the observational estimates of fluxes in staircases we can conclude that (i) double-diffusion is the dominant mixing agent, (ii) the vertical *T–S* transport can only be evaluated using methods that are fully compatible with the double-diffusive nature of the phenomenon, and (iii) the active oceanic environment adversely affects double-diffusive mixing in staircases.

And to conclude this chapter, here is a final thought for not-so-mature readers like myself. Why is finger transport lower in the ocean than in the laboratory? Because the ocean always keeps its fingers crossed.

11

Large-scale consequences

The global role of double-diffusion in the ocean and its impact on climate is perhaps the most contentious topic in the theory of double-diffusion. To say that opinions vary would be a huge understatement. The controversy is caused not so much by the incomplete understanding of double-diffusion, although certainly it does not help, but rather by the lack of firm criteria for what might be important and what is secondary for the general circulation of the ocean. In relative terms, oceanography is still a young science. As it goes through a period of teenage ambivalence, various problems go in and out of fashion and perceptions of what should be considered significant rapidly evolve. It certainly makes for an interesting and exciting period. But, if pressed for a definite opinion, many double-diffusers could respond in all honesty, "I am sure double-diffusion is important but I am not quite sure why."

The following discussion is an attempt to summarize a wide spectrum of arguments in favor of the tangible influence of double-diffusion on large-scale (i.e., comparable to the basin size in this context) patterns. Some of the rather extreme early views, ascribing double-diffusion an exclusive role in maintenance of the thermocline, are mentioned for reasons of historical interest. Overall though, this chapter is focused on more recent, more restrained and demonstrably more relevant arguments. We first examine the effects of fingering, which reflects both its higher incidence and the amount of attention it has received so far, and then proceed to discuss the potential large-scale impact of diffusive convection.

11.1 Effects of salt fingers

Density stratification and the Meridional Overturning Circulation

It all started as a fortunate misconception. The prevailing concern in physical oceanography during the sixties and seventies was the structure of the thermocline. The mainstream view on its dynamics at that time was expressed by the diffusive thermocline model (Robinson and Stommel, 1959; Munk, 1966) based on the assumed balance between vertical diffusion and the advection of density:

$$w\frac{\partial \rho}{\partial z} \sim K_V \frac{\partial^2 \rho}{\partial z^2}. \tag{11.1}$$

Analogous balances could be assumed for temperature, salinity or, for that matter, any other quasi-conservative quantity. The innocuous looking equation (11.1) held the promise of explaining ocean stratification using very basic and intuitive physical arguments. Furthermore, when supplemented by the geostrophic momentum balance, (11.1) made it possible to predict the strength of the Meridional Overturning Circulation (MOC) – one of the main components of the Earth's climate system.

The biggest unknown in the advective–diffusive balance (11.1) was, and still remains, K_V. If the molecular diffusivity is used, either $k_T \approx 1.4 \cdot 10^{-7}\,\mathrm{m^2\,s^{-1}}$ for temperature or $k_S \approx 1.5 \cdot 10^{-9}\,\mathrm{m^2\,s^{-1}}$ for salinity, then (11.1) can be satisfied only at vertical scales of 1 m and 1 cm respectively, which renders the whole model completely irrelevant for the thermocline problem. For the proponents of the diffusive theory, this inconsistency simply meant that the mixing coefficient K_V should be interpreted as the eddy-induced, rather than molecular, diffusivity. At this point, the fate of the diffusive theory hinged on the ability of ocean mixing to support the advective–diffusive balance – in short, on the average value of K_V. However, the task of evaluating the much-needed mixing coefficient on the basis of oceanographic observations proved to be a major challenge. Numerous obstacles, conceptual and technical, demanded no less than the birth of a new branch of oceanography, the small-scale ocean mixing program. Its ultimate mission was to test and hopefully verify the diffusive model of the thermocline and meridional overturning.

In the lower thermocline, vertical advection in mid-latitudes is expected to be generally upward ($w > 0$) in order to compensate for high-latitude sinking. The strength of upwelling can be estimated from the rates of deep water production ($w \sim 10^{-7}$ m s^{-1}). If (11.1) is valid, then K_V/w should be of the same order as the stratification scale (H). Inspection of vertical density profiles in the ocean suggests scales of $H \sim 10^3$ m and therefore the diffusivity required to maintain the advective–diffusive balance (Munk, 1966) is

$$K_V \sim wH \sim 10^{-4}\,\mathrm{m^2\,s^{-1}}. \tag{11.2}$$

The early large-scale numerical simulations lent some support to the diffusive model: In order to reproduce an MOC of realistic strength, the Atlantic-only models (e.g., Bryan, 1987) have to assume a vertical diffusivity that is commensurate with

Munk's canonical value (11.2). The diffusive argument can be further bolstered by noting that if w and K_V are taken to be uniform, (11.1) leads to

$$\rho = C \exp\left(\frac{w}{K_V} z\right) + \rho_0. \tag{11.3}$$

For $K_V > 0$, as expected for predominantly turbulent mixing, (11.3) predicts a gradual reduction in stratification with increasing depth. It could be argued that, qualitatively, the actual density distribution is not inconsistent with the predicted exponential pattern.

Now enter double-diffusion. The double-diffusive flux of density is necessarily counter-gradient (i.e., downward) and therefore opposes turbulent mixing. Thus, the presence of double-diffusion reduces the net diffusivity of density (K_V) and thereby affects the stratification in the upwelling ($w > 0$) regions, as evident, for instance, from (11.3). For downwelling ($w < 0$) conditions, typical of the upper subtropical thermocline, the potential role of double-diffusion could be even more profound. In this case, it is possible to balance the downward density advection by negative diffusion, provided that finger-induced mixing dominates the turbulent transport. This scenario can be readily recognized as a double-diffusive counterpart of the classical diffusive thermocline theory. The possibility that double-diffusion plays a major role in maintenance of the thermocline was also supported by arguments based on the energetics of the general circulation (Stern, 1968, 1969; Schmitt and Evans, 1978). These early attempts, imaginative and internally consistent, suffered from only one major flaw. They were based on crude and indirect estimates of vertical mixing, the only kind available at that time.

The only cure for the uncertainty was an improvement in measurements. However, when the technical advancements of the mid eighties finally opened the stream of microstructure-based data, they brought rather disappointing news for the believers in the mixing-controlled ocean. As microscale measurements became more accurate and widespread, it became increasingly clear that mixing in the ocean interior, regardless of its origin, is generally too weak to support the vertical advective–diffusive balance. Thermocline diffusivity, inferred from microstructure and tracer dispersion measurements, is typically on the order of $K_V \sim 0.1 \cdot 10^{-4}\,\mathrm{m^2\,s^{-1}}$ (Ledwell e*t al.*, 1993; Toole *et al.*, 1994; St. Laurent and Schmitt, 1999), which falls significantly below Munk's canonical value (11.2). Fortunately, these findings did not curb the interest in small-scale mixing. By the nineties, microscale oceanography had expanded into an independent and vibrant field. Numerous new applications were found, including mixing of nutrients, carbon sequestration, dynamics of the bottom waters and naval functions. However, the original motivation of the mixing program as a support system for the thermocline theory had to be revised and early expectations downscaled.

On the theoretical level, recent years have witnessed a general retreat from the predominantly diffusive thermocline models to the adiabatic view, ascribing thermocline structure to the ventilation of water masses at the sea-surface. Diapycnal thermocline mixing was shown to be essential in isolated, dynamically distinct regions, such as the diffusive internal thermocline front (Samelson and Vallis, 1997; Vallis, 2000), but it plays a secondary role in much of the upper ocean. It was also argued that the MOC can be driven by adiabatic mechanisms (Toggweiler and Samuels, 1998; Marshall and Radko, 2003; Radko and Kamenkovich, 2011) associated with inter-hemispheric asymmetries in the ocean/landmass distribution. Webb and Suginohara (2001) estimate that only a third of the volume of deep waters originating in high-latitude North Atlantic regions is upwelled in the interior in response to diabatic mixing – the rest is brought to the surface in the Southern Ocean by Ekman suction.

In view of these findings, it is not too surprising that the geometry of the model ocean can constrain the global influence of double-diffusion. Let us first examine a few closed-basin models. Figure 11.1 shows a set of numerical simulations of the large-scale circulation in the Northern-Hemisphere basin (Gargett and Holloway, 1992). The upper panel presents the overturning streamfunction in a salt-finger experiment. This calculation was performed using uniform but unequal diffusivities of heat and salt ($r = K_S/K_T = 2$), which is a fairly reasonable, if somewhat crude, parameterization of finger-induced mixing. The outcome of this experiment is nothing short of astounding. The flow structure differs dramatically from the classical picture of an overturning cell, characterized by localized high-latitude sinking and broad upwelling. The central panel in Figure 11.1 shows the circulation pattern realized in the simulation with equal heat/salt diffusivities, which is meant to represent conventional turbulence. The magnitudes of the overturning in finger-driven and turbulence-driven experiments differ by an order of magnitude and even the sense of circulation is reversed. The reduction of r below unity (lower panel in Fig. 11.1) leads to more limited, but still clearly visible, modification of the turbulent flow pattern. The dramatic variation of the overturning patterns in Figure 11.1 is an indication of the sensitivity of large scales to the ratio $r = K_S/K_T$, which appears to be even more important than the magnitude of diffusion. In this regard, it should be kept in mind that, among various mixing mechanisms, salt fingering is unique in its ability to raise r above unity.

Qualitative changes in the variability and patterns of thermohaline circulation associated with the inclusion of salt fingers have been noted in many other modeling studies. Ruddick and Zhang (1989) examined box models of the ocean circulation driven by heating and evaporation at the sea-surface. They discovered, for instance, that self-sustained oscillations, common in the classical turbulent mixing models (Stommel, 1986; Welander, 1982) do not arise for finger-driven mixing. Gargett

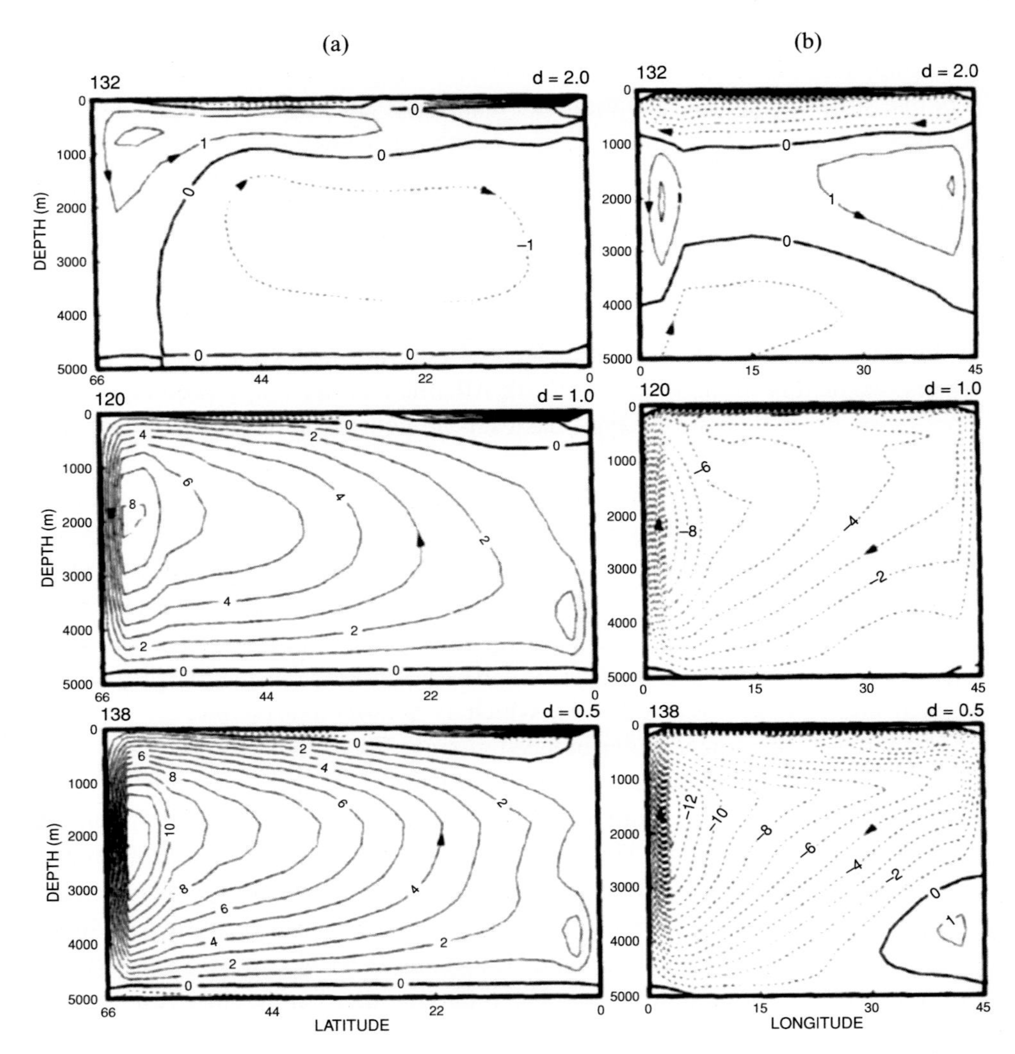

Figure 11.1 Meridional (a) and zonal (b) overturning streamfunctions for experiments with various values of $r = K_S/K_T$. Upper panels $r = 2$, middle panels $r = 1$, lower panels $r = 0.5$. From Gargett and Holloway (1992).

and Ferron (1996) examined multi-box models of thermohaline circulation and found that use of unequal diffusivities of heat and salt substantially modified both the equilibrium solutions and the transition points (in terms of forcing magnitude) between distinct states. An important step was made by Zhang *et al.* (1998), who incorporated a more realistic R_ρ-dependent double-diffusive parameterization into a numerical model. While in other respects the setup was similar to that of Gargett and Holloway (1992; Fig. 11.1), the more physical parameterization led to more

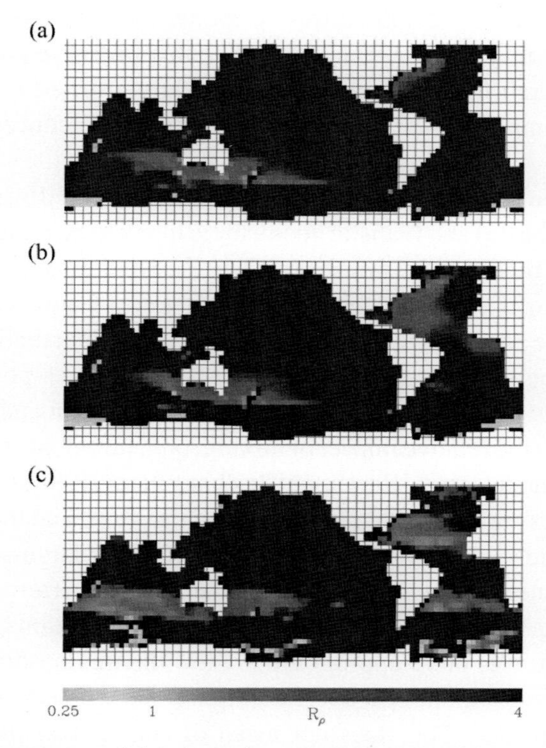

Figure 11.2 The pattern of density ratio at the 700 m depth in (a) the non-double-diffusive experiment, (b) double-diffusive experiment and (c) the annual-mean climatology. From Merryfield *et al.* (1999). See color plates section.

credible solutions. The effects of salt fingers were still pronounced, as reflected by a 22% reduction in the strength of the MOC. However, the flow structure was quite different from Gargett and Holloway's prediction: inclusion of salt fingers was not accompanied by the cardinal and implausible reorganization of the entire overturning circulation pattern.

In global models with realistic geometry and surface forcing, the effects of double-diffusion are even less prominent but still detectable (Merryfield *et al.*, 1999; Glessmer *et al.*, 2008). The inclusion of salt fingers brought modeled temperature and salinity closer to observations, increasing their values in the lower thermocline and in the deep ocean. Figure 11.2 shows the distribution of the density ratio at 700 m depth for the model runs (Merryfield *et al.*, 1999) in which double-diffusion was excluded (a) and included (b). Somewhat counterintuitively, taking double-diffusion into account created more favorable conditions for its occurrence (i.e., lower density ratio) particularly in the Atlantic. The distribution of R_ρ in the double-diffusive model (Fig. 11.2b) matched observations (Fig. 11.2c) much better than the non-double-diffusive pattern (Fig. 11.2a). However, the overall flow pattern was only slightly affected by double-diffusion and all major components of thermohaline circulation reduced by just a small percentage.

The cause of the dissimilar large-scale impact of double-diffusion in models with idealized (single-basin) and realistic geometry is likely to be related to the influence of the Antarctic Circumpolar Current (ACC) in the Southern Ocean. The absence of land barriers along its path results in distinct dynamical features that have no direct counterparts in the zonally blocked oceans regions (e.g., Marshall and Radko, 2003, 2006). The unique dynamics of the reentrant ACC engages powerful and fundamentally adiabatic mechanisms controlling the stratification and overturning. This inevitably reduces the relative impact of mixing, both turbulent and finger-driven. In the single-basin models (e.g., Bryan, 1987; Zhang *et al.*, 1998) diapycnal mixing is essential. It carries the full burden of supporting the MOC and therefore changes in the magnitude and type of mixing project directly onto large-scale patterns. It is possible to eliminate or even to reverse the sense of overturning circulation by simply modifying the mixing model. In realistic configurations, where alternative mechanisms exist, it is more appropriate to view the mixing-driven circulation as a correction – a substantial one but a correction nonetheless – to stronger adiabatically driven modes. This does not mean of course, that the diabatic interior processes can be discounted. Given the involvement of thermohaline circulation in issues of societal relevance, such as climate and ecosystem dynamics, its diffusively driven component should be better understood and better represented in models. However, in doing so, it is imperative to have a clear dynamical picture and realistic expectations for the interior mixing.

In view of apparent differences in model-based predictions of the impact of double-diffusion, a question could be raised whether much can be learnt from numerous idealized models (e.g., Gargett and Holloway, 1992) – models that focus on a selected subset of processes and ignore others in order to maintain physical transparency. I believe that the answer is yes and the reasons are three-fold. First of all, the general tendencies revealed and explained by the idealized models are invariably reflected in more realistic and comprehensive configurations (e.g., Merryfield *et al.*, 1999), albeit in a much reduced and harder-to-identify form. The second reason is related to the assessment of the relative roles of double-diffusion and turbulent mixing. These two types of microstructure affect large-scale patterns very differently and the dynamic dissimilarities can be interpreted more clearly when mixing effects are isolated. Finally, it should be realized that the integral measures and general circulation patterns tend to mask interesting regional effects. In areas where mixing is elevated and/or adiabatic advection is weak, flow can be governed by predominantly diffusive dynamics, brought to the fore by idealized mixing models.

Overall, the extant modeling studies suggest a detectable but moderate impact of salt fingers on the density stratification and global thermohaline circulation. This, however, does not preclude the possibility that double-diffusion could play a significant role in the context of global change (e.g., Oschlies *et al.*, 2003). Unlike small-scale turbulence and mesoscale variability, which are controlled by density distribution, the intensity of double-diffusion depends very strongly on temperature and salinity patterns. Therefore, the response of double-diffusion to climatic change can be qualitatively different from that of other mixing processes. Salt fingers are likely to introduce new powerful feedback mechanisms into the ecosystem dynamics, which should be – but seldom are – taken into account in the projections of future climate. Another potentially significant caveat was noted by McDougall (1987). He argued that effects associated with the nonlinearity of the equation of state can amplify double-diffusive transformation of water masses by as much as a factor of four. McDougall's proposition warrants further analysis, validation and, possibly, reevaluation of the large-scale impact of salt fingers. On this intriguing and unorthodox suggestion we leave the thermohaline circulation problem and proceed to discuss more subtle implications of double-diffusive convection.

The T–S relation and the pattern of density ratio

Some of the oldest and most fundamental questions in classical ocean dynamics are related to the form of the *T–S* relation in the central thermocline. Two aspects should be emphasized: (i) its remarkable tightness and (ii) the presence of broad regions with uniform density ratio in areas susceptible to fingering. Figure 11.3 shows representative vertical profiles of the density ratio in the central waters of the North Atlantic and North Pacific. In the Atlantic, which is characterized by lower density ratios and therefore more active salt fingering, regions of uniform density ratio are more pronounced and vertically extended. Figure 11.3 also presents a histogram of the Atlantic density ratios, which exhibits a well-defined peak at $R_\rho \approx 2$ and a sharp reduction in probability at lower density ratios. These observations immediately raise a set of related dynamical questions. What mechanisms are responsible for aligning the *T–S* relation along the density ratio curve? Why is the density ratio effectively precluded from drifting significantly below $R_\rho \approx 2$ and what sets this very specific limit?

So far, the most promising attempts to address these questions have involved salt fingers (Schmitt, 1981, 1990). Since the intensity of double-diffusion rapidly increases with decreasing R_ρ, regions with anomalously low values of density ratio

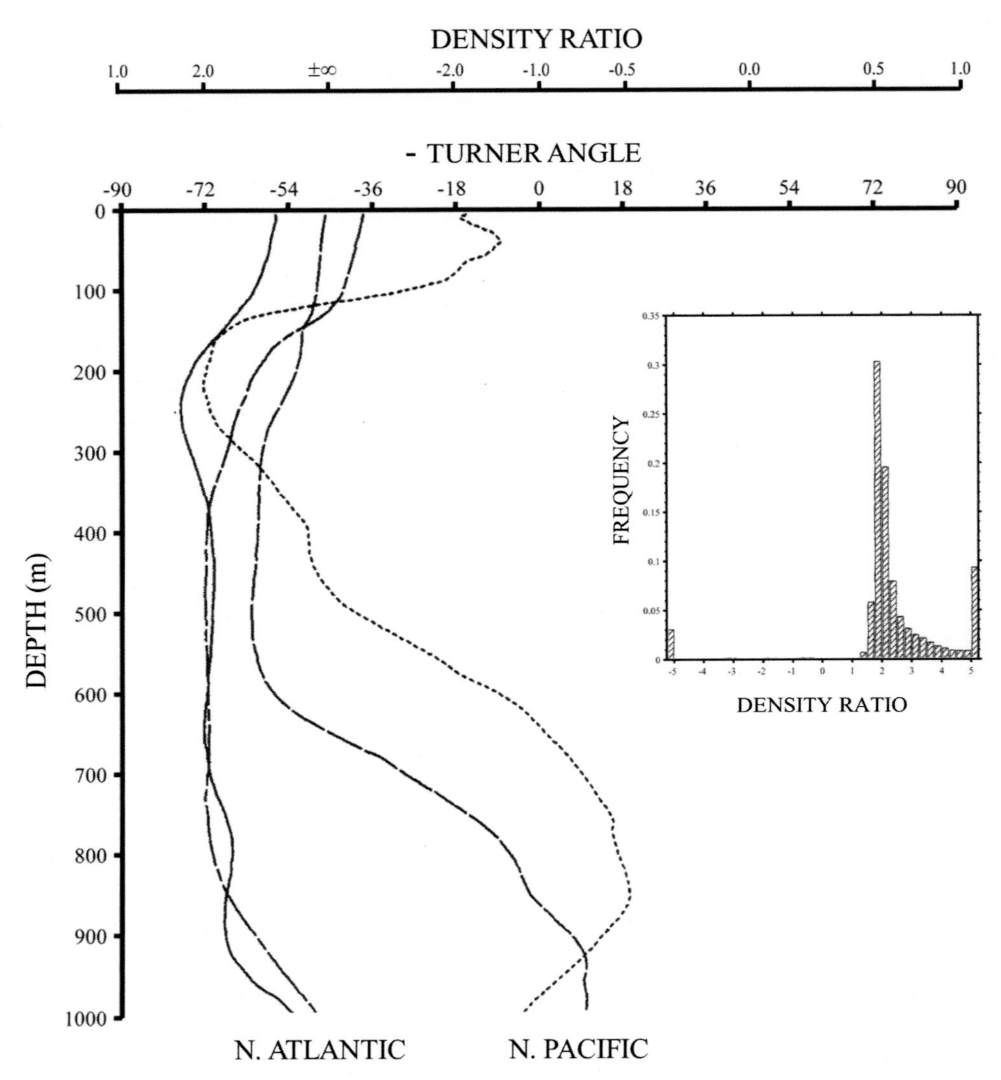

Figure 11.3 Density ratio computed over 100 m vertical intervals for four CTD stations in the eastern North Atlantic (solid), western North Atlantic (short-long dash), eastern North Pacific (short dash) and western North Pacific (dotted). The insert shows the histogram of occurrence of R_ρ in the Atlantic thermocline along 24° N. From Schmitt (1990).

inevitably become the sites of intense mixing. Fingers transport salinity faster than heat. As a result, they tend to reduce the salinity gradient more efficiently than the temperature gradient and thereby increase the density ratio $R_\rho = \frac{\alpha \bar{T}_z}{\beta \bar{S}_z}$. Thus, it could be argued that the density ratio is set by a balance between double-diffusive mixing acting to increase the density ratio and the large-scale forcing counteracting this tendency. It is generally accepted that salt fingers are most effective in modifying the large-scale *T–S* patterns for density ratios less than two. As discussed in Chapter 10, double-diffusive fluxes in the ocean reduce precipitously at $R_\rho > 2$, where fingers become weaker and more vulnerable to the adverse action of internal waves and intermittent turbulence. The nearly discontinuous on/off character of finger-induced mixing is essential; it implies that as soon as R_ρ reduces below the critical value of two, salt fingers get to work on increasing the density ratio, but they effectively disengage when $R_\rho > 2$. In order to produce the permanent low density ratio ($R_\rho < 2$) conditions, external forcing – which may be associated either with sea-surface heat/salinity fluxes or large-scale advection – needs to be strong enough to break the resistance of these tiny regulators. Of course, there are several locations in the world ocean where it does happen. In the Atlantic, the density ratio drops significantly below the critical value in the Mediterranean outflow region and in the Caribbean (C-SALT area), but these conditions are rather exceptional. Over most of the central thermocline, the salt-finger regulation of density ratio is highly effective.

To be completely fair, we should mention the possibility of non-double-diffusive regulation of the *T–S* relation. Stommel (1993) hypothesized that the density ratio could be constrained by the response of the upper mixed layer to atmospheric forcing. His temperature–salinity regulator model suggested the preferred horizontal mixed layer density ratio of two. In the ventilated regions of the thermocline, mixed layer properties are advected downward along the sloping isopycnals. This way, the surface density ratio can project onto its vertical distribution, which would explain the observed interior *T–S* pattern. What seriously undermines this argument – or, for that matter, any argument based on subsurface dynamics – is that the temperature and salinity patterns in the mixed layer are often characterized by density compensation on a wide range of scales (Rudnick and Ferrari, 1999). The projection of such patterns onto the vertical *T–S* profiles would produce a series of density-compensated segments with the density ratio close to unity. The absence of such features in most observations implies the existence of an internal mixing process acting to homogenize the density ratio.

It is interesting that it is possible to discriminate between the effects of turbulence and fingering on the *T–S* relation by exploring the nonlinearity of the equation of state (Schmitt, 1981). Since the expansion/contraction coefficients (α, β) in the ocean are non-uniform, curves of constant density ratio $R_\rho = \frac{\alpha T_z}{\beta S_z}$ in the *T–S* parameter space are distinct from the straight lines representing the constant gradient ratio $R_g = \frac{T_z}{S_z}$. Turbulence mixes temperature and salinity at equal rates, and therefore its effect on the *T–S* relation would be reflected in the tendency to homogenize the gradient ratio R_g. Salt fingers are different. The intensity of fingering is controlled by the density ratio and therefore they tend to align *T–S* values along the uniform R_ρ curves. An inspection of oceanographic data (Ingram, 1966; Schmitt, 1981) leaves no doubt with regard to the preferred pattern of the *T–S* relation. Observations (Fig. 11.4a) are obviously much better described by the uniform density ratio model. In fact, the standard error of the linear fit to the measured *T–S* values exceeds that of the $R_\rho = const$ model by an order of magnitude.

To illustrate the mechanics of density ratio regulation by salt fingers, Schmitt (1981) numerically integrated the one-dimensional (z) large-scale temperature and salinity equations (8.8) in time. The vertical fluxes were parameterized using the flux-gradient laws appropriate for salt fingers. Typical evolutionary patterns of this system are shown in Figures 11.4b,c. The experiment in Figure 11.4b was initiated by a uniformly low density ratio ($R_\rho = 1.1$), which evolved, within a few years, to the more typical value $R_\rho \approx 2$. The dotted line in Figure 11.4b represents the evolution of the density ratio pattern in a similar model for turbulent rather than finger-driven mixing, demonstrating just how ineffective turbulence is in terms of removing the density ratio anomalies. The experiment in Figure 11.4c was initiated by the variable density ratio, which in time developed extended quasi-uniform regions. Once again, the anomalously low values of R_ρ drifted towards the pervasive value of two.

The obvious limitation of Schmitt's (1981) model is related to the questionable ability of any one-dimensional system to reflect the dynamics of oceanic circulation. However, it was later shown that the key ideas of the model are sufficiently general. Using a fully three-dimensional framework, Schmitt (1990) went on to demonstrate that the $R_\rho = const$ character of the central waters can be plausibly attributed to double-diffusion but is inconsistent with either turbulent diapycnal or isopycnal mixing.

Another compelling argument in support of the double-diffusive control of the *T–S* relation is based on its tightness. The ocean is full of mesoscale eddies that actively stir the adjacent water masses predominantly along isopycnal surfaces. It is intuitively clear that the variation in eddy-transfer rates at different isopycnals is bound to produce strong vertical *T–S* variability, unless this tendency is balanced by some active internal mixing process. While such density-compensated variability has been detected in observations (Shcherbina *et al.*, 2009; Ferrari and Polzin, 2005), it is surprisingly limited in most of the ocean, particularly at large vertical scales (Fig. 11.3) exceeding that of thermohaline intrusions ($\sim$10 m). The density-compensated variability is most conveniently represented by "spiciness" – the quantity orthogonal to density in terms of its thermal and haline components – see (7.25). While mesoscale eddies and surface forcing are the obvious sources of spiciness, double-diffusion might be its major consumer (Schmitt, 1999). This expectation is consistent with the generic bias of large-scale non-double-diffusive numerical models towards unrealistic growth of spiciness (McWilliams, 1998).

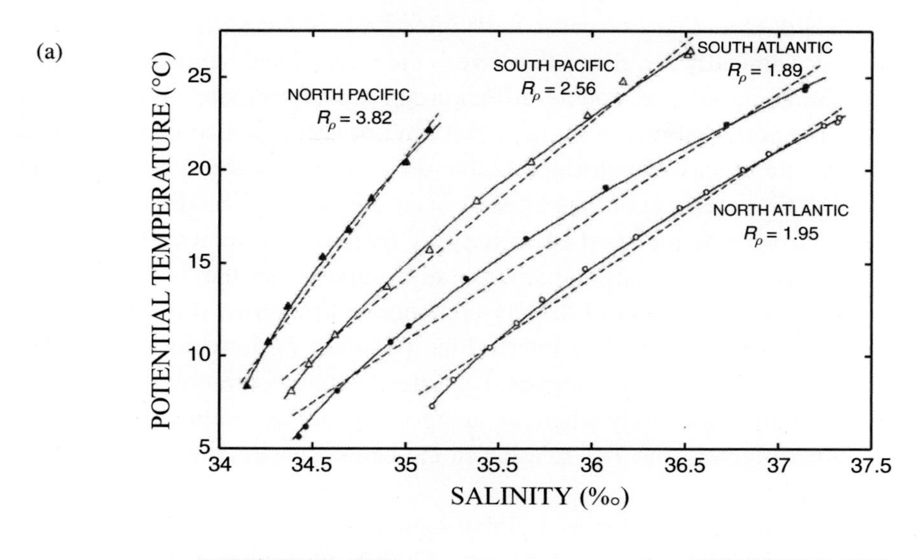

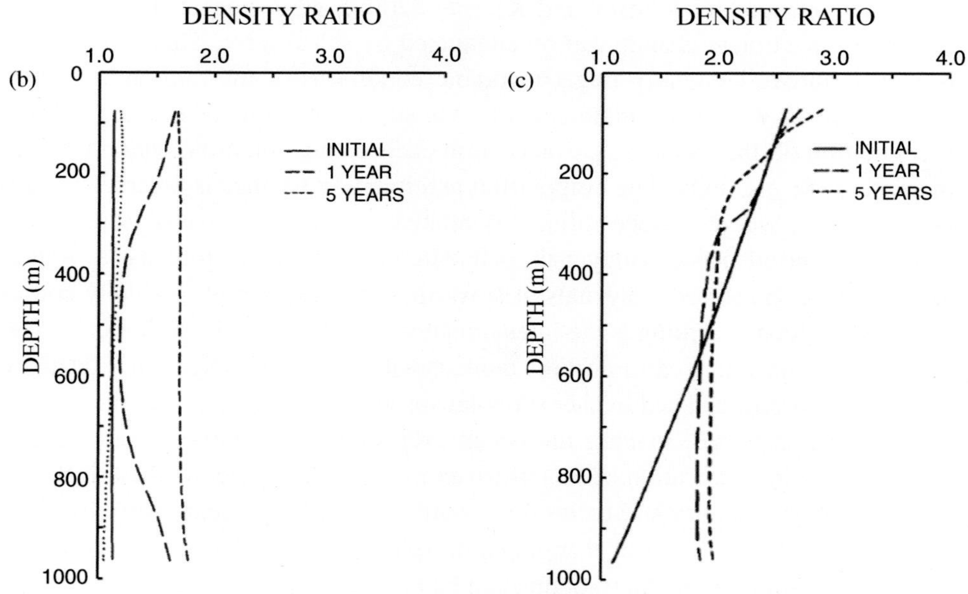

Figure 11.4 (a) Potential temperature versus salinity plot for four hydrographic stations in the Atlantic and Pacific. The linear fit (dashed) and a curve of constant density ratio (solid) are shown for each station. (b) Solution of the one-dimensional model. The density ratio pattern is shown initially (solid curve), after one year (long dash) and after five years (short dash). The profile after five years of turbulent mixing is indicated by a dotted curve. (c) The same as in (b) but for the non-uniform initial distribution of the density ratio. From Schmitt (1981).

What makes double-diffusion particularly effective in reducing ocean spiciness is its ability to dynamically discriminate between the thermal and haline components of density. Unlike turbulence, double-diffusion can actively respond, by varying its intensity, to the perturbations in T and S that are not accompanied by changes in density. There are at least two distinct double-diffusive mechanisms for removing spiciness. The first mechanism has already been alluded to. The direct effect of the preferential vertical transport of salinity by fingers is to remove the density ratio anomalies, which heavily constrains the excursions of the T–S curve and thereby suppresses the growth of density-compensated perturbations. The second mechanism involves thermohaline interleaving (Chapter 7). Intrusions are driven by – and act to diffuse – the isopycnal gradients of temperature and salinity. Active interleaving ceases only when the along-isopycnal variability is effectively removed, which tightens the T–S relation in the central waters, turning the spicy ocean into a mild one.

The view of double-diffusion as a global regulator of the T–S distribution was recently reinforced by Johnson and Kearney (2009), who suggested that oceanic signatures of climate change can be attenuated by salt fingers. Their proposition is very intuitive. Generally finger-favorable conditions of the mid-latitude thermoclines in all oceans are maintained by the surface patterns of evaporation and precipitation. In the subtropics, evaporation exceeds precipitation, which tends to increase surface salinity. The evaporation/precipitation balance is reversed in subpolar regions, where surface salinity is relatively low. Thus, water is relatively fresh and cold at denser isopycnals that outcrop in subpolar regions; at lighter and therefore shallower isopycnals, it is warm and salty. One of the likely consequences of global warming is the intensification of the Earth's hydrological cycle. A warmer atmosphere can transport more moisture meridionally, which leads to even saltier subtropical and fresher subpolar surface waters. As a result, the salinity stratification increases, making the ocean even more susceptible to salt fingers. However, the intensification of finger-driven mixing can accelerate the transfer of temperature and salinity anomalies downward into the deep ocean, partially compensating the direct effects of changing atmospheric forcing. This way, the imprint of global warming in the thermocline can be reduced. Johnson and Kearney warn of the potential dangers of ignoring the salt-finger attenuation: climate studies that focus on the variability of the upper ocean and that do not take double-diffusion into account could significantly underestimate the ocean change brought about by global warming.

The hypothesis of attenuation has been supported by the analysis of data from the repeated trans-Indian Ocean surveys along 32° S from 1987 to 2009. The Indian Ocean becomes progressively more susceptible to fingering from west to east. Johnson and Kearney (2009) argued that this pattern is responsible for qualitative differences in the evolution of the *T–S* anomalies in the eastern, central and western sectors of the basin. In the western sector, the density ratio systematically decreased and anomalies spread into deeper regions of the thermocline, which can be viewed as the natural response of the ocean that is only slightly affected by double-diffusion. In the central segment – the region characterized by moderate fingering – the downward spread of anomalies was still noticeable but the density ratio values did not change significantly. Finally, in the east, where fingering is always most intense, the *T–S* distribution remained largely invariant in time. These evolutionary patterns were interpreted as a manifestation of spatially non-uniform (more effective from west to east) moderation of changes in the water-mass properties by double-diffusion. The observed temperature and salinity tendencies were consistent with the prediction based on the vertical convergence of parameterized salt-finger fluxes, lending further credence to the idea of the finger-induced attenuation of the climate signal.

Regional effects

While the global impact of double-diffusion is difficult to evaluate and extant estimates are much debated, no serious doubts have been expressed with regard to the ability of salt fingers to profoundly influence regional climate and dynamics. The intensity of double-diffusive mixing is spatially inhomogeneous, with thermohaline staircases being the obvious mixing hot spots. A case in point is the Caribbean staircase. A fairly detailed description of this staircase and its transport characteristics resulted from major field programs, including C-SALT and the Salt Finger Tracer Release Experiment (SFTRE). The diffusivities of heat and salt in this area ($K_T \approx 0.45 \cdot 10^{-4}\,\mathrm{m^2\,s^{-1}}$ and $K_S \approx 0.9 \cdot 10^{-4}\,\mathrm{m^2\,s^{-1}}$) approach Munk's canonical value (11.2), which implies that mixing is sufficiently intense to effectively control the local flow patterns. One of the critical dynamical consequences of the elevated mixing rates in the C-SALT area is the injection of salinity into the Antarctic Intermediate Water, which preconditions waters for later sinking in the high-latitude Atlantic, thus affecting thermohaline circulation.

In contrast to the mixing environment of C-SALT, smooth-gradient regions are characterized by very low diffusivities ($\sim 10^{-5}\,\mathrm{m^2\,s^{-1}}$ or less). This disparity puts the C-SALT staircase into a league of its own in terms of its ability to affect water-mass transformation in the North Atlantic thermocline. An instructive illustration of the mixing intensity of the C-SALT staircase is based on the comparison of the staircase-driven and turbulent mixing (Ray Schmitt, private communication). The net salt flux in the C-SALT staircase can be estimated as a product of the salt diffusivity, C-SALT area and the vertical salinity gradient, resulting in

$$F_{\text{C-SALT}} \sim (0.9 \cdot 10^{-4}\,\text{m}^2\,\text{s}^{-1}) \times (1.3 \cdot 10^{12}\,\text{m}^2) \times (3 \cdot 10^{-3}\,\text{psu}\,\text{m}^{-1}) \\ \sim 3.5 \cdot 10^6\,\text{psu}\,\text{m}^3\,\text{s}^{-1}. \quad (11.4)$$

A similar estimate can be made for the turbulent (wave-induced) mixing over the whole area of the North Atlantic subtropical gyre:

$$F_{\text{turb}} \sim (0.05 \cdot 10^{-4}\,\text{m}^2\,\text{s}^{-1}) \times (20 \cdot 10^{12}\,\text{m}^2) \times (1.5 \cdot 10^{-3}\,\text{psu}\,\text{m}^{-1}) \\ \sim 1.5 \cdot 10^6\,\text{psu}\,\text{m}^3\,\text{s}^{-1}. \quad (11.5)$$

The comparison of (11.4) and (11.5) is astonishing. The salt flux over a relatively small C-SALT region, less than a tenth of the North Atlantic subtropics, significantly exceeds the net turbulent transport over the whole gyre. What the C-SALT staircase lacks in area, it more than makes up for in mixing intensity. The turbulent diffusivity inferred from mixing parameterizations of the wave field in the thermocline ($\sim 0.05 \cdot 10^{-4}\,\text{m}^2\,\text{s}^{-1}$) is simply incommensurate with the staircase diffusivity suggested by tracer release measurements ($\sim 0.9 \cdot 10^{-4}\,\text{m}^2\,\text{s}^{-1}$). Note also that most of the subtropical North Atlantic thermocline is susceptible to both double-diffusion and turbulence; in regions lacking staircases their mixing contributions are comparable. But even if we generously ignore the finger-driven transport everywhere outside of the C-SALT staircase, double-diffusion still beats turbulence hands down in terms of water-mass transformation.

Another example of the profound impact of double-diffusion on regional dynamics is given by the Tyrrhenian staircase in the western Mediterranean (Chapter 7). Its extraordinarily large steps (up to 500 m) and low values of the density ratio ($R_\rho \sim 1.2$) create favorable conditions for some of the most vigorous finger-driven mixing in the world ocean. The vertical transport of heat and salt by the Tyrrhenian staircase is recognized as the dominant mechanism for the exchange of properties between the Lewantine Intermediate Water and the Mediterranean Deep Water (Zodiatis and Gasparini, 1996). Because of its interactive character and ability to adjust to evolving environmental conditions, the Tyrrhenian staircase could prove to play a major role in controlling the Mediterranean climate. As the upper Mediterranean waters become warmer and saltier, the staircase responds by systematically increasing the layer heights and *T–S* variation across the interfaces. The increase in step size is accompanied by a corresponding intensification of mixing, which, in turn, acts to moderate climatic changes in the stratification of the Tyrrhenian Sea.

While permanent staircases are the most obvious candidates for the substantial modification of water masses, regional changes caused by double-diffusion have also been detected in smooth-gradient regions. Double-diffusion is a major mixing process between the intruding layers of subpolar and subtropical waters. Talley and Yun (2001) examined water-mass transformation in the Mixed Water Region of the North Pacific, where the Oyashio waters mix with more saline Kuroshio waters east of Japan. They showed that double-diffusion (mostly salt fingering) significantly affects the Oyashio water-mass properties and increases the density of the salinity minimum at the top of the North Pacific Intermediate Water. A rare example of an analysis of the role of fingering in the dynamics of deep/bottom water masses was presented by McDougall and Whitehead (1984). In this study, salt fingering was invoked to explain the variation in the Antarctic Bottom Water properties in the Atlantic.

Several interesting case studies of finger-induced large-scale effects focus on the so-called mode waters. The term mode water describes water masses with exceptionally uniform properties, usually formed during wintertime convection in regions of strong heat loss from the ocean to the atmosphere. The role of double-diffusion in the dynamics of South Pacific Eastern Subtropical Mode Water was explored in detail by Johnson (2006). Johnson's analysis was based on the observed evolution of large positive isopycnal T–S anomalies produced by unusually strong air–sea heat fluxes in austral winter 2004. The warm and salty water was then advected along the isopycnal surfaces. Within six months after injection, the magnitude of the perturbation was halved on the isopycnals at which it was introduced, but increased at higher densities. Johnson suggested that the observed evolution of the upper-ocean structure can be largely attributed to finger-driven mixing, which intensified when the T–S anomalies spread into the main thermocline. The model based on the parameterization of finger-driven mixing successfully reproduced the spatial and evolutionary patterns of both temperature and salinity, whereas a turbulence-based parameterization led to inconsistent predictions. Similarly, double-diffusion was shown to cause the rapid downstream modification of certain types of mode waters in the North Pacific (Saito *et al.*, 2011; Toyama and Suga, 2012). Salt fingers have also been implicated in the downstream variability of the Atlantic mode waters, transferring the excess salt, and to a lesser extent temperature, to deep layers (e.g., Gordon, 1981; Tsuchiya, 1986).

Biogeochemical applications

A different and often overlooked application of double-diffusion comes from a somewhat unexpected source – biological oceanography. Biologists first set their sights on salt fingers after a controversial attempt by Lewis *et al.* (1986) to use microstructure measurements to estimate the rate of supply of nutrients to the euphotic zone in the eastern subtropical North Atlantic. The inferred nitrate fluxes were about six times too small to balance the biological uptake of nitrate measured by incubation techniques and were also much lower than the estimates based on observed changes in upper-ocean oxygen over annual scales.

The apparent inconsistency between the microstructure-based estimates and the standard biogeochemical techniques demanded urgent resolution. By that time, the problem of evaluating the nitrate flux had already transcended the conventional boundaries of chemical and biological oceanography and entered into the discussion of climate change. The nitrate budget of the upper ocean affects photosynthetic incorporation of carbon dioxide and constrains the export of organic carbon from the ocean surface layer, which is the primary oceanic sink for atmospheric carbon dioxide. Thus, the uncertainty in the nitrate transport affects our ability to model carbon dioxide concentrations in the atmosphere and ultimately hinders climate prediction efforts.

A plausible resolution of the nitrate flux conundrum was proposed by Hamilton *et al.* (1989). These authors noted that the observations of Lewis *et al.* (1986) were taken in an ocean region susceptible to active salt fingering. In estimating the vertical flux from microstructure measurements of viscous dissipation, Lewis *et al.* used the pure-turbulence model and made no attempt to account for fingers. However, as we discussed in Section 10.4, the relationships between fluxes and kinetic energy dissipation are very different for finger-dominated and turbulence-dominated environments; use of an inappropriate diagnostic model inevitably leads to major errors.

In order to constrain the range of fluxes potentially supported by the dissipation measurements, Hamilton and collaborators examined the opposite limit of pure finger-driven mixing. They assumed that the eddy diffusivity of nitrate is equal to salt diffusivity and predicted a nitrate flux that was six times as large as previously inferred, for the same data, using the turbulence model. The finger-based calculation was in much better agreement with the incubation and geochemical estimates. The success of the finger model, combined with the failure of the turbulent one, suggested that the origin of microstructure analyzed by Lewis *et al.* was indeed double-diffusive. This proposition is quite sensible, given that the density ratio at the location of their measurements was consistently less than two. Admittedly, the finger model can only predict the upper limit for the vertical transport of properties. A more precise estimate of fluxes from viscous dissipation data requires a-priori knowledge of the microstructure composition. Nevertheless, the results reported by Hamilton *et al.* were highly significant. For the first time, the biogeochemical community realized the importance of discriminating between the two major mixing processes and the potential of double-diffusion to dramatically elevate biological productivity in the upper ocean.

Hamilton's ideas instigated a series of inquiries, data-based and modeling, into the contributions of salt fingers and turbulence to the vertical transport of nutrients into the euphotic zone. Dietze *et al.* (2004) analyzed hydrographic measurements taken along the meridional section at 30° W in the subtropical North Atlantic. Turbulent diffusivities were deduced from the finescale shear data according to the mixing parameterizations of the internal wave field (Gregg, 1989). For salt fingers, Dietze *et al.* used the density ratio based model of Zhang *et al.* (1998). The results unequivocally demonstrated the critical role of finger-induced mixing for ecosystem dynamics. Transport of nitrates into the nutrient-consuming layer by salt fingering was more than five times higher than transport due to wave-induced turbulence and substantially exceeded the combined effects of horizontal diffusion, eddy pumping and advection (vertical and horizontal).

Modeling studies have also been highly effective in illustrating the biogeochemical consequences of fingering. One of the perceived weaknesses of data-based analyses is related to their regional emphasis. Since most such studies focus on a specific location or a hydrographic section, the possibility exists that the inferences they provide are not representative of general oceanic conditions. In this regard, modeling can nicely complement and reinforce the observational message by providing net, basin-integrated estimates of nutrient fluxes and quantifying their spatial variability. An example is given by Oschlies *et al.* (2003), who used a coupled circulation–ecosystem model to quantify the contribution of finger-induced mixing to nutrient fluxes in the North Atlantic. The domain-averaged contribution of salt fingers was necessarily less than in the selected particularly finger-favorable locations (e.g., Dietze *et al.*, 2004) but still essential. Taking finger-induced mixing into account approximately doubled the flux of nitrates in the eastern part of the oligotrophic subtropical gyre. It was shown that the nutrient supply by fingering is at least comparable to the contribution from mesoscale eddies which, so far, have received considerably more attention (McGillicuddy *et al.*, 1998).

An even more definitive analysis of the double-diffusive influence on biogeochemical cycles was performed by Glessmer *et al.* (2008). Using a global ocean model, these authors argued that, despite the relatively modest impact of double-diffusion on upper-ocean temperatures and salinities, its effects on marine biology and biogeochemistry are dramatic and widespread. Taking salt fingers into account increased nutrient supply to the upper ocean, resulting in the enhancement of primary production by up to 100% over broad areas of the subtropical gyres. Higher biological productivity, in turn, led to substantial changes in the carbon dioxide and oxygen uptake of the ocean. To test the sensitivity of their simulations to the assumed parameterizations, Glessmer *et al.* considered two different double-diffusive flux laws (Zhang *et al.*, 1998; Large *et al.*, 1994). The differences in simulated nutrient supply between the two versions of the model turned out to be smaller than the differences caused by inclusion of salt fingers, which to some extent validated qualitative inferences from the model.

Although the inclusion of double-diffusive mixing clearly improved the correspondence with biogeochemical data for the upper ocean, particularly when used in conjunction with the Zhang *et al.* (1998) parameterization, Glessmer's model still underestimated the nitrate concentration in the subtropics and tropics. The residual error could be, at least partially, attributable to the uncertainties in the double-diffusive parameterizations. The key complication here – and the common thread running through all biogeochemical studies involving double-diffusion – is that even the smallest changes in mixing can have a substantial impact on the organic composition of the upper ocean. Such sensitivity reflects a highly nonlinear response of biology to the variation in nutrient supply, motivating future efforts to better understand and better parameterize double-diffusion.

Salt fountains in the ocean

To complement the discussion of biogeochemical applications, it is worth touching upon one more issue – an issue that most readers would probably classify now as entertaining rather than hard-core scientific or even serious, but things could change. Given the benefits of double-diffusion for ocean biota and our dependence on fisheries, the question arises whether the principles of double-diffusion could be exploited to artificially increase ocean productivity. It is becoming increasingly clear that fish supply is unlikely to keep up with growing demand without an intervention of some sort. However, if we give the ocean a helping hand by installing artificial double-diffusive systems, ameliorating the low productivity zones of the world ocean, then the problem could be alleviated. This proposition appears to be the stuff of sci-fi novels, but is it that unfeasible?

Inspired by the landmark discovery of "an oceanic curiosity: the perpetual salt fountain" (Stommel *et al.*, 1956), a group of Japanese researchers (Maruyama *et al.*, 2004) performed a curious experiment of their own. In August of 2002, they deployed a 300-meter long free-floating pipe (Fig. 11.5) in the main thermocline (60–360 m depth range) west of the Mariana Islands in the tropical Pacific, a region characterized by finger-favorable stratification. The purpose of the experiment was to explore the potential of salt fountains for fish-farming (something that Stommel *et al.* described as an "improbable application"). The salt fountain can draw up deep nutrient-rich water without an external energy source and thereby fertilize the euphotic zone with minimal investment of resources. Unlike salt fingers, which are disorganized and vulnerable to adverse ambient forcing (Chapter 10), the flow in the pipe is unidirectional and unimpeded by oceanic shears and turbulence. Therefore, the salt fountain mechanism is expected to be much more effective than fingering in transporting nutrients into the near-surface layer. An associated proposal has been made for the cultivation of ocean deserts by an array of salt fountains (Fig. 11.5a).

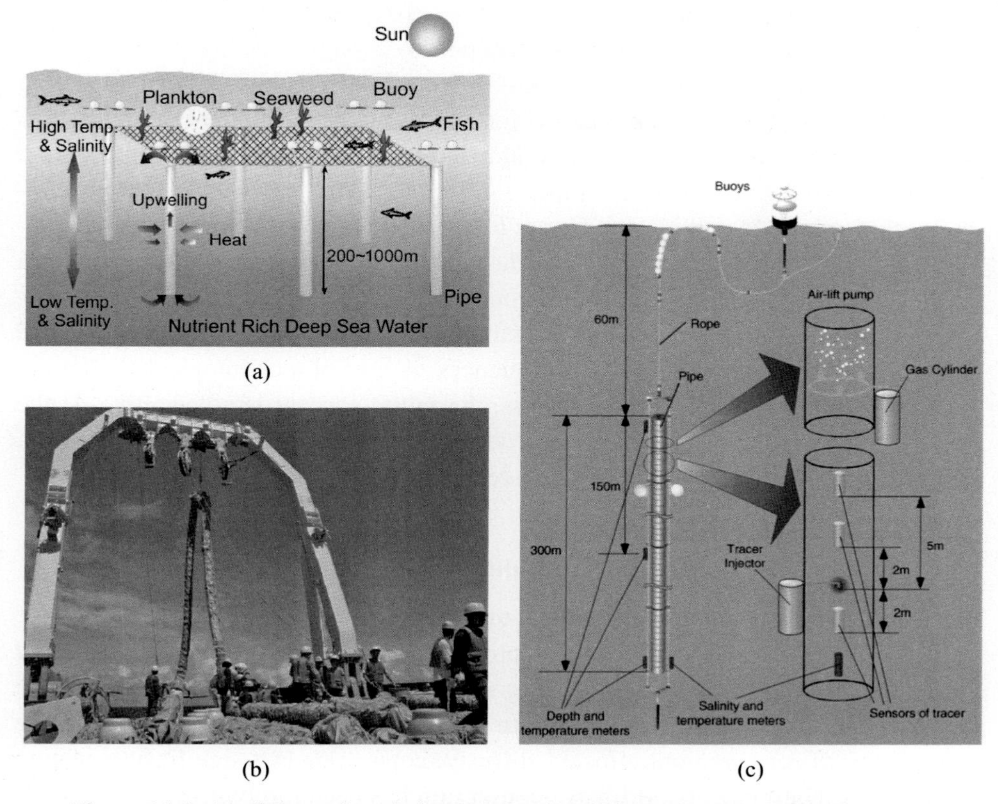

Figure 11.5. (a) Schematic of the proposed salt-fountain ocean farm (Laputa project). From Maruyama *et al.* (2004). (b) Deployment of the salt fountain. (c) Diagram of the experimental apparatus. From Tsubaki *et al.* (2007).

In a neat reference to a floating island in Jonathan Swift's *Gulliver's Travels*, the proposed undertaking was named the Laputa Project and the Mariana experiment was the first step in determining its feasibility.

Of course, neither the idea nor even the execution attempt is completely original. In 1971, Luis Howard and Henry Stommel performed a similar experiment in the Caribbean, but the outcome was deemed ambiguous because of the uncertain origin of the flow through the pipe. However, the 2002 Mariana experiment was a clear success. Throughout the duration of deployment, the system operated in accord with expectations based on salt-fountain physics and with the corresponding numerical predictions. Following the initial trial, similar salt-fountain experiments have been performed several times (Tsubaki *et al.*, 2007; Maruyama *et al.*, 2011), consistently generating artificial upwelling. Transport characteristics of the salt fountain were quite impressive, with upwelling velocities of up to 593 m/day, corresponding to a pumping rate of 43 m^3/day. To put these numbers in perspective, let us recall that diapycnal velocities in the ocean ($w \sim 10^{-7} - 10^{-6}\,\mathrm{m\,s^{-1}}$) are four to five orders of magnitudes less. Thus, the volume transport by a single pipe is equivalent to the natural upwelling over an area of ~500–5000 m^2. The signal of artificial upwelling was clearly visible in satellite ocean-color images taken during the Philippines Sea experiment (Maruyama *et al.*, 2011), indicating an increased surface chlorophyll distribution in the vicinity of the pipe.

At this point it is perhaps premature to speculate about the prospects of implementing salt-fountain farms on the industrial level. Before embarking on such a project, all its aspects – physical, environmental and even social – have to be considered to ensure that the benefits of artificial upwelling for the ocean ecosystem outweigh any potential adverse consequences. A host of technological challenges will need to be addressed. Nevertheless, the initial success is noteworthy. At the very least, it offers the long-awaited proof of concept for Stommel's curious idea, the idea that ultimately changed our understanding of ocean mixing.

11.2 Effects of diffusive convection

The extensive evidence accumulated during half-a-century of double-diffusive research, only a fraction of which is presented here, leaves little doubt that salt fingering is fully capable of influencing the large-scale dynamics. But what about its little sister, diffusive convection? In terms of both mixing intensity and the net volume of ocean regions affected, diffusive convection lags significantly behind salt fingers. Until recently, diffusive convection has been rarely invoked in discussions of global circulation patterns and the ocean climate. However, two factors could make a difference. First, diffusive convection is most common in high-latitude regions, in dangerous proximity to the formation sites of deep and bottom waters of the world ocean. Low preexisting density stratification in such locations tends to amplify water-mass transformation effects and even a modest amount of diffusively driven mixing can have a substantial impact on thermohaline circulation, the ocean's great conveyor belt. The second argument, particularly relevant for the diffusive layer in the Arctic, emphasizes the large contribution of diffusive convection relative to other mixing mechanisms. In the pristine regions of the central Arctic, wave energy is much less – by as much as a factor of fifty (Levine, 1990) – than typically observed in mid-latitude oceans. The anomalously low wave activity can support very limited levels of turbulent mixing ($K_V \sim 10^{-6}\ \mathrm{m^2\ s^{-1}}$). This by default makes diffusive convection the dominant interior mixing process, potentially governing key water-mass balances in the Arctic.

One of the first attempts to quantify the role of diffusive convection in high-latitude water-mass transformation was made by McDougall (1983), who argued that diffusive convection controls the formation rate of Greenland Sea bottom water and thereby influences global thermohaline circulation. A simple model was developed based on a balance between (i) vertical T–S fluxes driven by diffusive convection and (ii) the horizontal advection of the Atlantic water towards the center of the Greenland Gyre. McDougall's model was remarkably consistent with the observed formation rates of bottom water and its properties. The role of diffusive convection was also shown to be far more significant than the role of cabbeling – another mixing process known to affect high-latitude water-mass transformation, arising from variation in the thermal expansion coefficient. Given the involvement of Greenland bottom water in the maintenance of thermohaline circulation, McDougall's suggestion places diffusive convection right at the heart of the abyssal mechanics of the world ocean.

One of the most rapidly growing environmental concerns is related to the systematic reduction of the Arctic summer-time sea-ice coverage. The sea-ice cover in the Arctic Ocean has experienced dramatic changes in both extent and volume and these trends have notably accelerated in recent years. Observations reveal a decrease of 40% in the volume of sea ice in the Arctic Ocean over the past three decades, which corresponds to an average melt rate of 1% per year (Deser and Teng, 2008). The Arctic sea-ice decline is undoubtedly controlled by a combination of several processes, including solar input at the surface and albedo feedback effects – see the schematic in Figure 11.6. Recently, however, a suggestion has been made (Turner, 2010) that the ice melt could also be influenced by heat transport through the diffusive layer separating the cold and fresh waters of Pacific origin and the cold halocline region from warmer and saltier Atlantic waters.

The Atlantic water contains enough heat to melt all the ice in the Arctic, but the question is how this heat can reach the surface of the ocean. The most direct route is of course straight up, from the warm Atlantic layer through the diffusive staircases and eventually into the ice (Fig. 11.6). In this scenario, the rate of upward heat transport is controlled by the diffusive staircase, which plays the role of the transport bottleneck in the dynamics of the Arctic halocline. However, it is not immediately clear (i) whether this mechanism is physically viable and (ii) even if heat does penetrate the halocline and reach the surface, that the fluxes are large enough to significantly affect the melting rates. The exact dynamics of the heat transfer across the cold halocline region where the vertical temperature gradient changes sign is uncertain, but it likely involves a combination of lateral and vertical mixing processes. To address the first concern, Turner and Veronis (2000, 2004) performed a series of laboratory experiments designed to represent the Arctic configuration. These experiments demonstrated that extra heating at depth (the laboratory analogue of the Atlantic layer) leads to faster melting of floating ice.

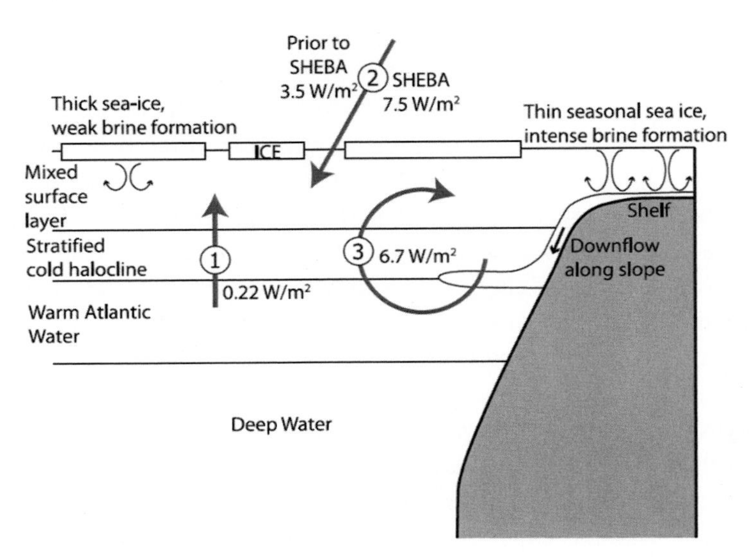

Figure 11.6 Mechanisms that can contribute to the melting of ice in the Arctic Ocean include: (1) upward double-diffusive transport from the warm layer of Atlantic water, (2) heat flux from the atmosphere and (3) downflows on the slope. From Turner (2010).

If we assume that the Arctic dynamics are adequately captured by the laboratory models, then it becomes rather straightforward to estimate the impact of diffusive convection on the ice content. Using the laboratory calibrated four-thirds flux law, Turner (2010) arrived at the vertical staircase-driven heat flux of $F_H = 7.1 \cdot 10^6\,\mathrm{J\,m^{-2}\,yr^{-1}}$. This flux is sufficient to melt 1% of sea ice in one year, which is of the same order as the observed melt rates (Deser and Teng, 2008). For the purpose of this discussion it is also important to keep in mind that the heat flux through the present-day staircase has increased by a factor of three to four within the past three decades, most likely in response to the advance of anomalously warm Atlantic water (e.g., Timmermans *et al.*, 2008). Thus, the dynamics suggested by the double-diffusive model are generally consistent with the sea-ice trend and the predicted melt rates come tantalizingly close to the observed values. Unless the agreement is coincidental, a significant fraction of the reduction in sea-ice coverage can be attributed to the increase in double-diffusive transport. Note also that Turner's argument is based on the laboratory-calibrated heat fluxes. Our ongoing research suggests that, in application to Arctic staircases, these estimates may require upward revision by as much as a factor of two to three, in which case the impact of diffusive convection could be even more substantial than Turner (2010) originally visualized.

The waters surrounding Antarctica are also generally susceptible to diffusive convection. One such region is the Weddell Sea, which has been studied most extensively as the main source of the Antarctic Bottom Water. It is interesting to consider the dynamics of this region with an eye to double-diffusion, but let us first review some basic background. Most of the Weddell Deep Water either forms at the continental shelves or is produced by intermittent open-ocean convection. A particularly energetic convective event resulted in the appearance of the so-called Weddell Polynya – a large area of ice-free water that formed in the eastern Weddell near the Maud Rise seamount and persisted through the winters of 1974–6. The spectacular spatial extent and duration of the Weddell Polynya indicated that the deep-ocean convection occurred on a massive scale. The associated deep-water formation rates greatly exceeded the shelf-derived production (Gordon, 1982). Major convective events of such scale produce a detectable impact on the lower limb of the thermohaline circulation and profoundly influence the regional climate. The extraordinary large amount of heat transferred from the ocean to the atmosphere during the Weddell Polynya years resulted in substantial cooling of the Antarctic Bottom Water and it took decades for the Weddell stratification to fully recover. While events of such scale have not occurred since 1976, more localized single-winter ice-free openings are common.

Despite the dramatic and climatically significant character of deep convection in the Weddell Sea, the specific conditions and mechanisms for its initiation are still uncertain. The winter stratification of the Weddell is much weaker than in the most of the world ocean. The marginally stable state is maintained by a subtle balance between (at least) two competing processes. The first one is the surface-layer densification driven by brine rejection during winter sea-ice growth, which tends to reduce the stratification. The destabilizing brine rejection is countered by the increasing upward heat flux from the warmer deep waters, which ultimately arrests the sea-ice growth, an effect often referred to as the "thermal barrier" mechanism. Thus, the precarious balance of the Weddell stratification is ultimately controlled by the processes involved in transporting heat through the main pycnocline. In order to predict the response of such a system – and to explain its resilience – to a wide a range of forcing conditions, it is essential to identify the dominant mixing agent. The list of usual suspects includes wave-induced turbulent mixing, cabbeling, double-diffusion and the mesoscale eddies impinging on Maud Rise. Their relative contributions still remain a subject of ongoing debate. However, a persistent stream of evidence (reviewed in Chapter 9) suggests that diffusive convection in the Weddell Sea is a significant, and possibly even the dominant, transport mechanism in the permanent pycnocline.

The possibility of predominantly diffusive control was reinforced by recent microstructure measurements made during the 2005 Maud Rise Nonlinear Equation of State Study expedition to the eastern Weddell Sea (Shaw *et al.*, 2013). One of the main objectives of this program was to quantify the variability and parameter

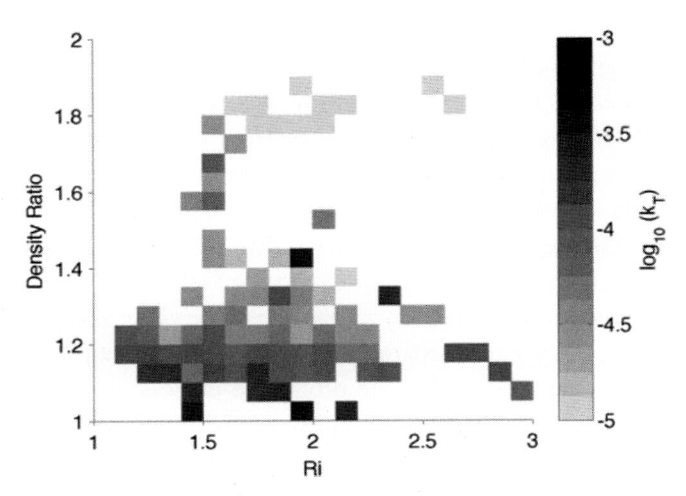

Figure 11.7 Thermal diffusivities in the Weddell pycnocline as a function of Richardson number and density ratio. Modified from Shaw *et al.* (2013).

dependence of thermal diffusivity (K_T) in the vicinity of Maud Rise. The most striking conclusion following from this analysis is the strong dependence of K_T on the finescale density ratio and the lack of statistically significant dependence on the strength of background shear, measured by the Richardson number (Fig. 11.7). Since turbulence is known to be sensitive to finescale shear, these results argue against the turbulent origin of mixing in the Weddell Sea. At the same time, typical diffusivity values and their rapidly decreasing dependence on density ratio are very much in line with the expectations for diffusive convection, as is the abundance of layers in the Weddell pycnocline.

Let us now consider the possible dynamic implications of diffusive mixing in the Weddell Sea. One of the distinguishing characteristics of diffusive convection is the counter-gradient (downward) transport of density. Recall also that this flux rapidly increases as the system drifts towards the marginally stable state ($R_\rho^* \to 1$). Combined, these properties produce a robust mechanism for the prevention of convective overturns. The wintertime brine rejection decreases R_ρ^*, intensifying double-diffusion, which in turn transports more density downward and thereby stabilizes the water column. This double-diffusion-specific negative feedback acts to strengthen the thermal barrier effect, making the Weddell stratification dynamically more resilient to external forcing. So why have we not observed convective overturns comparable to the Weddell Polynya for the last three-and-a-half decades? A definitive answer to this question clearly requires further research, observational and theoretical. However, the possibility of diffusive regulation appears very likely. Without double-diffusion, deep convection would be more common and energetic, making the Antarctic Bottom Water colder and more voluminous. This, in turn, could have a substantial impact on the global thermohaline circulation.

Diffusive convection has also been implicated in numerous regional processes. For instance, a numerical modeling study of sea-ice patterns in the West Antarctic Peninsula regions (Smith and Klinck, 2002) revealed strong sensitivity of wintertime ice thickness to double-diffusion. Doubling the parameterized diffusive flux reduced the ice thickness by a factor of two, tripling the flux-produced ice-free conditions in mid-winter, and removing double-diffusion increased the ice thickness by ~50% towards less realistic values. On the other hand, replacing diffusive convection by the equivalent turbulent transport (of comparable intensity but with equal heat and salt diffusivities) produced surface conditions that are inconsistent with regional observations.

While high-latitude examples are common and, to some extent, expected, diffusive convection can also affect large-scale dynamics in more moderate conditions. For an illustration, consider the mixing environment of the Black Sea. Diffusive convection is supported in much of its interior, below the upper pycnocline, and constitutes the main vertical transport mode for heat and salt in the deep waters. However, the most prominent feature of the Black Sea stratification is its bottom convecting layer. With a thickness of 300–450 m and spreading across the whole basin, it is the largest known diffusive layer in the world's ocean. Here, convection is maintained by geothermal heat flux, which is balanced by the upward flux across the diffusive interface at the top. This structure profoundly affects the deep waters of the Black Sea by laterally homogenizing the water-mass properties inside the convective layer and thus setting a laterally uniform lower boundary condition for the interior stratification. The lateral homogenization of the bottom layer has also been implicated in the uniform distribution of bottom sediments across the basin (Ozsoy *et al.*, 1991; Murray *et al.*, 1991; Ozsoy and Besiktepe, 1995).

On this peculiar example we conclude the discussion of the potential large-scale implications of double-diffusive convection. The list of topics is not meant to be exhaustive. The choice of illustrative examples was governed by three factors: (i) interesting dynamics, (ii) representativeness and (iii) potentially serious impact on climate. However, even within these limits, one subject is notably missing – thermohaline interleaving. This exception is intentional and it does not imply that interleaving is unimportant. On the contrary, it is our belief that interleaving is one of the major mechanisms for lateral mixing, affecting a wide range of large-scale processes. The key difficulty in this regard is that, at present, the transfer characteristics of intrusions are poorly constrained. Quantitative estimates of their large-scale consequences are rare and speculative. Unfortunately, a meaningful discussion of this topic will have to await the development of acceptable models for intrusive mixing. The next chapter, however, attempts an even more ambitious undertaking – analysis of double-diffusion in environments that are far more mysterious and inaccessible than the ocean interior. Let us talk about stars, planets, liquid metals and magma chambers.

12

Beyond oceanography

After Melvin Stern let the double-diffusive genie out of the bottle in 1960, it did not take long for his ideas to spread from oceanography to fields as diverse as astrophysics, geology, chemistry and numerous other physical and engineering sciences. As we think of it, conditions for double-diffusion are not particularly restrictive. All that is needed is a fluid with two or more density components that diffuse at different rates – a very common configuration that is perhaps even more widespread in nature than pure one-component media. At least one of these components should be stratified in the unstable sense (why not?) and there is the potential application for double-diffusion. The following review of double-diffusive applications is intentionally focused on several illustrative examples. Our objective is to reflect the breadth of the subject but at the same time we strive to avoid redundancy. In various applications, under the outer shell of different nomenclature, parameters and scales, lies the universal dynamical core. The key questions and challenges are similar and the knowledge acquired in one discipline is often transferable to others.

The variability of double-diffusive patterns in different media is largely caused by a wide range of two key non-dimensional parameters – the Prandtl number (Pr) and the diffusivity ratio (τ). The typical values of (Pr, τ) relevant for the most common applications are indicated in Figure 12.1. The Prandtl number varies by at least ten orders of magnitude and the diffusivity ratio varies by at least eight. Nevertheless, several basic double-diffusive characteristics are fairly stable. Figure 12.1 presents the non-dimensional linear salt-finger growth rate (a) and the flux ratio (b) as functions of τ and Pr for $R_\rho = 2$. As previously, we use the standard system of non-dimensionalization (1.11). Neither the growth rate (λ) nor the flux ratio (γ) are particularly sensitive to the diffusivity ratio, provided that $\tau \ll 1$. The dependence on Pr is bimodal. For high and moderate Prandtl numbers ($Pr \geq 1$), growth rates and flux ratios are constrained to the relatively narrow intervals of $0.1 < \lambda < 0.3$ and $0.5 < \gamma < 0.85$. However, both λ and γ become small in the limit $Pr \to 0$.

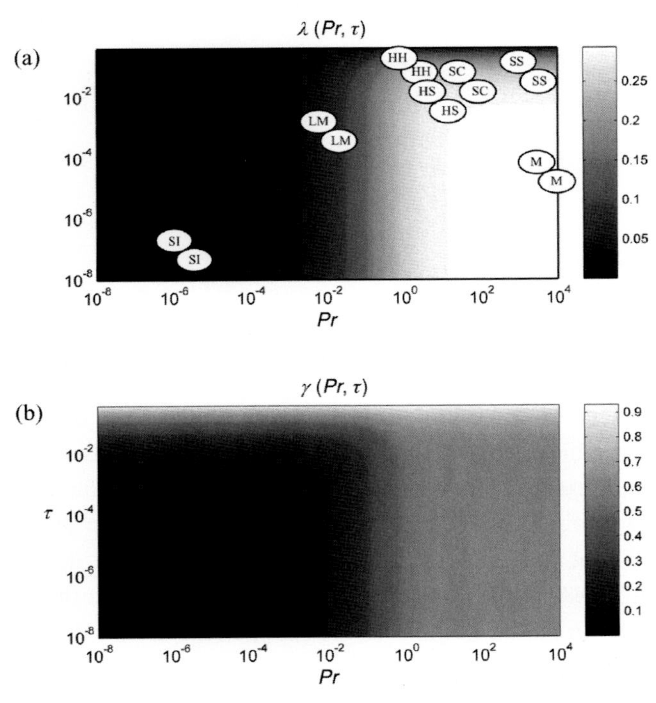

Figure 12.1 Linear characteristics of vertical fastest growing fingers at a density ratio of $R_\rho = 2$ as functions of the Prandtl number and the diffusivity ratio. (a) The non-dimensional growth rate. Regions in the parameter space occupied by various double-diffusive systems are indicated as follows: SI – stellar interiors, LM – liquid metals, HS – heat–salt, SC – semiconductor oxides, HH – humidity–heat, SS – salt–sugar and M – magmas. (b) The flux ratio. After Schmitt (1983).

The fundamentally nonlinear quantities are more difficult to evaluate. For instance, we are unaware of any fully validated unified theory of double-diffusive transport for arbitrary (τ, Pr). The great majority of studies are regime-specific and some regions of the parameter space have been explored better than others. The oceanographic heat–salt case has been subjected to the most intense scrutiny (although the author could be biased), but other fields are rapidly catching up. Several non-oceanographic examples are discussed below, in no particular order, and the first item on our list is the astrophysical case of extremely low Prandtl number and diffusivity ratio.

12.1 Astrophysics

Semiconvection

As astrophysicists have suspected for quite some time (Kato, 1966), double-diffusion is likely to play a fundamental role in the dynamics and evolution of massive stars and giant planets. Stellar and planetary systems are phenomenologically different and will be discussed separately. Both forms of double-diffusion, diffusive and fingering, are active and can lead to observable consequences. However, so far, more attention has been paid to diffusive convection, which in astrophysical literature is usually referred to as semiconvection (Spiegel, 1969, 1972). Semiconvective regions are most commonly found outside the convective core of stars. These regions contain a mixture of hydrogen and helium; heavier elements are also present but in most applications play a lesser dynamical role. There are several scenarios for the development of semiconvective zones (e.g., Merryfield, 1995). For instance, in the main-sequence stars with masses between roughly $10\ M_\odot$ and $30\ M_\odot$ ($M_\odot$ denotes the solar mass) the hydrogen-burning convective region shrinks. This process leaves behind a radially increasing hydrogen mass fraction, leading to an unstable chemical composition. Temperature, on the other hand, decreases and therefore has a destabilizing effect, which mirrors the diffusive configuration in high-latitude oceans. In astrophysical terms, the semiconvective region is Ledoux stable but Schwarzschild unstable – the Ledoux criterion quantifies the stability of radial adiabatic displacements; the Schwarzschild criterion is the analogous condition based only on the thermal density component.

It is interesting and perhaps surprising that stellar dynamics can be adequately modeled using the Boussinesq equations (1.1). The Boussinesq approximation is justified for phenomena operating on scales that are less than the pressure scale height, which for stars is typically one-tenth of their radius. Semiconvection – a relatively small-scale phenomenon – is expected to fall into this category. Another common approximation neglects the local nuclear energy generation, since the nuclear core spans only a small fraction of the interior. While it is comforting to know that geophysical and astrophysical systems are described by the same set of equations, the difference in governing parameters does affect the physical properties of double-diffusion. The compositional diffusivity in stars is extremely low in comparison with the thermal diffusivity ($\tau \sim 10^{-6} - 10^{-8}$); viscosity is predominantly radiative and also small ($Pr \sim 10^{-7} - 10^{-4}$). Thus, the hierarchical orders of governing parameters in stellar ($\tau < Pr \ll 1$) and oceanic ($\tau \ll 1 < Pr$) contexts are different. The immediate consequence of low Pr is that the range of density ratios for the linearly unstable stratification (2.5) is very wide:

$$1 < R_\rho^* < R_{\rho \mathrm{cr}}^*,\ R_{\rho \mathrm{cr}}^* = \frac{Pr + 1}{Pr + \tau} \sim \frac{1}{Pr} \gg 1. \tag{12.1}$$

In this regard, the astrophysical case is simpler than the geophysical. In the ocean, where $R_{\rho \mathrm{cr}}^* \sim 1.1$, the initiation of diffusive convection is attributed to fundamentally nonlinear mechanisms (Veronis, 1965, 1968). In contrast, the origin of semiconvection in stars and planets is clear; it is the linear instability of large-scale gradients. Hence, the relevant spatial and temporal scales can be determined by considering the linearly fastest growing modes. The corresponding wavelength of primary instabilities is on the order of $l_{fg} \sim 500$ km and the e-folding time is $t_{fg} \sim 10^5$ s (Merryfield, 1995). Note that l_{fg} is still much less than the thickness of the stellar semiconvective zone ($\sim 10^5$–10^6 km) and therefore diffusive mixing is usually treated as a local process operating in the effectively unbounded gradient.

Numerous pieces of indirect evidence suggest that semiconvection is critical for the internal structure and evolution of core-convective stars of various sizes and its inclusion in large-scale stellar models helps to meet the observational constraints (e.g., Crowe and Mitalas, 1982; Langer, 1991). More quantitative estimates in astrophysics are difficult to come by. The major impasse is the lack of reliable flux laws. Not only the numerical values of transport but even the functional forms of such laws are highly uncertain. The situation is exacerbated by the absence of terrestrial fluids satisfying the condition $(Pr, \tau) \ll 1$, which precludes the laboratory modeling of semiconvection. One of the big unknowns is the stratification pattern: it is not clear whether the semiconvective region remains smoothly stratified or breaks into a series of mixed layers, as one could anticipate from the oceanographic studies (Chapter 8). Both possibilities have been considered. Theory developed by Langer *et al.* (1983) assumes that mixing is driven by destabilization of growing oscillatory modes, which collapse without forming permanent staircases. An alternative view was articulated by Spruit (1992), who proposed the staircase model of semiconvection. In order to determine the heat and compositional transport, several ad hoc assumptions were made about the thickness of layers and interfaces. Progress in this direction was stalled by the lack of effective means for testing the proposed transport models.

Given the lack of viable alternatives, the approach based on direct numerical simulations (DNS) has become increasingly popular. The standard Fourier-based spectral model appears to be best suited for studies of semiconvection (e.g., Mirouh *et al.*, 2012). The obvious limitation of the numerical approach is that the dissipation scales of heat, composition and momentum in stars are dramatically different – by as much as four orders of magnitude. Even modern state-of-the-art models cannot simultaneously resolve such a range of scales and the choice of the Prandtl number and diffusivity ratio is dictated by the feasibility of simulations. Transport estimates in such circumstances can only be obtained through extrapolation towards the realistic parameters.

Nevertheless, simulations have brought much needed insight into the dynamics of semiconvection and particularly into the conditions of layering (Rosenblum *et al.*, 2011; Mirouh *et al.*, 2012). Overall, the astrophysical situation has proven to be quite similar to the oceanographic case. Extensive exploration of the numerically accessible parameter range – small but not too small (Pr, τ) – revealed that layering

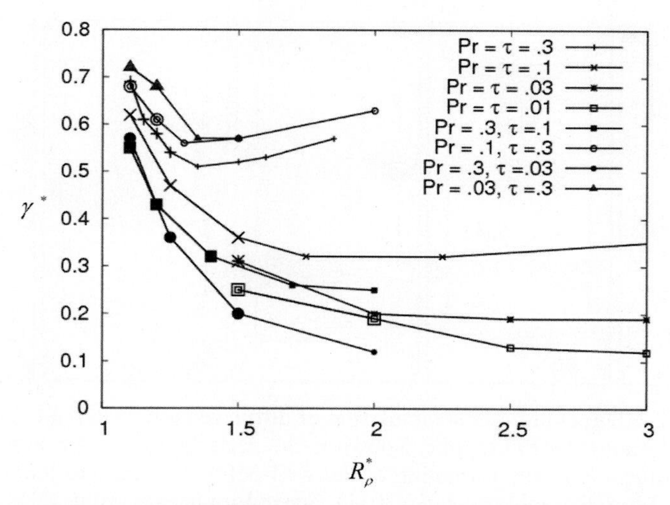

Figure 12.2 The equilibrium diffusive flux ratio as a function of the diffusive density ratio in a series of DNS. The experiments resulting in layering are marked by larger symbols. From Mirouh *et al.* (2012).

can occur in low-Pr systems and that it is caused by the γ-instability (Chapter 8). Typical patterns of the flux ratio, estimated from small-domain simulations, are shown in Figure 12.2; the runs resulting in spontaneous layering are indicated by larger symbols. These experiments indicate that the decrease of the flux ratio with the density ratio is a necessary and sufficient condition for the formation of semiconvective staircases. In all cases considered, the layering condition $\left(\frac{\partial \gamma^*}{\partial R_\rho^*} < 0\right)$ is met for sufficiently low diffusive density ratios ($1 < R_\rho^* < R_{\min}^*$). Note that in the astrophysical ($Pr \ll 1$) case, it is critical to incorporate into the flux ratio γ^* the contribution from the molecular heat flux (a secondary effect for high Pr applications). The γ-instability modes developing in this parameter range grow at a rate consistent with the theoretical prediction (8.11), eventually transforming the background stratification into a well-defined staircase.

However, the analogies between the astrophysical and oceanographic cases (Chapter 8) do not end here. Layers that form first are thin and unsteady (Fig. 12.3). They undergo a series of merging events, in which strong interfaces grow further at the expense of weaker ones. Weak interfaces gradually erode and eventually disappear, following the B-merger scenario described in Section 8.4. The characteristic size of steps correspondingly increases in time until there is only one interface left within the limits of the computational domain (Fig. 12.3). Such evolutionary similarities of layering across different forms of double-diffusion (fingering/diffusive) and parameter ranges (low/high Pr) are truly striking. They are indicative of the robust and universal nature of thermohaline layering.

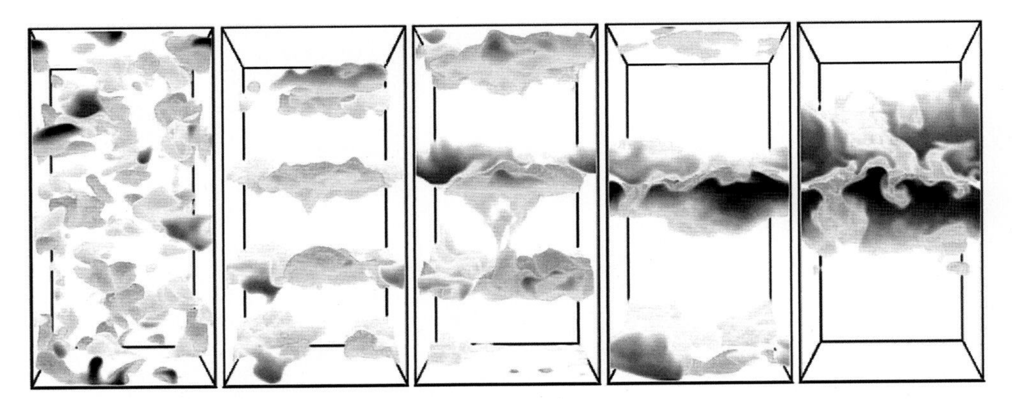

Figure 12.3 Direct numerical simulation of diffusive convection in the astrophysically relevant (low *Pr*) regime. Shown is the perturbation of the composition at various times. Note the formation of the well-defined layers and their sequential mergers. From Rosenblum *et al.* (2011). See color plates section.

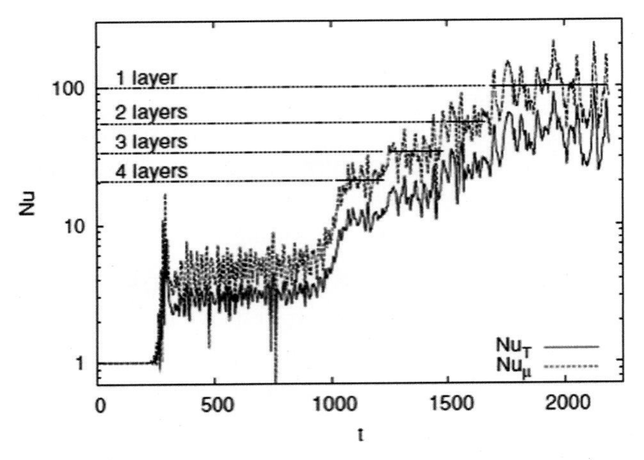

Figure 12.4 Time record of the thermal (solid line) and compositional (dashed line) fluxes. Note the systematic increase in the fluxes associated with layer-merging events. From Rosenblum *et al.* (2011).

As in the ocean (Schmitt *et al.*, 2005; Radko, 2005), formation of a staircase and the sequential layer-merging events dramatically increase the vertical transport (Fig. 12.4), which poses a major challenge for parameterizing semiconvection. In the non-layer-forming regime ($R_\rho^* > R_{\min}^*$), transport could be deduced from small-domain DNS. For instance, Mirouh *et al.* (2012) notes that the available simulations (Fig. 12.5) can be adequately described by a universal analytical expression:

$$Nu - 1 = (0.75 \pm 0.05)\left(\frac{Pr}{\tau}\right)^{0.25\pm0.15} \frac{1-\tau}{R_\rho^* - 1}(1-r), \qquad (12.2)$$

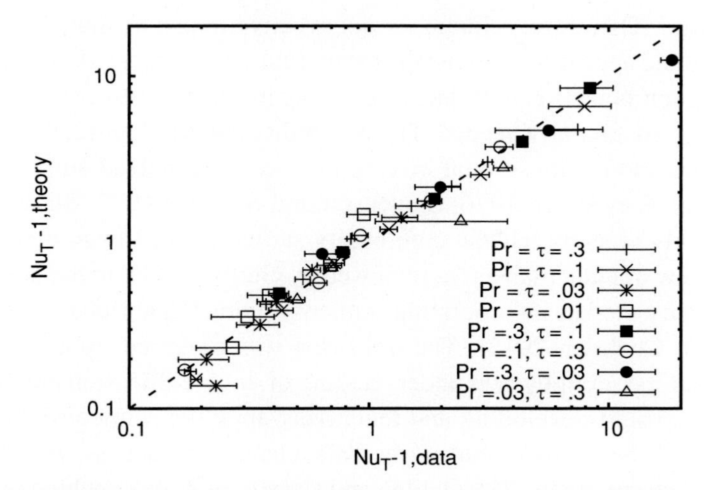

Figure 12.5 Comparison between the theoretical prediction of heat transport (12.2) and the corresponding numerical data. From Mirouh *et al.* (2012).

where

$$r = \frac{R_\rho^* - 1}{R_{\rho\mathrm{cr}}^* - 1}. \tag{12.3}$$

The analysis in Mirouh *et al.* offers a promising starting point for the non-layering part of the problem. Future simulations, better resolved and capturing a wider range of (Pr, τ) will undoubtedly refine these flux laws and improve the accuracy of extrapolations, but the direction such research should take is well defined.

The way forward in the layering regime ($R_\rho^* < R_{\mathrm{min}}^*$) is less clear. It is safe to assume that heat and compositional fluxes in the staircase exceed the smooth-gradient values, but by how much? These fluxes sensitively depend on the equilibrium height of steps – the point where mergers are arrested (Fig. 12.4). However, the theory of layer equilibration is currently at a highly qualitative stage (Radko, 2005). Its applicability to the astrophysical regime is particularly uncertain, given the lack of observational guidance. Layer-merging significantly slows down as a staircase coarsens. The mergers of thick steps occur on time scales that are much longer than the growth period of primary instabilities. Therefore, the prospects of DNS-modeling the entire merging sequence and the ultimate stable state are problematic. On the positive side, limited numerical accessibility is likely to stimulate new theoretical developments. As we have seen so often in the past, the failure of brute-force methods inevitably creates fertile grounds for the birth of highly original ideas and unorthodox approaches.

Compared to the stellar case, modeling semiconvection in giant planets poses a lesser challenge. Planetary Prandtl numbers and diffusivity ratios are moderate (Pr, $\tau \sim 0.01$) and the dissipation scales of heat, composition and momentum are not too dissimilar. Therefore, reliable semiconvective transport models for planetary dynamics could appear in the not-too-distant future (Mirouh *et al.*, 2012). Semiconvection in giant planets is a subject of growing interest in astrophysics and models of this nature are urgently needed. The possibility that semiconvection can regulate both thermal and compositional mixing has been recognized since the seventies (Salpeter and Stevenson, 1976; Stevenson and Salpeter 1977; Stevenson, 1985). However, only recently has the community started to shift from the crude purely adiabatic view of giant planets (as reviewed by Guillot, 2005) to the new generation of evolutionary models incorporating semiconvection (Chabrier and Baraffe, 2008; Leconte and Chabrier, 2012). The transition was triggered by a combination of two factors. First, our physical understanding of double-diffusion reached the level where we can start developing and tentatively applying semiconvective transport models with a better than shot-in-the-dark chance of success. Another impetus comes from observations. The Galileo and Cassini missions enabled detailed measurements of the compositions of the atmospheres of Jupiter and Saturn. Extrasolar planet surveys provided statistically significant data on their composition and global characteristics, and future missions will undoubtedly expand the already impressive body of information. The possibilities are now opening to firmly connect semiconvection theory with observations. It is a win–win proposition that will concurrently improve physical interpretation of data and afford the observational validation of mixing models.

The impact of semiconvective mixing on the structure and evolution of giant planets could be profound and multifold. Semiconvection has been considered as the plausible origin of the low observed heat flux of Uranus (Gierasch and Conrath, 1987), of the high metallicity of the envelopes of Uranus and Neptune, and of the erosion of the core of Jupiter and Saturn (Guillot *et al.*, 2004). Non-adiabatic models have been proposed to explain the uncharacteristically large observed radii of some transiting exo-planets (Chabrier and Barraffe, 2008). The inclusion of semiconvective dynamics in evolutionary models (Leconte and Chabrier, 2012) improves the agreement with observations and leads to more reliable estimates of the size of the convective layers. While the full description of Jovian (gas giant) dynamics has to await development of accurate mixing parameterizations, it is possible to draw several general conclusions (Leconte and Chabrier, 2012) that are robust and not sensitive to the details of a semiconvective model. In particular: (i) Jovian planets might be significantly more enriched in heavy elements (by 30–60%) than previously thought; (ii) their interior heat content might be much larger; and (iii) the inner temperature profile significantly departs from the adiabatic profile.

Leconte and Chabrier (2012) emphasize that the significance of semiconvection is not limited to its effects on the current compositional and thermal structures of giant planets. Taking semiconvection into account can impact our understanding of planet formation and their cooling properties. For instance, if new semiconvective models substantially modify the estimates of heavy material content, the planet formation theories should be revised accordingly. New results suggest an early and more efficient capture of planetesimals in the protoplanetary nebulae of giant planets. Another revision of the conventional chronology of giant planets is required to reflect the less efficient heat transport, a consequence of the semiconvective inhibition of convective mixing. The associated larger heat content and the possibility of significant core erosion also directly impact the planet cooling histories.

Fingering convection

While astrophysicists are more concerned with semiconvection, it is generally accepted that stellar stratification often becomes susceptible to fingering as well. It is less clear whether its mixing intensity is sufficient to substantially change the composition of stars on the evolutionary time scale. The answer requires knowledge of the relevant flux laws. Early theoretical attempts to predict fingering transport in the astrophysical context (Ulrich, 1972; Kippenhahn *et al.*, 1980) were based on uncertain assumptions and, furthermore, produced mutually inconsistent estimates of the transport. The oceanographic flux laws are in much better shape but, due to the differences in molecular characteristics, their application to astrophysical phenomena is unwarranted. Numerical simulations of fingering are also problematic due to the disparity of the dissipation scales of heat, momentum and composition. Thus, the obstacles for the development of a consistent theory (or at least a simple recipe) for finger-driven transport largely mirror the complications in dealing with semiconvection in stars.

On the positive side, similar problems can be tackled by similar methods. For instance, fingering convection has been successfully modeled in the numerically accessible range, which was followed by imaginative extrapolations of results to realistic values of (Pr, τ). As in semiconvection, the Fourier-based spectral model is a weapon of choice for astrophysicists, which makes it possible to avoid artificial contamination of the results by inappropriate boundary conditions (Denissenkov, 2010; Denissenkov and Merryfield, 2011b). One of the success stories was reported by Traxler *et al.* (2011b), who noted that simulations in the numerically accessible parameter range can be described using the universal expression for the eddy diffusivity of the composition K_μ:

$$K_\mu = 101\sqrt{\nu k_\mu}\exp(-3.6r)(1-r)^{1.1}, \tag{12.4}$$

where k_μ is the molecular compositional diffusivity, ν is viscosity and

$$r = \frac{R_\rho - 1}{\tau^{-1} - 1}. \tag{12.5}$$

An important caveat is that the flux law (12.4) was obtained for the smooth-gradient model. The assumed absence of staircases is consistent with the existing numerical simulations and theoretical arguments. For instance, the flux ratio monotonically increases with the density ratio implying the absence of γ-instability, the most common cause of step formation (Chapter 8). This argument, however, has to be considered with caution since the linear stability of a uniform gradient does not completely preclude spontaneous layering, which could be driven by fundamentally nonlinear processes (diffusive layering in high-latitude oceans being a prime example). Until the absence of fingering staircases in stellar atmospheres is proven, it is perhaps prudent to interpret (12.4) as a lower bound for the finger-driven transport. Overall though, despite all their current limitations and uncertainties, numerical studies have significantly advanced our understanding of the importance of fingering in astrophysics. In particular, all simulations consistently support two major qualitative conclusions: (i) the heat transport by fingers in stellar atmospheres is negligible but (ii) the compositional transport is substantial and can affect the evolution of stars and their observable characteristics.

One of the examples of finger-induced effects in stellar dynamics involves the planetary pollution conundrum. Central stars of planetary systems tend to exhibit surprisingly high metallicity – an intriguing observation, begging for a mechanistic explanation. Two distinct proposals have been considered. It is possible that metallicity is primordial: the planetary systems could be formed more easily in a higher metallicity disk. The accretion hypothesis, on the other hand, assumes that the planet formation mechanism is independent of the disk metallicity and the enhancement is caused by subsequent planetary infall. The most common criticism of the accretion idea is based on the lack of correlation between the size of the stellar convection zones and their metallicity. If the accretion mechanism is dominant, stars with thinner convection zones should be polluted more. The size of the outer convection zone rapidly decreases with stellar mass and therefore the relative metallicity enhancement should be larger for heavy stars (Laughlin and Adams, 1997). However, such correlation has not been detected. It is an interesting counter-argument for the accretion theory but is it sufficiently rigorous? Recent studies suggest that the mechanics of planetary pollution could be more complicated (Vauclair, 2004; Garaud, 2011; Theado and Vauclair, 2012). The added metallicity does not necessarily stay trapped in the outer convection zone, but could be drained into the interior of a star. The likely mixing mechanism is double-diffusive, by metallic fingers.

Using the flux law (12.4), Garaud (2011) successfully modeled the radial large-scale distribution of composition and demonstrated that fingering serves as an effective regulator of surface metallicity. The first post-infall stage is convective mixing, triggered by the accretion of planetary material in a thin surface layer of a star. This brief period is followed by a much longer fingering phase, during which the metallicity is drained deep into the interior. Garaud notes the sensitive dependence of the finger-induced dilution rates on the star size: the larger the star, the faster metallicity is removed from the surface. The enhanced post-impact metallicity in large stars could be compensated by faster dilution into the interior. This fundamentally double-diffusive effect could explain the lack of systematic variation of metallicity with star size – the original argument against the accretion hypothesis. In view of this result, the debate over accretion and the primordial dynamics of planetary pollution is bound to shift in a different direction. While the size–metallicity correlation argument no longer appears convincing, taking into account double-diffusive mixing poses a problem of a different kind. The double-diffusive draining of the metal-rich material into the stellar interior leaves only a fraction of the initial metallicity excess in the surface convective layer – much smaller than the average observed overmetallicity (Theado and Vauclair, 2012). Thus, the question is still open of how to explain the enhanced metallicity in central stars of planetary systems. Another suggestive finding of Theado and Vauclair (2012) and Garaud (2011) is that the fingering region can extend downward to the lithium-burning region, which could help to rationalize recent observations (Gonzalez 2008; Israelian *et al.*, 2009) of lower lithium abundances in planet-bearing stars.

There are several other scenarios of stellar evolution in which fingering could play a major role. For instance, red giants are known to experience extra mixing in their convectively stable radiative zones, separating the hydrogen-burning shell from the bottom of the convective envelope. A promising physical mechanism for this extra mixing is fingering convection. The evolutionary model that takes into account parameterized finger-induced mixing (Charbonnel and Zahn, 2007) appears to be consistent with numerous spectroscopic measurements of surface composition in low-mass red giants after they reach bump luminosity. On the other hand, the estimates of transport suggested by recent finger-resolving DNS (Denissenkov, 2010; Traxler *et al.*, 2011b) fall by at least an order of magnitude below the levels required to explain observationally inferred extra mixing. The inconsistency can be caused by two factors (or their combination). The DNS could underestimate mixing due to their present inability to operate in the realistic parameter range. Likewise, one cannot rule out the possibility that the extra mixing in red giants is associated with a yet undetermined non-double-diffusive transport mechanism and that the agreement of the finger-based model of Charbonnel and Zahn (2007) with observations is coincidental.

Fingering convection has also been implicated in the mixing of carbon-rich material deposited at the surface of a star in a mass-transferring binary (Stancliffe *et al.*, 2007). It is generally assumed that the material accreted from a no longer visible companion of such stars remains on the surface and does not mix with the interior. However, the new material has higher mean molecular weight than the original stellar composition and therefore is likely to drive fingering convection. Stancliffe *et al.* (2007) suggests that the ensuing mixing rapidly transfers the accreted material deep into the star interior, affecting up to 90% of its volume. As a result, the surface abundance of carbon decreases by nearly an order of magnitude relative to the non-fingering estimates. Some of the accreted carbon-rich material mixes into the high-temperature interior region where CN-cycling occurs. Therefore, the finger-based model also displays a substantial increase in the abundance of nitrogen in the period after the surface convection zone extends to the layers that have gone through nuclear fusion (the first dredge-up). Taking into account finger-driven transport makes it possible to bring the predicted values and evolutionary patterns of carbon/iron and nitrogen/iron ratios into close agreement with observations, something that the non-double-diffusive version of the model fails to accomplish.

12.2 Geology and geophysics

I hope we all enjoyed this brief voyage to space – time to go back to the Earth and take a look at signs of double-diffusion in its interior. There are several geophysical systems that have been considered as likely sources of double-diffusive phenomena. The most frequently discussed examples include magma chambers (Huppert and Sparks, 1984; Clark *et al.*, 1987; Toramaru and Matsumoto, 2012) and the core–mantle boundary (Kellogg, 1991; Hansen and Yuen, 1994; Buffett and Seagle, 2010). In addition, diffusive convection driven by hydrothermal forcing can occur in individual groundwater wells (Love *et al.*, 2007) and, on much larger scales, throughout the Earth's crust (Schoofs *et al.*, 1999, 2000). Magma chambers have been studied more extensively for reasons of observational accessibility and will be discussed here first.

The term magma chamber refers to an underground pool that contains molten rock under high pressure. The liquid material in a magma chamber can fracture the solid material around it and create outlets, occasionally reaching the surface of the Earth and resulting in a volcanic eruption. Magma chambers are located 1–10 km below the Earth's surface and their obvious locations are directly under active volcanoes. The most natural thermal structure of a magma chamber is the one in which it is cooled at the upper surface. Thus, the thermal component of density is unstable. At the same time, numerous studies of volcanic rocks suggest that magma chambers are often characterized by stable compositional gradients. The major components of silicate melts are only weakly diffusive ($\tau \sim 10^{-3}-10^{-8}$), which creates favorable conditions for diffusive convection (e.g., Huppert and Sparks, 1984; Hansen and Yuen, 1995).

The early studies of double-diffusion in a geological context were motivated by observations of well-defined layers commonly seen in igneous rocks (Fig. 12.6). Their similarity to thermohaline staircases in the ocean has led to a suggestion that layers were formed in the liquid state of rocks by double-diffusive processes (Huppert and Turner, 1981; Kerr and Turner, 1982). However, closer inspection of igneous layering reveals signatures of several distinct mechanisms (reviewed by Naslund and McBirney, 1996), and hopes for a universal explanation for all or even most incidences of layering are likely to be disappointed. It also remains unclear how the layered liquid can transform into a layered solid without major disruptions of regularity. Nevertheless, the double-diffusive hypothesis remains among the most popular explanations of layering. The major challenge for igneous petrologists is to develop techniques for identification, by the composition and textural patterns of layered rocks, the likely origin of layering. Meeting such a challenge requires familiarity with the basic (and possibly not-too-basic) physics of double-diffusive convection.

There are several features that distinguish the dynamics of geological double-diffusive systems from their oceanographic and astrophysical counterparts. One peculiarity is that molten rocks are highly viscous. Viscosity depends on the magma temperature and its silica content. In different types of magma, viscosity can vary by up to ten orders of magnitude – a remarkably wide range for any physical parameter. However, even for the least viscous magmas, viscosity exceeds thermal diffusivity by more than four orders of magnitude ($Pr \geq 10^4$). Generally, the disparity in the dissipation coefficients leads to significant complications for analytical and numerical modeling, as we have seen in the astrophysical examples. However, the effectively infinite Prandtl number of geological systems appears to simplify the dynamics. Non-dimensionalizing Boussinesq equations (1.1) using the standard system (1.11) and taking the limit of high Prandtl number ($Pr \to \infty$) reduces the governing equations to

$$\begin{cases} -\nabla p + (T - S)\vec{k} + \nabla^2 \vec{v} = 0, \\ \dfrac{\partial T}{\partial t} + \vec{v} \cdot \nabla T = \nabla^2 T, \\ \dfrac{\partial S}{\partial t} + \vec{v} \cdot \nabla S = \tau \nabla^2 S, \\ \nabla \cdot \vec{v} = 0. \end{cases} \tag{12.6}$$

Figure 12.6 (a) Typical variability in layer thickness (m) in observations. (b)–(e) Selected examples of cyclic layering. From Toramaru and Matsumoto (2012).

Thus, the nonlinearities in this limit are retained in the equations for temperature (T) and composition (S) but not in the momentum equations, which can facilitate analytical and numerical developments. The high Prandtl number version of the standard Boussinesq model (12.6) adequately reflects the qualitative properties of geological flows. However, if precise quantitative information is sought, geologists often employ the so-called extended Boussinesq version, in which the temperature equation is modified to include viscous and adiabatic heating.

The asymptotic high Prandtl number system has been successfully modeled numerically. For instance, Hansen and Yuen (1989, 1995) used it to examine the propensity of double-diffusive magmas to form layers. Analogous to the oceanographic case, the ability of such systems to maintain preexisting layers and to create new ones from un-layered states is controlled by the density ratio, as are the cooling rates of magma. The numerical solutions obtained by Hansen and Yuen exhibited rich phenomenology. Interesting effects include the coexistence of stratified and homogeneous layers, as well as transient layering followed by convective overturns. While layering was readily developing in the diffusive case, it did not occur in the analogous fingering systems. Viscous heating was shown to have a significant impact on double-diffusive convection. The variation in viscosity across the magma chamber could also be of major importance (Turner and Campbell, 1986) but is rarely incorporated in modeling studies.

Aside from high viscosity (a simplifying factor), another distinguishing characteristic of geological fluid dynamics is the possibility of crystallization (a complicating factor), which occurs when the melt cools. Crystallization is one of the major mechanisms for the initiation and maintenance of double-diffusive convection. On the most fundamental level, crystallization affects the fluid dynamics of magma by the so-called differentiation effect – the production of compositional gradients in the initially homogeneous media. There are at least two common mechanisms for crystallization-driven differentiation. The first one is the crystal settling, in which stratification is caused by the displacement of crystals relative to a melt. The second, and perhaps more likely (Turner and Campbell, 1986) mechanism involves the growth of crystals at the boundaries of a magma chamber, which selectively depletes the fluid adjacent to newly formed crystals of certain chemical components. For instance, the crystallization of dense olivine after cooling generates residual liquids of lesser density. Crystallization in magmas causes much larger changes in melt density than the associated temperature changes. Thus, if cooling is applied at the upper surface, the residual melts in the upper layers become colder and lighter, due to a different composition, than the magma in the lower part of the chamber. This creates favorable conditions for diffusive convection. Since crystallization and boundary cooling are generic attributes of magma dynamics, double-diffusive processes of one form or another should be ubiquitous in magma

(a)

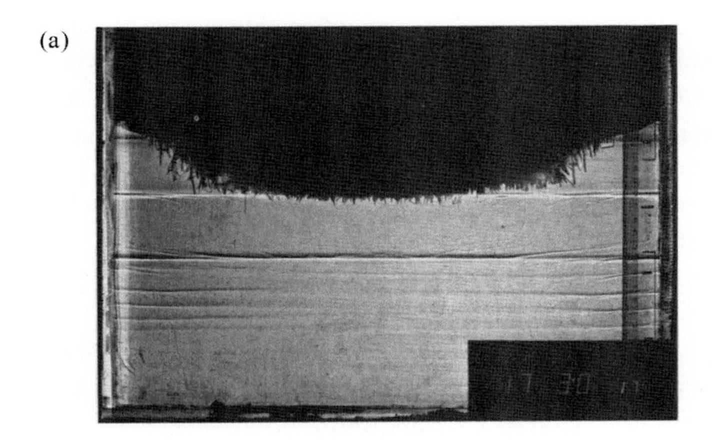

(b)

Figure 12.7 Shadowgraphs of crystallization. (a) Crystallization of Na_2CO_3 in a constant vertical concentration gradient cooled from above. (b) A vertical concentration gradient cooled from the side. From Chen and Turner (1980).

chambers. Melting and assimilation of the solid boundaries of magma chambers can lead to analogous dynamic sequences – differentiation, double-diffusion and, ultimately, formation of layers.

Much of our understanding of the combined systems involving crystallization and double-diffusive processes comes from the analogous laboratory experiments with crystallizing aqueous solutions. Figure 12.7a presents an experiment in which a tank containing sodium carbonate (Na_2CO_3) solution was cooled at the top (Chen and Turner, 1980). Regardless of its initial stratification pattern (e.g., gradient or homogeneous) the system eventually developed a well-defined set of diffusive layers. Laboratory experiments have been highly effective in terms of emphasizing the coupled nature of the convective/crystallization dynamics. The term "convective fractionation" is commonly used in geological literature to describe processes involving convection of residual fluids away from fixed growing crystals. Figure 12.7a presents one of many examples of such coupled phenomena – crystallization induces double-diffusive layers and these layers, in turn, influence the subsequent growth of the crystals. As a result, layers in the crystallized matter are well defined and characterized by a significant variation in properties. Several interesting attempts have been made to represent the coupled dynamics of magma chambers in the numerical models (Oldenburg *et al.*, 1989; Oldenburg and Spera, 1991) using simple parameterizations of crystallization. Most recently, Toramaru and Matsumoto (2012) developed a one-dimensional numerical model based on double-diffusive dynamics as well as the kinetics of diffusion-limited crystallization. They were able to reproduce major features of layering observed in a number of locations, including such subtle characteristics as the spatial trends in layer thickness.

Two-dimensional effects play an important role in double-diffusive systems with crystallization. The crystallization can happen on one or several surfaces simultaneously and so can erosion of solid boundaries. In some cases, heat escapes through the top of a chamber, in others through the sidewalls and the floor. Figure 12.7b presents an experiment in which cooling and crystallization occur at the vertical boundary. The fluid released by crystallization in this case is less dense than the original mixture and, since the density is dominated by the compositional content, the light and cold fluid rises along the wall. Both laboratory experiments and field observations suggest that boundary currents of this type are laminar over their whole depth for a wide range of chamber sizes and magma types. The boundary current continuously supplies cold and light fluid upward, slowly building up diffusive stratification in the chamber. The combination of vertical and horizontal destabilizing temperature gradients leads to the formation of a series of double-diffusive layers clearly visible in Figure 12.7b. In a certain parameter range, crystallization leads to a denser residual fluid, which can flow downward forming double-diffusively stable stratification (cold and compositionally dense fluid below the warm and compositionally light). Nevertheless, layering in this case is still possible due to thermohaline interleaving (Chapter 7) driven by the lateral gradients introduced at the cooled boundary (Leitch, 1990). The interleaving can be particularly spectacular in the case of preexisting stratification. The presence of sloping boundaries adds a new element of complexity to the problem and affects the patterns of secondary double-diffusive structures (Martin and Campbell, 1988). Another set of double-diffusive processes is triggered by the formation of convective chimneys, which focus upwelling currents into narrow vertical channels (Tait and Jaupart, 1992).

A series of dynamically interesting effects in magma chambers are associated with the time dependence of the filling pattern. It is believed that large magma chambers are not filled by a single large pulse but by several smaller pulses over an extended period of time, which by itself could lead to layering. Most significantly, the non-uniform filling can produce a rich array of thermal/compositional configurations, many of which are likely to develop active double-diffusive convection. For instance, even the timing of a new pulse can determine the convective pattern. Fractionation of tholeiitic magma has a non-monotonic effect on density. First, the density decreases due to crystallization of olivine and bronzite. However, as lighter elements start to crystallize, density increases again beyond the original density of the primitive magma. Finally, crystallization of iron–titanium oxides leads to another reversal of the density trend. Thus, depending on the cooling history of the magma, the new material could be lighter or denser than the old fractionated magma. As a result, magma can convect, double-diffuse or cool off without mixing.

The periodic replenishment of magma chambers leads to several plausible scenarios in which finger-favorable stratification is produced (e.g., Campbell, 1996). For instance, the new pulse of magma that enters the chamber may be significantly colder and therefore denser than the magma already present in the chamber. But if the new magma is compositionally lighter, fingering convection ensues. In view of this possibility, double-diffusion has been considered as a cause of columnar finger-like structures occasionally observed in basaltic rock formations (Kantha, 1980, 1981). Double-diffusion may also occur when the chemical content of the adjacent magmas is dissimilar and individual components diffuse at different rates. In this case, the circulation is controlled by the diffusion of two or more chemical components, rather than by temperature and composition, giving rise to a wide range of processes that can be grouped under the heading of multicomponent convection (Huppert and Sparks, 1984).

Geometry is also important for circulation in the replenished chambers. If the new material is light and enters the chamber through a relatively narrow inlet at the bottom boundary, then "filling box" mechanics (Baines and Turner, 1969) will be engaged. The replenishing magma will first float towards the surface of the chamber and then gradually spread downward, creating finger-favorable stratification. Turner and Campbell (1986) point out that different dynamics are expected for funnel-shaped and inverted-funnel chambers. The gravitationally stable but possibly double-diffusive stratification is likely to be better developed in the case of the normal funnel due to the channeling effect of the inward-sloping walls. The aspect ratio of a chamber fundamentally affects the fluid dynamics of magma, as do numerous other factors not mentioned in this review. Overall, the variety of possible configurations in magma chambers is astounding, and each scenario is associated with a distinct set of physical process and circulation patterns. Being primarily an oceanographer, I may be running the risk of sounding a bit disloyal to my chosen field, but it seems that geological systems offer an extraordinarily wide spectrum of intriguing and largely unresolved double-diffusive problems. The sheer number of possibilities may be unmatched by any other application.

We should also add that the interest in double-diffusion of magma goes far beyond intellectual curiosity. Understanding the fluid dynamics of magma chambers aids in the recognition of precursors to volcanic eruptions, assists with mineral exploration, and addresses the associated environmental problems. The mixing processes that occur between different types of magma coming into direct contact are particularly significant in this regard. Knowledge of the mixing rates is important for designing petrogenetic models, which are used to infer the origin and structure of igneous rocks. The especially intense mixing of mafic and felsic magmas could cause an increase in pressure and trigger eruptions. However, the mixing rates in magma chambers are highly variable, which makes their prediction difficult. There are numerous examples of effective mixing of different types of magma. A parallel stream of evidence suggests that distinct magmas frequently coexist in close proximity to each other, maintaining their individual characteristics for long periods of time (e.g., Turner and Campbell, 1986). The systematic analysis of the conditions that control mixing of magmas, particularly double-diffusive mixing, is a prerequisite for development of quantitative evolutionary models and is the subject of ongoing research.

In addition to magma dynamics, there are several other geological systems in which double-diffusion is expected to arise and affect circulation patterns. In particular, the dynamics of the core–mantle boundary (CMB) has been a focus of much interest and research activity (Hansen and Yuen, 1994; Montague and Kellog, 2000; Lay *et al.*, 2004). The CMB is involved in the thermo-chemical coupling of the mantle and the core and thus plays a major role in the planetary thermo-chemical evolution. One of the key variables controlling planetary heating is the CMB heat flux. Current estimates of this flux are poorly constrained, partially because of the complex and not fully understood fluid mechanics of this region. Seismic studies find evidence of a seismic discontinuity 150–450 km above the CMB, which suggests the presence of a thermo-chemically distinct boundary layer, known as the D″ layer. This region is highly heterogeneous in chemical composition and temperature, which provides a favorable environment for double-diffusive instabilities.

The influence of diffusive convection on the structure, heat flux and chemical transport characteristics of the D″ layer has been emphasized by numerous studies. Without the stabilizing compositional gradients, this region would be strongly unstable and therefore susceptible to rapid destruction and entrainment into the lower mantle. Several modeling studies (e.g., Montague *et al.*, 1998, Montague and Kellogg, 2000) have examined the stability of the D″ layer and concluded that order one (or higher) values of the diffusive density ratio are required to maintain its structural integrity. Hansen and Yuen (1988) argued that the observed pattern of topographic variations of the CMB could be attributed the fundamentally double-diffusive instabilities of the D″ layer. Samuel and Farnetani (2003) showed how the presence of a dense diffusive layer at the base of the mantle can profoundly affect the convective pattern in its interior. For instance, it provides a plausible explanation for the observationally inferred persistence of geochemical heterogeneities in a vigorously convecting mantle for billions of years. Inclusion of the thin diffusively convecting layer, acting as a thermo-chemical barrier, in the numerical models of the mantle (Farnetani, 1997) reduces the typical excess temperature of the plumes reaching the surface, bringing it into agreement with the inferences based on petrological models.

Buffett and Seagle (2010) suggest that diffusive processes should be active in a stratified layer at the top of the core, which is maintained by diffusion of light elements through the CMB. These authors develop a physical model for the evolution of this diffusive layer. Stratification is assumed to be close to density neutral but staircase-like in temperature and composition. The overall thickness of the stratified region is set by a balance between its diffusive spreading and the adverse tendency associated with the inner core growth and nucleation. The model predicts thickness of 60 to 70 km, although the accuracy of this estimate is difficult to ascertain. The significance of the diffusive layer is related to its control of the transfer rates of the light elements (oxygen and silicon) into the core interior.

Double-diffusion in porous media has already developed into an actively developing discipline in its own right (Nield, 1968; Taunton *et al.*, 1972; Griffiths, 1981; Green, 1984; Chen and Chen, 1993; Diersch and Kolditz, 2002; Zhao *et al.*, 2008; Trevelyan *et al.*, 2011; among many others) but the primary applications of such studies are geophysical. Porous double-diffusive convection is likely to occur in the dendritic mushy zones, formed during the cooling and crystallization of magmas. Another application, on much larger scales, is related to the motion of hydrothermal flows in the Earth's crust. The hydrothermal circulation extends to depths of about 10 km and is affected by both thermal and haline chemical buoyancy. The temperature stratification is maintained by geothermal heating and regional variations in salinity are inevitable due to active geochemical processes, seawater intrusion or contaminant disposal. There are a number of key physical differences between the homogeneous and porous media. Unlike heat, solute cannot diffuse through a solid matrix. As a result, temperature perturbations disperse more rapidly in porous media, which greatly reduces the effective diffusivity ratio. Even more unusual are the advective effects: chemical components are advected by the velocity of the liquid component but heat is effectively transported by the much slower average velocity of the media. Thus, the fluid dynamics of porous systems can be both double-diffusive and double-advective. Mechanical dispersion is a new and significant form of mixing that appears in porous material due to the obstructions a fluid element experiences on a particular flow path. Momentum balance is better represented by Darcy's law (Darcy, 1856) and its extensions, rather than by the Boussinesq model commonly used for homogeneous media. The porosity values (the fraction of volume occupied by fluid) in the crust are generally low ($\phi \sim 10^{-4} - 10^{-1}$) and therefore the circulation patterns and the consequences of porous double-diffusive processes differ from their homogeneous counterparts.

The most natural double-diffusive configuration for fluids circulating in the Earth's crust is diffusive convection – geothermal heating from below destabilizes the system whereas the chemical gradient is stabilizing – and this case has been studied more extensively. As expected, numerical simulations (Schoofs *et al.*, 1999, 2000) reveal that diffusive effects often lead to the formation of well-defined horizontal layers with nearly uniform temperature and composition. However, for typical geological parameters, diffusive convection in low-porosity media appears to be more disorganized than in pure-fluid systems. Chaotic features include irregular transitions from layered to non-layered stratification patterns. Phase (liquid/vapor) separation is another common and important process in hydrothermal settings and the inclusion of multi-phase effects can considerably affect double-diffusive flow patterns and transport characteristics (Geiger *et al.*, 2005). In this complicated and highly nonlinear system, the conventional governing parameters (such as the density ratio and the Rayleigh numbers) tend to lose some of their controlling influence on the dynamics of diffusive convection, giving way to parameters characterizing phase transitions and permeability of the media. Diffusive convection in the Earth's crust has significant implications for ore deposit formation and mineralization in hydrothermal systems, for sedimental history of subsiding basins, for groundwater pollution, and for heat transport at mid-ocean ridges. Fingering convection in porous media has also been considered (e.g., Imhoff and Green, 1988) as one of the possible scenarios of groundwater contamination in which warm, chemically laden solute overlies cooler, fresher and denser groundwater.

12.3 Chemistry

Numerous fluid dynamical processes in nature, including several geophysical and astrophysical phenomena, are accompanied by chemical reactions. These reactions add a new level of complexity to double-diffusive convection. The double-diffusive system ceases to be conservative; not only do density components diffuse and advect but they can be produced internally. The possibility of strong interactions between chemical kinetics and fluid dynamics has been recognized for quite some time (e.g., Dewel *et al.*, 1983; Pojman and Epstein, 1990). However, research activities in this field have greatly intensified in the last decade or so. A new lease on life for chemical double-diffusion was brought by a fuller appreciation of the depth and richness of coupled hydro-chemical systems.

Some effects are straightforward and could be easily anticipated. If an exothermic autocatalytic reaction results in a product that is compositionally denser than the reactant solution then compositional and thermal changes have opposing influences on density – the antagonism that lies at the heart of double-diffusive dynamics. Such configurations arise in the nitric acid–iron(II), chlorite–thiosulfate, chlorite–tetrathionate, chlorate–sulfite, chlorite–thiourea, and bromate–sulfite autocatalytic fronts: the products contract while the reaction is exothermic. Free radical polymerizations also fall into the same category: they are very exothermic and form polymer products that are denser than the original monomers (Pojman *et al.*, 1992). Thus, if a chemical front is spreading downward (product above the reactant) then the system becomes susceptible to fingering. In the case of an ascending front (product below the reactant), the conditions could favor diffusive convection. Of course, the occurrence of double-diffusive convection in such systems is not guaranteed. Depending on the values of governing parameters, the system could also become buoyantly convective (top-heavy) or completely stable, but our main interest lies in the classical double-diffusive configurations (bottom-heavy and unstable). If the front spreads very slowly, allowing enough time for double-diffusive convection to fully establish, and if advective-diffusive effects are more significant than the chemical forcing, then the dynamics reduces to that of classical non-reactive systems. The role of chemical reactions in such cases is limited to creating the background stratification that supports double-diffusive convection.

However, most frequently, chemistry introduces a series of new and fundamentally coupled phenomena. Pojman *et al.* (1991) emphasize the impact of double-diffusive processes on the structure and propagation velocity of the chemical front, supporting their arguments by experiments with the iron (II)–nitric acid system. The rate of propagation of a chemical front is set by a balance between the diffusion of the reacting material and the intensity of the chemical reaction. The double-diffusive processes, salt-finger or diffusive, dramatically accelerate the transfer of properties across the front relative to molecular diffusion, which increases the speed of the front by up to two orders of magnitude. The descending (fingering) fronts are more distributed and generally propagate at higher speeds than the ascending ones. The motion of the front, in turn, can affect the double-diffusive circulation patterns. A number of studies have modeled, experimentally and numerically, the fluid dynamics of chlorite–tetrathionate reactive fronts (Bansagi *et al.*, 2003, 2004; Toth *et al.*, 2007), and examples of fingering are shown in Figure 12.8. Kalliadasis *et al.* (2004) noted that fingers formed in the descending fronts are frequently characterized by a “frozen” cellular structure, and – unlike non-reactive fingers – remain

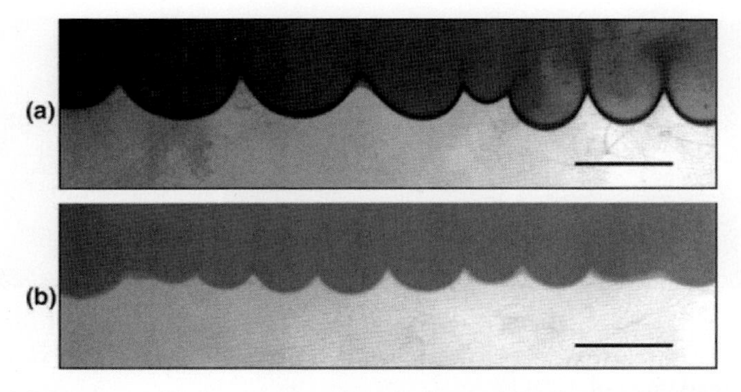

Figure 12.8 Images of fronts propagating downward. Darker regions represent the product solution and lighter the reactant. The black bar corresponds to 5 mm. From Bansagi *et al.* (2004).

spatially constrained to a narrow high-gradient interface and do not penetrate into the homogeneous interior.

Even more subtle and intriguing phenomena have been reported by D'Hernoncourt *et al.* (2006, 2007a,b), who argued that double-diffusive destabilization of the ascending exothermic chemical fronts is possible even when both temperature and compositional gradients are overall stabilizing. This counterintuitive effect is driven by the interplay between chemistry and fluid dynamics. When a fluid particle is displaced from the upper (warm and light) region occupied by the product-rich material into the reacting zone, it rapidly adjusts its temperature but not its compositional content. Thus, it is less engaged in chemical reactions than the surrounding reactant-rich fluid. The particle receives less heat chemically, becomes denser, and continues to sink. The positive feedback implies the existence of a new type of fingering instability, which does not arise in non-reactive media. More detailed analysis reveals that the chemical instability takes the form of small-scale cellular structures located in the reacting front and that it is highly sensitive to the diffusivity ratio. This instability is most obvious in doubly stable configurations. However, the chemical modes are also present in double-diffusive regimes, where they can substantially broaden the parameter space of instability. Of particular theoretical interest are the mixed instabilities that critically depend on both chemical and classical double-diffusive mechanisms.

Of course, temperature need not be the main diffusive component, and innumerable possibilities exist for double-diffusive effects driven by the differences in diffusivities of individual chemical components. For instance, Rica *et al.* (2010) performed a series of experiments with acid-catalyzed chlorite–tetrathionate reactions in a horizontally propagating front. Thermal effects on buoyancy were largely negligible, yet the purely compositional double-diffusion was sufficiently vigorous

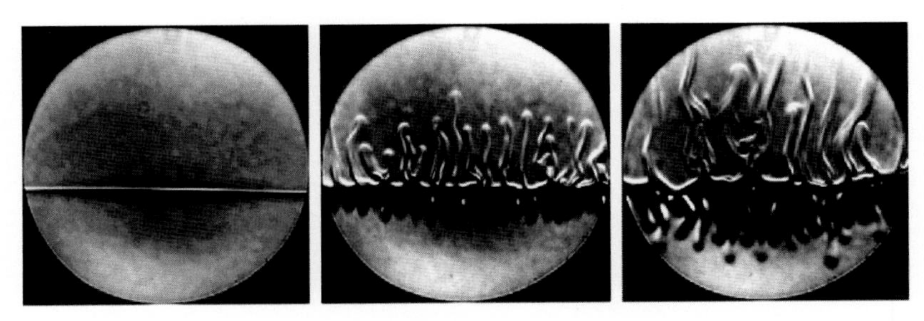

Figure 12.9 Evolution of the interface between a solution of HCl on top of a solution of CsOH. A snapshot was taken every minute. The field of view is 3 cm. From Almarcha *et al.* (2011).

and profoundly affected the patterns and evolution of the system. A number of studies have focused on the instabilities of the interfaces between solutions containing different reactants (Zalts *et al.*, 2008; Almarcha *et al.*, 2010a,b). Such configurations are more complicated and phenomenologically rich than the systems with an initially homogeneous reactant and typically are more unstable. The dynamics of such systems are controlled by the relative densities and diffusivities of both reactants and their product.

Interesting examples were presented by Almarcha *et al.* (2011), who performed a series of experiments on the neutralization of a strong acid by a strong base in an aqueous solution. By systematically varying the initial concentrations and the type of base reactant, various forms of buoyancy-driven circulation were produced and systematically examined in the Hele-Shaw cell. Figure 12.9 presents an experiment in which a solution of hydrogen chloride (HCl) was placed in the upper layer and caesium hydroxide (CsOH) in the lower layer. Since H^+ is the fastest diffusing cation and OH^- the fastest diffusing anion, the product of the reaction diffuses slower than both of these reactants. Thus, for suitably chosen experimental parameters (density decreases upward but gently enough not to quench double-diffusion) it is possible to induce the circulation pattern that is concurrently controlled by diffusive convection in the upper layer and by fingering in the lower layer. These dynamics are easily identifiable in Figure 12.9 by irregular plumes rising above the interface and narrow fingers spreading downward. Thermal effects in this experiment are of marginal significance. The evolutionary pattern shown here is one of several possible outcomes; depending on the reactants and their concentration, circulation can be limited to diffusive plumes in the upper solution only or to fingers in the lower solution. However, the hybrid system (Fig. 12.9) is of particular interest since it conjoins in a single experiment the adversary diffusive and fingering phenomena. The simplicity and elegance of the experimental setup in this case is afforded by the clever use of chemical reactions to control fluid dynamics – a nice example of an effective and truly multidisciplinary approach.

12.4 Materials science and engineering

Over the years, double-diffusion has become recognized as an important process in quite a few industrial applications and is currently the subject of intense research activity in these areas. The general philosophy of engineering studies is conspicuously dissimilar to that of more basic physical sciences. In most industrial problems, double-diffusion is regarded as an undesirable phenomenon: it compromises the quality of metal alloys, increases the chances of rollover in liquid gas tanks, and adversely affects the performance of solar ponds. Engineering research is motivated by the "know your enemy" principle, which actually can be an even more effective driver than the intellectual curiosity of a physicist or a mathematician. However, the pragmatic engineering attitude has introduced a strong bias towards the solution of immediate practical problems, leading the field a bit astray from analyzing the fundamentals of double-diffusive dynamics. In this section, we discuss two representative examples of industrial applications: solidification of metal alloys and double-diffusive effects in solar ponds.

Solidification of metal alloys

There are several ways in which double-diffusion comes into play in solidifying systems. On the most primary level, it affects the morphology of the liquid/solid interface (Coriell and Sekerka, 1981; Schaefer and Coriell, 1984; Lan and Tu, 2000). The geometric complexities of dendritic solidifying zones can be ultimately traced to small-scale instabilities of initially planar interfaces. The significance of such instabilities in the dynamics of solidification has long been known (Langer, 1980), and their ability to create intricate patterns is truly impressive. The exquisite shape of a snowflake, incredibly complex and precise, is perhaps the most common and striking example of the spontaneous evolution of a two-phase interface. In some cases, the instability is purely morphological; it is driven by capillary effects, diffusion kinetics and thermodynamic constraints. However, the interfacial instabilities become very different in the presence of buoyancy-driven circulation in the liquid layer. The fluid dynamical and morphological instabilities are intrinsically linked by the boundary conditions at the interface and their interaction becomes particularly intricate in doubly diffusive media (Coriell *et al.*, 1980; Nguyen Thi *et al.*, 1989; Anderson *et al.*, 2010). New morphological/double-diffusive unstable modes emerge and profoundly affect the evolution of an interface. Somewhat counterintuitively, these coupled instabilities can occur even when both diffusing substances in the liquid region are stably stratified. The double-diffusive version of the classical phase-change problem is phenomenologically rich and offers new physical routes to the breakdown of the interface and the formation of dendrites.

An example of a more industrially motivated type of double-diffusive research in metallurgy is given by studies of freckle formation in metal alloys. Freckle defects are the segregation channels that are formed during directional solidification of alloys, and their occurrence remains a major problem in the casting industry. For instance, freckles can severely weaken superalloy turbine blades for high-temperature applications. The rejection rate of turbine blades due to macrosegregation defects can be as high as 40%. It is generally accepted (Giamei and Kear, 1970; Sarazin and Hellawell, 1988; Hansen *et al.*, 1996) that freckles are caused by buoyancy-driven circulation – convection or double-diffusion. The latter is commonly referred to in metallurgical literature as thermosolutal convection. The dynamics of freckling is analogous to the convective fractionation effect that occurs in magma chambers (Section 12.2); it involves the interaction between the morphology of solidification and fluid dynamics.

When the alloy is cooled from below and liquid released during solidification is compositionally light, the zone immediately above the mushy solidifying layer becomes susceptible to fingering. The buoyancy-induced circulation plays a dual role in the dynamics of solidification: it directly causes the segregation of the material and also provides a mechanism for the transport of dendritic fragments from the mushy region into the bulk liquid. Surviving fragments become nuclei for solidification and block the parental dendritic front. The resulting inhomogeneities of the solid material can be substantial, profoundly affecting its mechanical characteristics. A series of such finger-driven freckles is clearly visible on the surface of a superalloy ingot shown in Figure 12.10. Similar adverse double-diffusive effects often compromise the production of chemically homogeneous crystals used in the electronics industry.

The closed character of casting systems poses obvious constraints on the direct measurement of solidification in metal alloys. To address the observational limitations, two lines of inquiry have been actively pursued. The older traditional approach (e.g., Copley *et al.*, 1970; Sample and Hellawell, 1984; Hansen *et al.*, 1996) involves the analysis of transparent aqueous analogs followed by the extrapolation of the results to opaque metallic systems based on the post-mortem examination of fully solidified samples. The major problem with this method is that the analogies between the metallic and aqueous systems are very qualitative. Prandtl numbers and diffusivity ratios differ by orders of magnitude and, consequently, so do the double-diffusive fluxes. Typical incubation periods in solidifying metals are very brief compared to crystallizing aqueous systems, which makes the

Figure 12.10 An example of freckling in a metal alloy. From Hansen *et al.* (1996).

extrapolation attempts even more questionable. Experimental uncertainties, combined with computational advancements over the past few decades, have led the casting community to become increasingly more interested in the numerical modeling of solidification (reviewed by Voller *et al.*, 1990; Beckermann, 2002; among others).

Numerical models of solidification have been developed since the 1960s (Flemings and Nereo, 1967, 1968a,b). However, the early models were principally deficient in their focus on the solute redistribution during solidification contraction; double-diffusive effects were not taken into account. As a result, these models failed to capture the essential mechanisms for the formation of freckles. The next generation of numerical models appeared much later (Bennon and Incropera, 1987; Beckermann and Viskanta, 1988). The entire casting system, including its liquid, mushy and solid components, was treated using the unified formalism of volume-averaged equations. The differences in types of media were taken into account by assuming spatially variable permeability, which removed the need to model the evolution of the liquid/mushy interface. The ability of these models to explicitly represent double-diffusive effects substantially improved their predictive capabilities.

However, several significant challenges remain. Mesoscale solidification models are only as good as the macroscopic parameterizations they assume for all essential microscopic processes, including nucleation, microsegregation, thermal/solutal undercooling and solid movement. The development of accurate parameterizations – currently an active area of research – is necessarily conditioned by progress in the physical understanding of solidification microdynamics (e.g., Heinrich and Poirier, 2004). In this regard, it is interesting to note that double-diffusion is also engaged on the microscale level. Double-diffusively induced currents inside interdendritic regions affect the dendrite growth and pattern formation. Taking the thermosolutal circulations into account reduces the spacing between the adjacent dendrites (Spinelli *et al.*, 2004, 2005), which impacts the macroscopic properties of the mushy zone.

Another major difficulty is associated with the wide range of dynamically important scales present in double-diffusive convection, from the solute dissipation scale to the characteristic size of the casting domain. The modeling of freckle segregation with mesh adaptation schemes (Sajja and Felicelli, 2011) appears to be particularly promising in this regard. A different set of complications arise in modeling multicomponent alloys. The analysis of generalized solidification models developed for multicomponent systems (e.g., Schneider and Beckermann, 1995) has revealed that additional elements induce substantial changes in the convective patterns and macrosegregation dynamics. This finding has brought into question earlier attempts to model multicomponent systems with binary models. Overall though, I am happy to report that, despite all the obstacles, the field of solidification modeling is rapidly advancing. In the not too distant future we can expect development of reliable computational models that can guide innovation in the casting industry. The ultimate goal of such models is to reduce or even eliminate the need for the expensive and time-consuming experimental trial-and-error stage in the design of ingots and castings.

Solar ponds

The first impression of a person trying to figure out the operating principle of a solar pond is that this device is nothing more than the diffusive convection experiment, reproduced on a large scale and industrially utilized. The essential dynamical component of solar ponds is the initially introduced stable salinity stratification. The relatively shallow layer of water (1–4 m) in solar ponds is nearly translucent to solar radiation and therefore much of the radiant solar heat is absorbed at the bottom. As a result, water in deep regions of a solar pond is much warmer than that near the surface. In the absence of salinity stratification, this would lead to convective overturns and rapid homogenization. However, salt maintains a stable

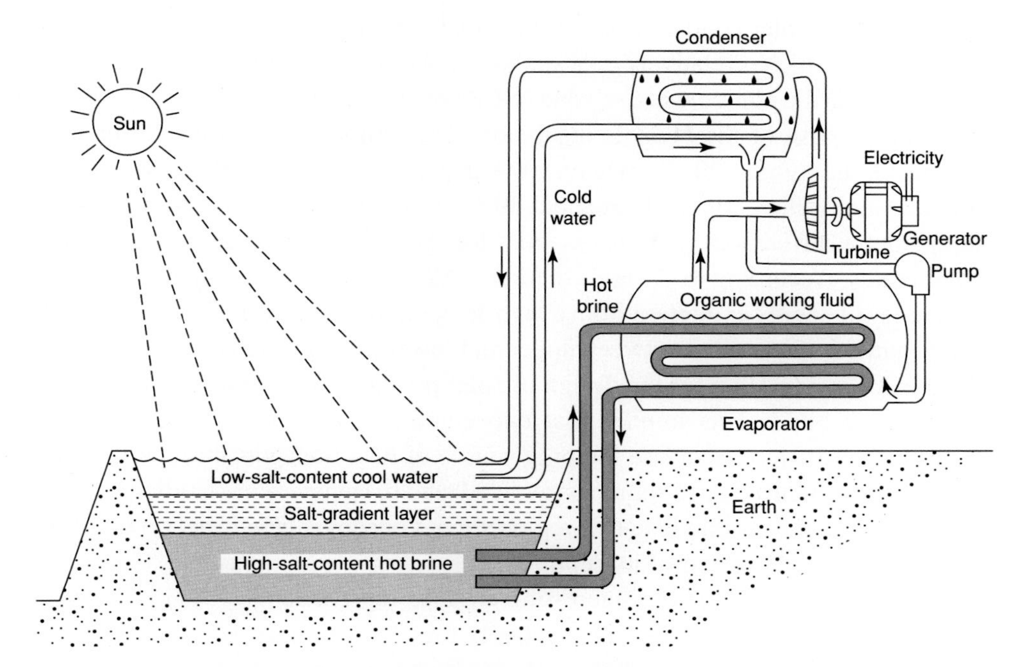

Figure 12.11 Schematic diagram of a power-generating solar pond.

density distribution, the circulation is minimal, and heat remains trapped in the lower layer. Thermal energy collected in this manner can be used for various applications, such as heating, desalination, refrigeration, drying and electric power generation (Fig. 12.11). While in this section we focus on artificial solar ponds, it should be mentioned that there are natural examples of solar pond dynamics, such as the saline Solar Lake in the Sinai Peninsula of Egypt.

Salty, warm and dense water that is located below cold, fresh and buoyant sounds very familiar – it is the configuration prone to diffusive convection. But does this mean that double-diffusion has found a new practical use in the form of solar ponds? Not exactly. The mixing induced by diffusive convection can transport heat to the surface, where it would be rapidly lost to the atmosphere. This could seriously compromise the dual purpose of a solar pond as a thermal energy collector and as a heat storage mechanism. Once again, double-diffusion shows its dark side and should be avoided in solar ponds by all means. This is not a trivial task since the medium (heat–salt) is fundamentally double-diffusive and therefore the solar pond problem has stimulated a fair amount of research. Designing solar ponds and making proper maintenance decisions requires clear understanding of the conditions that can trigger double-diffusive convection and of its consequences (Tabor, 1980, 1981).

Typically, a solar pond consists of three distinct regions (Fig. 12.11): (i) the top mixed layer, which is relatively cool and fresh, (ii) the warm and salty bottom mixed layer and (iii) the non-convective zone (NCZ) separating the two mixed layers. The primary purpose of the NCZ is insulation. The temperatures in the lower layer can reach as high as 90 °C, whereas the upper layer is in equilibrium with the ambient atmosphere, typically at about 30 °C. Maintenance of large top-to-bottom temperature differences (ΔT) is essential for the efficient extraction of energy in solar ponds. According to Carnot's theorem, ΔT controls the maximum theoretical efficiency of energy extraction from a high-temperature reservoir (Carnot, 1824). In this regard, the insulation between upper and lower mixed layers is critical and the foremost consideration for the design of solar ponds is to minimize heat transport through the NCZ. This implies that two conditions are met: (i) the inevitable molecular diffusion should be the only source of heat transport through the NCZ and any form of turbulence should be avoided; and (ii) temperature gradients should be low, which means that the NCZ region should occupy a large fraction of a solar pond, typically about half of its depth.

Both requirements lead to interesting engineering challenges. As discussed in Chapter 4, diffusive systems with relatively low density ratios ($R_\rho^* < \tau^{-1/2}$) tend to spontaneously evolve to a statistically steady configuration in which convecting layers are separated by a thin interface. In order to avoid this scenario, salinity concentration in the lower layer should be very high ($R_\rho^* > 10$). This complicates the maintenance of solar ponds, requiring continuous brine injection in the lower layer, the removal of salt crystals and surface washing (Ouni *et al.*, 2003). But even if the density ratio is kept high, convective motions in the lower mixed layer can still lead to the gradual erosion and/or vertical drift of the NCZ (Zangrando and Fernando, 1991; Karim *et al.*, 2011). The life expectancy of solar ponds due to breakdown of the NCZ is about 10–15 years, after which the solar pond has to be drained and refilled in order to continue its operation. The destruction of the gradient region occurs by means of local instabilities at the lower boundary of the NCZ, induced by bottom-layer convection. To some extent, the problem could be mitigated by introducing the initial salt stratification that minimizes the strength of such instabilities – an interesting fluid dynamical problem in its own right. Karim *et al.* (2011) also suggest that the durability of solar ponds can be substantially increased by introducing a grid that would suppress the convective vortices in the lower layer. Implementation of thin horizontal membranes in solar ponds to prevent the vertical transport of salt is another interesting idea, which is currently at the research and development stage.

Another pressing problem for the operation of solar ponds, which is also double-diffusively induced, involves the generation of lateral interleaving at the sidewalls (Akbarzadeh and Manins, 1988; Jubran *et al.*, 1999). The relatively small-scale

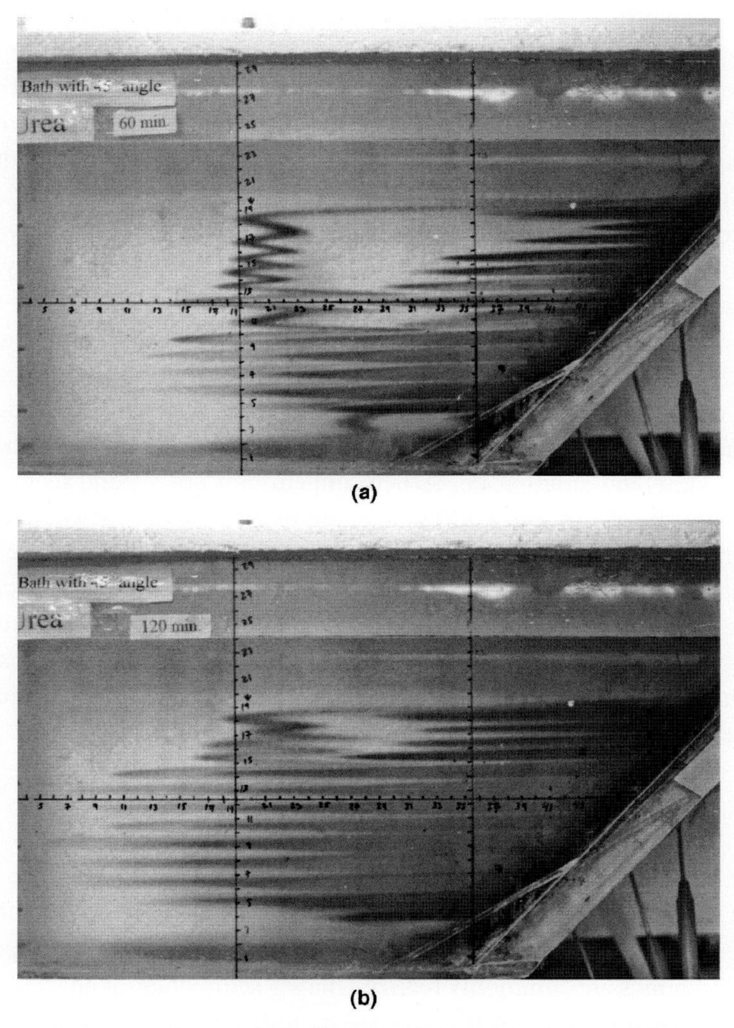

Figure 12.12 Interleaving in a solar pond with inclined sidewalls. From Jubran *et al.* (1999).

intrusive structures, driven by heat absorbed at the sidewalls, can lead to substantial changes in the overall stratification of the NCZ. If this effect is not accounted for, the gradient zone can become unstable, which would ultimately lead to the destruction of the NCZ. Sidewall interleaving is a generic property of diffusively stratified systems (Chapter 7). Intrusive currents could be particularly intense in solar ponds with inclined walls, which are often used to minimize the ground heat loss and to eliminate shadows. Figure 12.12 shows a series of nearly horizontal layers spreading from the sidewalls into the interior of a solar pond. Their progression is quite rapid and effects on the stability of the NCZ could be profound. As a way of reducing the adverse consequences of sidewall interleaving, Jubran *et al.* (1999) suggest several measures: (i) usage of high angles of sidewalls; (ii) experimentation with different types of salts – intrusions are sensitive to the molecular properties of the media; and (iii) increasing the roughness of sidewalls, which has a detrimental effect on intrusions.

Despite the numerous problems of this nature, solar pond technology is promising. The first solar pond experiments were made in the late 1950s, and by the 1970s significant progress had been made (Tabor, 1980). Solar ponds have been built and industrially operated in several countries (Israel, Australia, India and Tunisia, among others). For many years, solar ponds have been discussed as an attractive source of cheap energy in developing countries, most of which are located in warm low-latitude climates. However, the implementation of this idea has been disappointingly slow and small-scale. A new impetus for solar pond technology was brought about by the recent worldwide surge of interest in renewable energy. On the other hand, solar ponds now have to compete with other emerging technologies: solar panels, wind-driven turbines and bio-fuels. Whether solar ponds can survive and succeed in such a competitive environment depends on many factors, one of which is our ability to design reliable, low-cost but long-lasting systems. Progress in this direction requires complete understanding of the fluid dynamics of solar ponds and, particularly, of the ubiquitous double-diffusive effects.

12.5 Other applications

There are several applications of double-diffusive convection that are not easily classified into well-defined scientific disciplines and therefore are combined here under the hopelessly vague heading of "other." For instance, archetypal double-diffusive dynamics are commonly realized during sedimentation in stratified fluids. Stratified sedimentation in itself is a broad multidisciplinary subject, largely motivated by environmental concerns. Specific applications include a variety of natural and industrial processes: dynamics of dust particles in the air, sedimentation in rivers and oceans, collective evolution of swimming microorganisms, dispersal of atmospheric pollutants, dilution of undersea sewage and dredged materials. If the background density stratification is gravitationally stable and the particle concentration decreases downward, the sedimenting particles could play the role of a destabilizing agent. The resulting instability takes a form dynamically analogous to fingering convection (Drake, 1971; Chen, 1997). These particle-driven dirt fingers have been studied experimentally in two-layer configurations (Green, 1987; Hoyal *et al.*, 1999), in a gradient stratification (Houk and Green, 1973) and at the base of spreading gravity currents (Maxworthy, 1999; Parsons *et al.*, 2001).

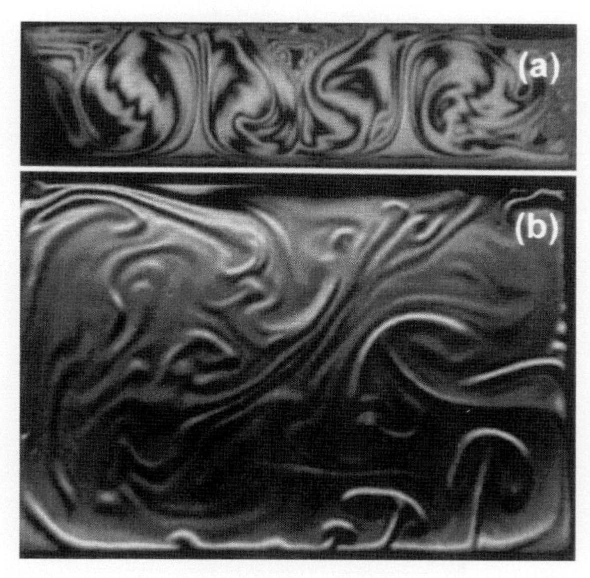

Figure 12.13 Flow patterns in soap films. (a) The steady circulation regime for the temperature variation of $\Delta T = 15$ °C. (b) Chaotic pattern realized for $\Delta T =$ 45 °C. From Martin and Wu (1998).

All this talk about dirty fingers makes us yearn for some soap. Some peculiar examples of double-diffusion can be found among experiments with freely suspended soap films (Martin and Wu, 1998). At first, such experiments come across as a bit esoteric and frivolous, although undeniably entertaining. In reality, however, they carry the significant function of isolating essentially two-dimensional aspects of double-diffusive convection in the laboratory setting. In this regard, soap films offer an attractive alternative to the thin-gap (Hele-Shaw) experiments, where solid boundaries strongly impede flow, resulting in fundamentally frictional dynamics. The setup of soap film experiments is simple and circulation patterns are easily observable. Figure 12.13 shows the diffusive convection that occurs when a soap film is heated from below. The competing double-diffusive mechanisms in this case are the density reduction due to heating and the film's tendency to increase its thickness downward under the action of gravity. As the temperature difference between the bottom and the top of the film is gradually increased, the system first transitions from a stable steady state to a time-dependent state characterized by periodic oscillations of a standing-wave pattern. When the temperature difference is further increased, the system develops the steady circulation pattern presented in Figure 12.13a. This regular pattern changes once again for even larger temperature differences, resulting in the chaotic and disorganized state in Figure 12.13b. Overall, the dynamics and transitions observed in the soap film experiment follow

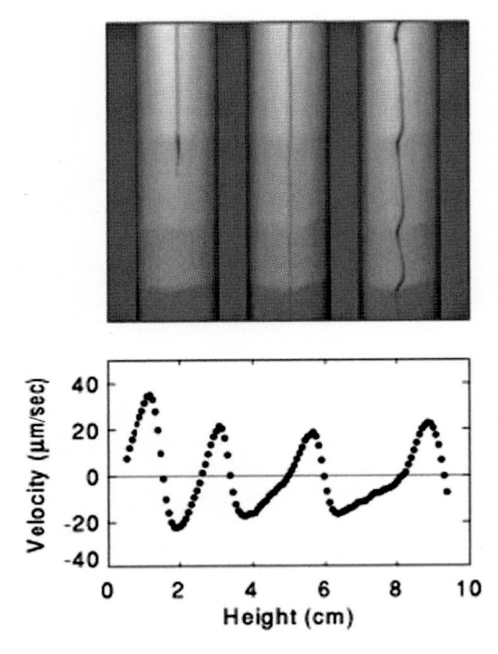

Figure 12.14 Direct visualization of convective flow in a stratified suspension. (Top) Evolution of stratified emulsion and formation of layers. The circulation pattern is revealed by the deformation of the initially vertical dye line. (Bottom) The horizontal flow velocity along the tube's center line as measured from the photographs above. From Mueth *et al.* (1996).

the scenario realized in the analytical and numerical studies of diffusive convection (Veronis, 1965, 1967).

An interesting example of double-diffusion at play is given by spontaneous layering of initially homogeneous colloidal suspensions in narrow containers. This effect has been known for more than a century (Brewer, 1884) but a clear physical explanation has eluded observers. Several mechanisms have been proposed to explain these observations, including spinodal decomposition, formation of streaming flows, and generation of Burgers shocks. However, a series of flow visualization experiments with creaming emulsions (Mueth *et al.*, 1996) finally provided strong evidence in favor of the double-diffusive mechanism of layering. These experiments demonstrated that each layer in a stratified suspension is associated with a convecting roll (Fig. 12.14). The circulation pattern was shown to be highly sensitive to the horizontal thermal gradient, which led Mueth *et al.* to suggest that the origin of the layers is the instability due to coupling between the vertical gradient in the concentration of suspended particles and the horizontal thermal gradient. This mechanism is dynamically analogous to the thermohaline interleaving in laterally bounded fronts discussed in Chapter 7. The lateral gradients required to engage such dynamics are surprisingly low ($\sim$10 mK m^{-1}) and could easily have been present in most, if not all, previous experiments with colloidal suspensions.

Double-diffusion in moist air is rarely invoked in the discussions of double-diffusive convection. The diffusivity ratio of humidity/heat is close to unity ($\tau \approx 0.8$) and therefore double-diffusive effects are usually assumed to be secondary. This assumption has been questioned by Bois and Kubicki (2002, 2003), who pointed out that the dynamics of saturated air can differ from that of unsaturated systems. In the saturated case, the buoyancy-driven circulation is inherently coupled with the liquid/vapor phase change. Since the diffusivities of the liquid and gaseous phases are different, double-diffusive processes are readily engaged. Bois and Kubicki argue that the resulting instabilities come in the form of moisture fingers, dynamically analogous to salt fingers in the ocean. These moisture fingers, among other processes, could be responsible for the specific texture of "fleecy clouds" in the statically stable moist-saturated air. The oscillatory diffusive regime is also possible in the atmosphere but should not be as common as fingering under typical meteorological conditions.

The list of applications could go on and on. There is an immensely broad range of problems that have formal mathematical analogies with double-diffusive convection, although the physics may be dissimilar. One such problem is the analysis of Langmuir circulation – a system of shallow, counter-rotating vortices, commonly manifested by parallel streaks of foam and debris at the surface of a natural body of water. While this phenomenon is unrelated to double-diffusion of heat and salt, the governing equations that are used to describe Langmuir circulation are identical to the standard double-diffusive system. In this case, velocity and temperature play the roles of T and S respectively (Leibovich, 1983). Another example of isomorphism is presented by the so-called GSF (Goldreich–Schubert–Fricke) instabilities associated with differential rotation in stably stratified stars (Goldreich and Schubert, 1967). The difference in the rate of energy and lepton diffusion in post-collapse stellar cores can drive the neutron-finger instability, which could play a role in supernova explosions (Wilson *et al.*, 1986; Wilson and Mayle, 1993). Convection in a magnetic field is yet another field that has direct parallels with double-diffusive convection (Proctor and Weiss, 1982). A whole class of problems emphasizes the dynamically distinct roles of buoyancy components even when their molecular diffusivities are the same – the fundamentally two-component effects may occur due to the nonlinearities of a mixing model or the differences in boundary conditions (Welander, 1989; Ferrari and Young, 1997). Another Pandora's Box opens when the difference in viscosities, rather than diffusivities, of mixing components is taken into account (Homsy, 1987). Such double-viscous systems are known to produce viscous fingering, a phenomenon that has apparent similarities to the buoyancy-driven fingering.

Undoubtedly, the exchange of knowledge between different fields, related by equivalent dynamics and/or by similar math, can be highly beneficial for all parties involved. But whether or not such analogue systems should be considered part of the core theory of double-diffusion is open to interpretation. In this monograph, we adopted a rather conservative approach – the "if it is not about heat and salt, then don't complain" principle. The whole grey area of related-to-double-diffusion studies has been mentioned in passing. The breadth of the subject is the author's only excuse. Considering that the number of articles published on the salt-finger subject alone already exceeds a quarter of a million (Google Scholar) it becomes apparent that any attempt to summarize the status of the field in a single book requires major sacrifices. The choices that have been made are unavoidably subjective and many of them might be unintentionally unfair. On this apologetic note, let us proceed towards the end. A couple of brief personal remarks in the next chapter will conclude our story.

13

Perspectives and challenges

Double-diffusion is a fun science. It is filled with unresolved dynamical problems and spiced up with controversial ideas. It challenges the imagination with counter-intuitive phenomena; it operates on vastly different scales and connects dissimilar disciplines. If this book conveys at least some of the beauty and depth of the subject, I am happy – the time invested in writing it was well spent.

As with any other extensive review, our discussion of the current state of double-diffusion inevitably leads to questions and speculations regarding its prospects. It was great for the last fifty years, how about the next fifty? Let us first state the obvious: the future of this field is secure. Interest in double-diffusion is on the rise, as reflected by the exponentially increasing frequency of publications, and there are no reasons to expect a slow-down. Many signs point towards much more rapid growth in years to come. Just by looking at the current playing field in fluid dynamics, one gets a sense of an impending change. Several classical disciplines are already beyond their prime; their central, most fundamental problems have been either resolved or abandoned, giving way to less exciting incremental developments. In this sense, double-diffusion is in excellent shape; we don't really understand a whole lot. The fall-out from aging sciences can only further invigorate double-diffusive research. The exploration of multicomponent phenomena is also the most natural evolutionary path for many established subjects (convective turbulence being the prime example), which is another reason to expect fluid dynamics to become progressively more double-diffusive. Regarding its real-world motivation, double-diffusion is involved in such a broad spectrum of applications that it is hard to imagine a more stable portfolio. In summary, we predict, as confidently as one can without a clairvoyant's crystal ball, that our field will continue to advance and its best days are yet to come. Perhaps it is more interesting and productive to consider whether some actions should be taken to accelerate double-diffusive research even more. Are there some artificial barriers that could be removed, applications emphasized or promising strategies explored? A serious conversation about the direction of double-diffusive research can benefit all parties involved and is long overdue. No firm prescriptions are given here, but as a starting point for the discussion I would like to offer a few observations and personal remarks.

13.1 Perceptions

Curious as it may be, double-diffusion does not exist in isolation. It is rooted in the traditionally defined sciences: oceanography, materials science and astrophysics, just to name a few. The strong connection with paternal disciplines is vital for its expansion and there is some room for improvement in this area. An important step involves raising the level of understanding and awareness of double-diffusion among non-experts. Most researchers in physical and environmental fields are familiar with the concept of double-diffusion, but important caveats tend to escape attention. Half-knowledge creates fertile ground for misconceptions that often distort the perception of our field. To illustrate the need for better communication with the outside world, let us review some evidently false notions that still occasionally surface in discussions with non-double-diffusers:

(i) *Double-diffusion is an esoteric science.* Any conceivable metric shows that double-diffusion has long graduated from oceanographic curiosity to a broad and vibrant field. For instance, the number of publications listing "salt finger" as a keyword exceeds that for "thermocline" or "meridional overturning circulation" – a couple of the most popular topics in oceanography. The significant and demonstrable impact of double-diffusion on climate and the biological productivity of the oceans is a reason in itself to invest more effort in its analysis. Double-diffusion is spread between several disciplines, which makes it difficult to fully grasp its overall importance, particularly for researchers reluctant to cross interdisciplinary boundaries. However, it should be realized that the cumulative influence of double-diffusion greatly exceeds its contribution to each subject. All the information required to dispel the "esoteric" myth is readily available; it only remains to make it widely known.

(ii) *Double-diffusion is similar to other mixing processes in the ocean.* Double-diffusion is unique in many ways. Unlike mechanically generated turbulence, it is highly interactive: double-diffusive fluxes are sensitive to the background stratification and the fluxes, in turn, affect stratification. The resulting feedback mechanisms trigger a series of secondary finescale instabilities (intrusions, collective and layering modes), which impart life and spontaneity into the ocean. Another essential double-diffusive characteristic is the counter-gradient density flux: double-diffusion makes light water lighter and dense water denser.

Its ability to "unmix" density leads to sharpening, rather than diffusing, of vertical gradients. In the absence of double-diffusion, density patterns in the ocean would be as smooth, predictable and hopelessly boring as a bald man's scalp. Combining double-diffusion with turbulence under the general heading of small-scale mixing trivializes and masks very unusual dynamics and can ultimately lead to significant conceptual misinterpretations.

(iii) *Double-diffusion is a small-scale phenomenon.* Diffusive layers spread across the Beaufort Gyre, remaining laterally coherent from coast to coast and maintaining their identity for years. Thermohaline intrusions can reach vertical scales of hundreds of meters and laterally mix water masses over hundreds of kilometers. The fingering staircase in the Tyrrhenian Sea splits the water column into homogeneous layers half a kilometer thick. The ability of double-diffusion to directly affect such an enormous range of scales – from the salt dissipation scale ($\sim 10^{-3}$ m) to the basin size ($\sim 10^{6}$ m) – may be unmatched by any other physical phenomenon.

For many double-diffusers, concerns of this nature appear so unreasonable that no attempt is made to address them – well, isn't it obvious? However, in order for our field to completely break the shell of isolation, basic facts about double-diffusion have to be patiently and pervasively conveyed to as wide an audience as possible. The payoff for publicizing double-diffusion could be substantial: new workers, fresh ideas and productive interactions. Double-diffusion is addictive and the newly initiated tend to become loyal and active contributors to the subject.

13.2 Barriers

In addition to the challenges of addressing the general scientific audience, double-diffusers also have to cope with internal communication barriers, many of which are caused by the multidisciplinary nature of our field. While key questions and goals in various applications are similar, very often practitioners are unaware of parallel developments in other disciplines. As a result, the transfer of ideas and methods can be random and slow. The problem is exacerbated by the highly specialized structure of education in the natural sciences. Graduate programs are fine-grained; we produce molecular biophysicists, physical oceanographers and extractive metallurgists. An alternative approach (the Eastern European model) would be to offer broadly defined graduate degrees in "physical sciences", which assume some common background in math and the ability to speak the same scientific language. Of course, specialization has its advantages and I am sure that many subjects have benefitted immensely from narrowly focused education. However double-diffusion, I am afraid, may have gotten a raw deal in this system. For instance, oceanography still remains the largest consumer and motivator of double-diffusive research. Yet, oceanographic education lacks several components that help in the pursuit of such quintessentially physical subjects as double-diffusion. Exposure to classical fluid mechanics and analytical techniques is limited. Courses in nonlinear instability – the heart and soul of double-diffusion – are notably absent in most oceanographic curricula. To alleviate the problems caused by excessive specialization, an argument could be made for (i) embracing flexibility in the type of research expected from faculty members in applied departments, (ii) stronger support of interdisciplinary activities, and (iii) placing more emphasis on fundamentals in graduate coursework.

Of course, interdisciplinary barriers are natural and, to some extent, inevitable: we read different journals, attend different conferences and may not be personally familiar with colleagues in other departments. However, disconnect can be detected even within the boundaries of the same discipline. In most fields, double-diffusion is actively studied by theoreticians, observationalists, experimentalists and modelers. There is yet another, and much larger, group of researchers whose primary focus is not on double-diffusion per se but on various phenomena that could be affected by it. The urgent request from this group is the accurate parameterization of double-diffusion in the simplest user-friendly format. Such pragmatic interests should also be known and taken into account when developing double-diffusive products. Unfortunately, the exchange of information between all these types of researchers is not as fluid and swift as it could be. Needless to say, the intra-disciplinary barriers are counterproductive and preventable.

It is perhaps most fitting to mention at this point that a precedent for an open and unifying approach to double-diffusive research was set by the founder of the field, Melvin Stern. Being a hard-core theoretician by trade, throughout his career he maintained keen interest in all aspects of double-diffusion. For him, new observational, experimental and modeling developments were a constant source of theoretical ideas and deeper insight into double-diffusion. In an interview for the Florida State University newsletter, Melvin recalled his participation in early laboratory fingering experiments:

> An experiment and collaboration with Turner led to the happiest professional day of my career. The day we did an experiment on salt-finger convection . . . we had been using mixes of heat and salt, but heat diffuses into the atmosphere too readily, so we had the idea of replacing heat with sugar. We were not sure it would work because salt and sugar have such a slight difference of molecular diffusivity that, we did not think anything would happen. But, there it was on the first try! It was an eye opener – the effect was large and very beautiful. That experiment aided my professional development in particular because I began to think in a different way.

What a truly scientific attitude, uncompromisingly objective and inclusive. This all-embracing mindset has to be emulated by the community in order to finish what Melvin started and finally unravel the salt-finger dynamics.

I end this monograph with a plea for collaboration. For an astrophysicist, the task of understanding a double-diffusive study in the oceanographic context may seem daunting at first, but a second look usually reveals that the differences in the way we approach the problem are rather superficial. It is not unfeasible to monitor key developments in double-diffusive research across several subjects and this experience can be most instructive and stimulating. The opportunities for collaboration are ample and should be pursued without hesitation. And there is really no excuse for the lack of interaction between experimentalists, modelers and theorists in the same discipline. The time has come for all of us to get our hands dirty and take care of these fingers.

References

Akbarzadeh, A., and P. Manins, 1988: Convective layers generated by side walls in solar ponds. *Solar Energy*, 41, 521–529.

Almarcha, C., P. M. J. Trevelyan, P. Grosfils, and A. De Wit, 2010a: Chemically driven hydrodynamic instabilities. *Phys. Rev. Lett.*, 104, 044501.

Almarcha, C., P. M. J. Trevelyan, L. A. Riolfo, *et al.*, 2010b: Active role of a color indicator in buoyancy-driven instabilities of chemical fronts. *J. Phys. Chem. Lett.*, 1, 752–757.

Almarcha, C., Y. R'Honi, Y. De Decker, *et al.*, 2011: Convective mixing induced by acid-base reactions. *J. Phys. Chem. B*, 115, 9739–9744.

Anderson, L. G., G. Björk, O. Holby, *et al.*, 1994: Water masses and circulation in the Eurasian Basin: results from the *Oden* 91 expedition. *J. Geophys. Res.*, 99, 3273–3283.

Anderson, D. M., G. B. McFadden, S. R. Coriell, and B. T. Murray, 2010: Convective instabilities during the solidification of an ideal ternary alloy in a mushy layer. *J. Fluid Mech.*, 647, 309–333.

Anschutz, P., and G. Blanc, 1996: Heat and salt fluxes in the Atlantis II Deep (Red Sea). *Earth Planet. Sci. Lett.*, 142, 147–159.

Armi, L., D. Hebert, N. Oakey, *et al.*, 1988: The movements and decay of a Mediterranean salt lens. *Nature*, 333, 649–651.

Armi, L., D. Hebert, N. Oakey, *et al.*, 1989: Two years in the life of a Mediterranean salt lens. *J. Phys. Oceanogr.*, 19, 354–370.

Baines, P. G., and A. E. Gill, 1969: On thermohaline convection with linear gradients. *J. Fluid Mech.*, 37, 289–306.

Baines, W. D., and J. S. Turner, 1969: Turbulent buoyant convection from a source in a confined region. *J. Fluid Mech.*, 37, 51–80.

Balmforth, N. J., and J. A. Biello, 1998: Double diffusive instability in a tall thin slot. *J. Fluid Mech.*, 375, 203–233.

Balmforth, N. J., S. G. Llewellyn Smith, and W. R. Young, 1998: Dynamics of interfaces and layers in a stratified turbulent fluid. *J. Fluid Mech.*, 355, 329–358.

Balmforth, N. J., S. A. Ghadge, A. Kettapun, and S. D. Mandre, 2006: Bounds on double-diffusive convection. *J. Fluid Mech.*, 569, 29–50.

Bansagi, Jr., T., D. Horvath, A. Toth, *et al.*, 2003: Density fingering of an exothermic autocatalytic reaction. *Phys. Rev. E*, 68, 055301(R).

Bansagi, Jr., T., D. Horvath, and A. Toth, 2004: Multicomponent convection in the chlorite-tetrathionate reaction. *Chem. Phys. Lett.*, 384, 153–156.

Barenblatt, G. I., 1996: *Scaling, Self-similarity, and Intermediate Asymptotics*. Cambridge University Press.

Barry, M. E., G. N. Ivey, K. B. Winters, and J. Imberger, 2001: Measurements of diapycnal diffusivity in stratified fluids. *J. Fluid Mech.*, 442, 267–291.

Batchelor, G. K., 1959: Small-scale variation of convected quantities like temperature in turbulent fluid. *J. Fluid Mech.*, 5, 113–133.

Batiste, O., E. Knobloch, A. Alonso, and I. Mercander, 2006: Spatially localized binary-fluid convection. *J. Fluid Mech.*, 560, 149–158.

Beal, L. M., 2007. Is interleaving in the Agulhas current driven by near-inertial velocity perturbations? *J. Phys. Oceanogr.*, 37, 932–945.

Beckermann, C., 2002: Modeling of macrosegregation: applications and future needs. *Int. Mater. Rev.*, 47, 243–261.

Beckermann, C., and R. Viskanta, 1988: Double-diffusive convection during dentritic solidification of a binary mixture. *PhysicoChem. Hydrodyn.*, 10, 195–213.

Belyaev, V. S., and Yu. D. Chashechkin, 1989: Free thermoconcentration convection above a localized heat source. *Fluid Dyn.*, 24, 184–190.

Bennett, A. F., 2002: *Inverse Modeling of the Ocean and Atmosphere*. Cambridge University Press.

Bennon, W. D., and F. P. Incropera, 1987: A continuum model for momentum, heat and solute transfer in binary solid-liquid phase change systems. I. Model formulation. *Int. J. Heat Mass Transfer*, 30, 2161–2170.

Bensoussan, A., J. Lions, and G. Papanicolaou, 1978: *Asymptotic Analysis for Periodic Structures*. North-Holland.

Bergeon, A., and E. Knobloch, 2008a: Spatially localized states in natural doubly diffusive convection. *Phys. Fluids*, 20, 034102.

Bergeon, A., and E. Knobloch, 2008b: Periodic and localized states in natural doubly diffusive convection. *Physica D*, 237, 1139–1150.

Bianchi, A. A., A. R. Piola, and G. J. Collino, 2002: Evidence of double diffusion in the Brazil-Malvinas Confluence. *Deep-Sea Res. I*, 49, 41–52.

Biescas, B., V. Sallares, J. L. Pelegri, *et al.*, 2008: Imaging meddy finestructure using multichannel seismic reflection data. *Geophys. Res. Lett.*, 35, L11609.

Biescas, B., L. Armi, V. Sallares, and E. Gracia, 2010: Seismic imaging of staircase layers below the Mediterranean Undercurrent. *Deep-Sea Res. I*, 57, 1345–1353.

Bois, P.-A., and A. Kubicki, 2002: Double diffusive aspects of the convection in moist-saturated air. In R. Drouot, F. Sidoroff, and G. Maugin, eds., *Continuum Thermodynamics*. Solid Mechanics and Its Applications, 76, Springer, pp. 29–42, doi:10.1007/0-306-46946-4_1.

Bois, P.-A., and A. Kubicki, 2003: A theoretical model for double diffusive phenomena in cloudy convection. *Ann. Geophys.*, 21, 2201–2218.

Boldrin, A., and S. Rabitti, 1990: Hydrography of the brines in the Bannock and Tyro anoxic basins (eastern Mediterranean). *Marine Chem.*, 31, 21–33.

Borue, V., and Orszag, S. A., 1996: Turbulent convection with constant temperature gradient. In *APS, Division of Fluid Dynamics Meeting*, 24–26 November 1996, abstract no. CG.01. pp. 1.

Brewer, W. H., 1884: On the subsidence of particles in liquids. *Mem. US Natl. Acad. Sci.*, 3, 165–175.

Bryan, F., 1987: Parameter sensitivity of primitive equation ocean general circulation models. *J. Phys. Oceanogr.*, 17, 970–985.

Buckingham, E., 1914: On physically similar systems: illustrations of the use of dimensional equations. *Phys. Rev.*, 4, 345–376.

Buffett, B. A., and C. T. Seagle, 2010: Stratification of the top of the core due to chemical interactions with the mantle. *J. Geophys. Res.*, 115, B04407, doi:10.1029/2009JB006751.

Calzavarini, E., C. R. Doering, J. D. Gibbon, D. Lohse, A. Tanabe, and F. Toschi, 2006: Exponentially growing solutions on homogeneous Rayleigh-Bénard convection. *Phys. Rev. E*, 73, 035301(R).

Campbell, I. H., 1996: Fluid dynamic processes in basaltic magma chambers. In R. G. Cawthorn, ed., *Layered Intrusions*. Developments in Petrology, 15, Elsevier, pp. 45–76.

Canuto, C., M. Y. Hussaini, A. Quarteroni, and T. A. Zang, 1987: *Spectral Methods in Fluid Dynamics*. Springer Series in Computational Physics, Springer-Verlag.

Canuto, V. M., A. M. Howard, Y. Cheng, and M. S. Dubovikov, 2002: Ocean turbulence. Part II: Vertical diffusivities of momentum, heat, salt, mass, and passive scalars. *J. Phys. Oceanogr.*, 32, 240–264.

Canuto, V. M., Y. Cheng, and A. M. Howard, 2008: A new model for Double Diffusion + Turbulence. *Geophys. Res. Lett.*, 35, L02613.

Caplan, S., 2008: Microstructure signatures of equilibrium double-diffusive convection. Thesis. Naval Postgraduate School, Monterey, California.

Carmack, E. C., K. Aagaard, J. H. Swift, *et al.*, 1997: Changes in temperature and tracer distributions within the Arctic Ocean: results from the 1994 Arctic Ocean section. *Deep-Sea Res.*, 44, 1487–1502.

Carnot, S., 1824: *Réflexions sur la puissance motrice du feu et sur les machines propres a d'evelopper cette puissance*. Paris: Chez Bachelier.

Chabrier, G., and I. Baraffe, 2008: Heat transport in giant (Exo)planets: a new perspective. *Astrophys. J. Lett.*, 661, L81, doi:10.1086/518473.

Chan, C. L., W.-Y. Chen, and C. F. Chen, 2002: Secondary motion in convection layers generated by lateral heating of a solute gradient. *J. Fluid Mech.*, 455, 1–19.

Chandrasekhar, S., 1961: *Hydrodynamic and Hydromagnetic Stability*. Clarendon Press.

Charbonnel, C., and J.-P. Zahn, 2007: Thermohaline mixing: a physical mechanism governing the photospheric composition of low-mass giants. *Astron. Astrophys.*, 467, L29–L32.

Chen, C. F., 1974: Onset of cellular convection in a salinity gradient due to a lateral temperature gradient. *J. Fluid Mech.*, 63, 563–576.

Chen, C. F., 1997: Particle flux through sediment fingers. *Deep-Sea Res.*, 44, 1645–1654.

Chen, F., and C. F. Chen, 1993: Double-diffusive fingering convection in a porous medium. *Int. J. Heat Mass Transfer*, 36, 793–807.

Chen, C. F., and F. Chen, 1997: Salt-finger convection generated by lateral heating of a solute gradient. *J. Fluid Mech.*, 352, 161–176.

Chen, C. F., and J. S. Turner, 1980: Crystallization in a double-diffusive system. *J. Geophys. Res.*, 85, 2573–2593.

Chen, C. F., D. G. Briggs, and R. A. Wirtz, 1971: Stability of thermal convection in a salinity gradient due to lateral heating. *Int. J. Heat Mass Transfer.*, 14, 57–65.

Chereskin, T. K., and P. F. Linden, 1986: The effect of rotation on intrusions produced by heating a salinity gradient. *Deep-Sea Res.*, 33, 305–322.

Clark, S., F. J. Spera, and D. A. Yuen, 1987: Steady-state double-diffusive convection in magma chambers heated from below. In B. O. Mysen, ed., *Magmatic Processes: Physicochemical Principles*. The Geochemical Society.

Cooper, J. W., and H. Stommel, 1968: Regularly spaced steps in the main thermohaline near Bermuda. *J. Geophys. Res.*, 73, 5849–5854.

Copley, S. M., A. F. Giamei, S. M. Johnson, and M. F. Hornbecker, 1970: The origin of freckles in unidirectionally solidified castings. *Metall. Trans.*, 1, 2193–2204.

Coriell, S. R., and R. F. Sekerka, 1981: Effects of convective flow on morphological stability. *PhysicoChem. Hydrodyn.*, 2, 281–293.

Coriell, S. R., M. R. Cordes, W. J. Boettinger, and R. F. Sekerka, 1980: Convective and interfacial instabilities during unidirectional solidification of a binary alloy. *J. Cryst. Growth*, 49, 13–28.

Crapper, P. F., 1975: Measurements across a diffusive interface. *Deep-Sea Res.*, 22, 537–545.

Crowe, R. A., and R. Mitalas, 1982: Semiconvection in low-mass main sequence stars. *Astron. Astrophys.*, 108, 55–60.

Cushman-Roisin, B., D. McLaughlin, and G. Papanicolaou, 1984: Interactions between mean flow and finite-amplitude mesoscale eddies in a barotropic ocean. *Geophys. Astrophys. Fluid Dyn.*, 29, 333–353.

Da Costa, L. N., E. Knobloch, and N. O.Weiss, 1981: Oscillations in double-diffusive convection. *J. Fluid Mech.*, 109, 25–43.

Darcy, H., 1856: *Les fontaines publiques de la ville de Dijon*. Paris: Dalmont.

Denissenkov, P. A., 2010: Numerical simulations of thermohaline convection: implications for extra-mixing in low-mass RGB stars. *Astrophys. J.*, 723, 563–579.

Denissenkov, P. A., and W. J. Merryfield, 2011: Thermohaline mixing: does it really govern the atmospheric chemical composition of low-mass red giants? *Astrophys. J. Lett.*, 727, L8.

Deser, C., and H. Teng, 2008: Recent trends in Arctic sea ice and the evolving role of atmospheric circulation forcing, 1979–2007. In E. T. DeWeaver, C. M. Bitz, and L.-B. Tremblay, eds., *Arctic Sea Ice Decline: Observations, Projections, Mechanisms and Implications*. Geophysical Monograph, 180, American Geophysical Union, pp. 7–26.

Dewel, G., P. Borckmans, and D. Walgraef, 1983: Spatial patterns and double diffusion in chemical reactions. *Proc. Natl. Acad. Sci. USA*, 80, 6429–6430.

D'Hernoncourt, J., A. Zebib, and A. De Wit, 2006: Reaction driven convection around a stably stratified chemical front. *Phys. Rev. Lett.*, 96, 154501.

D'Hernoncourt, J., A. Zebib, and A. De Wit, 2007a: On the classification of buoyancy-driven chemo-hydrodynamic instabilities of chemical fronts. *Chaos*, 17, 013109, doi:10.1063/1.2405129.

D'Hernoncourt, J., A. De Wit, and J. A. Zebib, 2007b: Double-diffusive instabilities of autocatalytic chemical fronts. *J. Fluid Mech.*, 576, 445–456.

Diersch, H.-J. G., and O. Kolditz, 2002: Variable-density flow and transport in porous media: approaches and challenges. *Adv. Water Resour. Res.*, 25, 899– 944.

Dietze, H., A. Oschlies, and P. Kahler, 2004: Internal-wave-induced and double-diffusive nutrient fluxes to the nutrient-consuming surface layer in the oligotrophic subtropical North Atlantic. *Ocean Dyn.*, 54, 1–7.

Dijkstra, H. A., and E. J. Kranenborg, 1998: On the evolution of double-diffusive intrusions into a stably stratified liquid: the physics of self-propagation. *Int. J. Heat Mass Transfer*, 41, 2113–2124.

Doering, C. R., and P. Constantin, 1996: Variational bounds on energy dissipation in incompressible flows. III. Convection. *Phys. Rev. E*, 53, 5957–5981.

D'Orgeville, M., B. L. Hua, R. Schopp, and L. Bunge, 2004: Extended deep equatorial layering as a possible imprint of inertial instability. *Geophys. Res. Lett.*, 31, L22303.

Drake, D. E., 1971: Suspended sediment and thermal stratification in Santa Barbara Channel, California. *Deep-Sea Res.*, 18, 763–769.

Drazin, P. G., and W. H. Reid, 1981: *Hydrodynamic Stability*. Cambridge University Press.

Eady, E. T., 1949: Long waves and cyclone waves. *Tellus*, 1, 33–52.

Edwards, N. R., and K. J. Richards, 1999: Linear double-diffusive-inertial instability at the equator. *J. Fluid Mech.*, 395, 295–319.

Edwards, N. R., and K. J. Richards, 2004: Nonlinear double-diffusive intrusions at the equator. *J. Mar. Res.*, 62, 233–259.

Farnetani, C. G., 1997: Excess temperature of mantle plumes: the role of chemical stratification across D″. *Geophys. Res. Lett.*, 24, 1583–1586.

Fedorov, K. N., 1988: Layer thickness and effective diffusivities in "diffusive" thermohaline convection in the ocean. In J. Nihoul and B. Jamart, eds., *Small-Scale Turbulence and Mixing in the Ocean*. Elsevier, pp. 471–480.

Fer, I., P. Nandi, W. S. Holbrook, R. W. Schmitt, and P. Paramo, 2010: Seismic imaging of a thermohaline staircase in the western tropical North Atlantic. *Ocean Sci.*, 6, 621–631.

Fernandes, A. M., and R. Krishnamurti, 2010: Salt finger fluxes in a laminar shear flow. *J. Fluid Mech.*, 658, 148–165.

Fernando, H. J. S., 1989: Buoyancy transfer across a diffusive interface. *J. Fluid Mech.*, 209, 1–34.

Fernando, H. J. S., 1990: Comments on 'interfacial migration in thermohaline staircases'. *J. Phys. Oceanogr.*, 20, 1994–1996.

Ferrari, R., and K. L. Polzin, 2005: Finescale structure of the *T–S* relation in the Eastern North Atlantic. *J. Phys. Oceanogr.*, 35, 1437–1454.

Ferrari, R., and W. R. Young, 1997: On the development of thermohaline correlations as a result of nonlinear diffusive parameterization. *J. Mar. Res.*, 55, 1069–1101.

Flemings M. C., and G. E. Nereo, 1967: Macrosegregation: Part I. *Trans. Metall. Soc. AIME*, 239, 1449–1461.

Flemings, M. C., and G. E. Nereo, 1968a: Macrosegregation: Part II. *Trans. Metall. Soc. AIME*, 242, 41–49.

Flemings M. C., and G. E. Nereo, 1968b: Macrosegregation: Part III. *Trans. Metall. Soc. AIME*, 242, 50–55.

Foster, T. D., and E. C. Carmack, 1976: Temperature and salinity structure in the Weddell Sea. *J. Phys. Oceanogr.*, 6, 36–44.

Gama, S., M. Vergassola, and U. Frisch, 1994: Negative eddy viscosity in isotropically forced 2-dimensional flow – linear and nonlinear dynamics. *J. Fluid Mech.* 260, 95–126.

Garaud, P., 2011: What happened to the other Mohicans? The case for a primordial origin to the planet–metallicity connection. *Astrophys. J. Lett.*, 728, L30.

Gargett, A. E., 1989: Ocean turbulence. *Annu. Rev. Fluid Mech.*, 21, 419–451.

Gargett, A. E., 2003: Differential diffusion: an oceanographic primer. *Prog. Oceanogr.*, 56, 559–570.

Gargett, A. E., and B. Ferron, 1996: The effect of differential vertical diffusion of *T* and *S* in a box model of thermohaline circulation. *J. Mar. Res.*, 54, 827–866.

Gargett, A. E., and G. Holloway, 1992: Sensitivity of the GFDL ocean model to different diffusivities for heat and salt. *J. Phys. Oceanogr.*, 22, 1158–1177.

Gargett, A. E., and R. W. Schmitt, 1982: Observations of salt fingers in the central waters of the eastern North Pacific. *J. Geophys. Res.*, 87, 8017–8092.

Garrett, C., 1982: On the parameterization of diapycnal fluxes due to double-diffusive intrusions. *J. Phys. Oceanogr.*, 12, 952–959.

Geiger, S., T. Driesner, C. A. Heinrich, and S. K. Matthai, 2005: On the dynamics of NaCl-H_2O fluid convection in the Earth's crust. *J. Geophys. Res.*, 110, B07101, doi:10.1029/2004JB003362.

Georgi, D. T., 1978: Fine structure in the Antarctic Polar Front Zone: its characteristics and possible relationship to internal waves. *J. Geophys. Res.*, 83, 4579–4588.

Giamei, A. F., and B. H. Kear, 1970: On the nature of freckles in nickel-base superalloys. *Metall. Trans.*, 1, 2185–2192.

Gierasch, P. L., and B. J. Conrath, 1987: Vertical temperature gradients on Uranus: implications for layered convection. *J. Geophys. Res.*, 92, 15019–15029.

Glessmer, M. S., A. Oschlies, and A. Yool, 2008: Simulated impact of double-diffusive mixing on physical and biogeochemical upper ocean properties. *J. Geophys. Res.*, 113, C08029, doi:10.1029/2007JC004455.

Godreche, C., and P. Manneville, 2005: *Hydrodynamics and Nonlinear Instabilities*. Cambridge University Press.

Goldreich, P., and G. Schubert, 1967: Differential rotation in stars. *Astrophys. J.*, 150, 571–587.

Golub G. H., and C. F. van Loan, 1983: *Matrix Computations*. Johns Hopkins University Press.

Gonzalez, G., 2008: Parent stars of extrasolar planets – IX. Lithium abundances. *Mon. Not. R. Astron. Soc.*, 386, 928–934.

Gordon, A. L., 1981: South Atlantic thermocline ventilation. *Deep-Sea Res.*, 28A, 1239–1264.

Gordon, A.L., 1982: Weddell Deep Water variability. *J. Mar. Res.*, 40, 199–217.

Grant, H. L., B. A. Hughes, W. M. Vogeland, and A. Molliet, 1968: The spectrum of temperature fluctuations in turbulent flow. *J. Fluid Mech.*, 34, 423–442.

Green, T., 1984: Scales for double-diffusive fingering in porous media. *Water Resour. Res.*, 20, 1225–1229.

Green, T., 1987: The importance of double diffusion to the setting of suspended material. *Sedimentology*, 34, 319–331.

Gregg, M. C., 1987: Diapycnal mixing in a thermocline: a review. *J. Geophys. Res.*, 92, 5249–5286.

Gregg, M. C., 1989: Scaling turbulent dissipation in the thermocline. *J. Geophys. Res.*, 94, 9686–9698.

Gregg, M. C., and T. B. Sanford, 1987: Shear and turbulence in thermohaline staircase. *Deep-Sea Res.*, 34, 1689–1696.

Griffiths, R. W., 1979: The transport of multiple components through thermohaline diffusive interfaces. *Deep-Sea Res.*, 26A, 383–397.

Griffiths, R. W., 1981: Layered double-diffusive convection in porous media. *J. Fluid Mech.*, 102, 221–248.

Griffiths, R. W., and A. A. Bidokhti, 2008: Interleaving intrusions produced by internal waves: a laboratory experiment. *J. Fluid Mech.*, 602, 219–239.

Griffiths, R. W., and B. R. Ruddick, 1980: Accurate fluxes across a salt-sugar finger interface deduced from direct density measurements. *J. Fluid Mech.*, 99, 85–95.

Grossmann, S., and D. Lohse, 2000: Scaling in thermal convection: a unifying theory. *J. Fluid Mech.*, 407, 27–56.

Guillot, T., 2005: The interiors of giant planets: models and outstanding questions. *Annu. Rev. Earth Planet. Sci.*, 33, 493–530.

Guillot, T., D. J. Stevenson, W. B. Hubbard, and D. Saumon, 2004: The interior of Jupiter. In F. Bagenal, T. E. Dowling, and W. B. McKinnon, eds., *Jupiter: The Planet, Satellites and Magnetosphere*. Cambridge University Press, pp. 35–57.

Hallock, Z., 1985: Variability of frontal structure in the Southern Norwegian Sea. *J. Phys. Oceanogr.*, 15, 1245–1254.

Hamilton, J. M., M. R. Lewis, and B. R. Ruddick, 1989: Vertical fluxes of nitrate associated with salt fingers in the world's oceans. *J. Geophys. Res.*, 94, 2137–2145.

Hansen, U., and D. A. Yuen, 1988: Numerical simulations of thermal-chemical instabilities at the core–mantle boundary. *Nature*, 334, 237–240.

Hansen, U., and D. A. Yuen, 1989: Dynamical influences from thermal-chemical instabilities at the core–mantle boundary. *Geophys. Res. Lett.*, 16, 629–632.

Hansen, U., and D. A. Yuen, 1994: Effects of depth-dependent thermal expansivity on the interaction of thermal-chemical plumes with a compositional boundary. *Phys. Earth Planet. Inter.*, 86, 205–221.

Hansen, U., and D. A. Yuen, 1995: Formation of layered structures in double-diffusive convection as applied to the geosciences. In A. Brandt and H. Fernando, eds., *Double-Diffusive Convection*. Geophysical Monograph, 94, American Geophysical Union, pp. 135–149.

Hansen, U., A. Hellawell, S. Z. Lu, and R. S. Steube, 1996: Some consequences of thermosolutal convection: the grain structure of castings. *Metall. Mater. Trans. A*, 27A, 569–581.

Hebert, D., 1988: Estimates of salt-finger fluxes. *Deep-Sea Res.*, 35, 1887–1901.

Hebert, D., 1999: Intrusions: what drives them? *J. Phys. Oceanogr.*, 29, 1382–1391.

Hebert, D., N. Oakey, and B. R. Ruddick, 1990: Evolution of a Mediterranean salt lens. *J. Phys. Oceanogr.*, 20, 1468–1483.

Hebert, M. A., 2011: Numerical simulations, mean field theory and modulational stability analysis of thermohaline intrusions. Thesis. Naval Postgraduate School, Monterey, California.

Heinrich, J. C., and D. R. Poirier, 2004: Convection modeling in directional solidification. *C. R. Mécanique*, 332, 429–449.

Hoare, R. A., 1966: Problem of heat transfer in Lake Vanda, a density stratified Antarctic lake. *Nature*, 10, 787–789.

Hoare, R. A., 1968: Thermohaline convection in Lake Vanda, Antarctica. *J. Geophys. Res.*, 73, 607–612.

Holbrook, W. S., P. Paramo, S. Pearse, and R. Schmitt, 2003: Thermohaline fine structure in an oceanographic front from seismic reflection profiling. *Science*, 301, 821–824.

Holloway, G., and A. Gargett, 1987: The inference of salt fingering from towed microstructure observations. *J. Geophys. Res.*, 92, 1963–1965.

Holyer, J. Y., 1981: On the collective instability of salt fingers. *J. Fluid Mech.*, 110, 195–207.

Holyer, J. Y., 1983: Double-diffusive interleaving due to horizontal gradients. *J. Fluid Mech.*, 137, 347–362.

Holyer, J. Y., 1984: The stability of long, steady, two-dimensional salt fingers. *J. Fluid Mech.*, 147, 169–185.

Holyer, J. Y., 1985: The stability of long steady three-dimensional salt fingers to long wavelength perturbations. *J. Fluid Mech.*, 156, 495–503.

Homsy, G., 1987: Viscous fingering in porous media. *Annu. Rev. Fluid Mech.*, 19, 271–311.

Houk, D., and T. Green, 1973: Descent rates of suspension fingers. *Deep-Sea Res.*, 20, 757–761.

Howard, L. N., 1961: Note on a paper of John W. Miles. *J. Fluid. Mech.*, 10, 509–512.

Howard, L. N., 1963: Heat transport by turbulent convection. *J. Fluid Mech.*, 17, 405–432.

Howard, L. N., 1964: Convection at high Rayleigh number. In H. Gortler, ed., *Proceedings Eleventh International Congress on Applied Mechanics, Munich*. Springer, pp. 1109–1115.

Hoyal, D. C. J. D., M. I. Bursik, and J. F. Atkinson, 1999: Setting-driven convection: a mechanism of sedimentation from stratified fluids. *J. Geophys. Res.*, 104, 7953–7966.

Huppert, H. E., 1971: On the stability of a series of double-diffusive layers. *Deep-Sea Res.* 18, 1005–1021.

Huppert, H. E., and P. F. Linden, 1979: On heating a stable salinity gradient from below. *J. Fluid Mech.*, 95, 431–464.

Huppert, H. E., and D. R. Moore, 1976: Nonlinear double-diffusive convection. *J. Fluid Mech.*, 78, 821–854.

Huppert, H. E., and R. S. J. Sparks, 1984: Double-diffusive convection due to crystallization in magmas. *Annu. Rev. Earth Planet. Sci.*, 12, 11–37.

Huppert, H. E., and J. S. Turner, 1972: Double-diffusive convection and its implications for the temperature and salinity structure of the ocean and Lake Vanda. *J. Phys. Oceanogr.*, 2, 456–461.

Huppert, H. E., and J. S. Turner, 1980: Ice blocks melting into a salinity gradient. *J. Fluid Mech.*, 100, 367–384.

Huppert, H. E., and J. S. Turner, 1981: Double-diffusive convection. *J. Fluid Mech.*, 106, 299–329.

Huppert, H. E., R. C. Kerr, and M. A. Hallworth, 1984: Heating or cooling a stable compositional gradient from the side. *Int. J. Heat Mass Transfer*, 27, 1395–1401.

Hurle, D. T. J., and E. Jakeman, 1971: Soret-driven thermosolutal convection. *J. Fluid Mech.*, 47, 667–687.

Imhoff, P. T., and T. Green, 1988: Experimental investigation of double-diffusive groundwater fingers. *J. Fluid Mech.*, 188, 363–382.

Ingram, M. C., 1966: The salinity extrema of the World Ocean. Ph.D. thesis. Oregon State University, Corvallis, Oregon.

Inoue, R., H. Yamazaki, F. Wolk, T. Kono, and J. Yoshida, 2007: An estimation of buoyancy flux for a mixture of turbulence and double diffusion. *J. Phys. Oceanogr.*, 37, 611–625.

Inoue, R., E. Kunze, L. St. Laurent, R. W. Schmitt, and J. M. Toole, 2008: Evaluating salt fingering theories. *J. Mar. Res.*, 66, 413–440.

Israelian, G., E. Delgado Mena, N. C., Santos, *et al.*, 2009: Enhanced lithium depletion in Sun-like stars with orbiting planets. *Nature*, 462, 189–191.

Itsweire, E. C., K. N. Helland, and C. W. Van Atta, 1986: The evolution of grid-generated turbulence in a stably stratified fluid. *J. Fluid Mech.*, 162, 299–338.

Itsweire, E. C., J. R. Koseff, D. A. Briggs, and J. H. Ferziger, 1993: Turbulence in stratified shear flows: implications for interpreting shear-induced mixing in the ocean. *J. Phys. Oceanogr.*, 23, 1508–1522.

Jacobs, S. S., H. E. Huppert, G. Holdsworth, and D. J. Drewry, 1981: Thermohaline steps induced by melting of the Erebus Glacier tongue. *J. Geophys. Res.*, 86, 6547–6555.

Jevons, W. S., 1857. On the cirrous form of cloud. *London, Edinburgh, and Dublin Philos. Mag. J. Sci.*, 4th Series, 14, 22–35.

Johnson, G. C., 2006: Generation and initial evolution of a mode water θ–S anomaly. *J. Phys. Oceanogr.*, 36, 739–751.

Johnson, G. C., and K. A. Kearney, 2009: Ocean climate change fingerprints attenuated by salt fingering? *Geophys. Res. Lett.*, 36, L21603, doi:10.1029/2009GL040697.

Joseph, D. D., 1976: *Stability of Fluid Motions, Vol. II*. Springer.

Joyce, T. M., 1976: Large-scale variations in small-scale temperature/salinity finestructure in the main thermocline of the Northwest Atlantic. *Deep-Sea Res.*, 23, 1175–1186.

Joyce, T. M., 1977: A note on the lateral mixing of water masses. *J. Phys. Oceanogr.*, 7, 626–629.

Joyce, T. M., W. Zenk, and J. M. Toole, 1978: The anatomy of the Antarctic polar front in the Drake Passage. *J. Geophys. Res.*, 83, 6093–6113.

Jubran, B. A., Kh. A. Ajlouni, and N. M. Haimour, 1999: Convective layers generated in solar ponds with fertilizer salts. *Solar Energy*, 65, 323–334.

Kalliadasis, S., J. Yang., and A. De Wit, 2004: Fingering instabilities of exothermic reaction-diffusion fronts in porous media. *Phys. Fluids*, 16, 1395–1409.

Kantha, L. H., 1980: A note on the effect of viscosity on double-diffusive processes. *J. Geophys. Res.*, 85, 4398–4404.

Kantha, L. H., 1981: 'Basalt finger' – origin of columnar joints? *Geol. Mag.*, 118, 251–264.

Karim, C., S. M. Jomaa, and A. Akbarzadeh, 2011: A laboratory experimental study of mixing the solar pond gradient zone. *Solar Energy*, 85, 404–417.

Kato, S., 1966: Overstable convection in a medium stratified in mean molecular weight. *Publ. Astron. Soc. Jpn.*, 18, 374–383.

Kelley, D. E., 1984: Effective diffusivities in ocean thermohaline staircases, *J. Geophys. Res.*, 89, 10484–10488.

Kelley, D. E., 1988: Explaining effective diffusivities within diffusive oceanic staircases. In J. C. J. Nihoul and B. M. Jamart, eds., *Small-Scale Turbulence and Mixing in the Ocean*. Elsevier, pp. 481–502.

Kelley, D. E., 1990: Fluxes through diffusive staircases: a new formulation. *J. Geophys. Res.*, 95, 3365–3371.

Kelley, D. E., H. J. S. Fernando, A. E. Gargett, J. Tanny, and E. Ozsoy, 2003: The diffusive regime of double-diffusive convection. *Prog. Oceanogr.*, 56, 461–481.

Kellogg, L. H., 1991: Interaction of plumes with a compositional boundary at 670 km. *Geophys. Res. Lett.*, 18, 865–868.

Kerr, O. S., 1989: Heating a salinity gradient from a vertical sidewall: linear theory. *J. Fluid Mech.*, 207, 323–352.

Kerr, O. S., 1990: Heating a salinity gradient from a vertical sidewall: nonlinear theory. *J. Fluid Mech.*, 217, 529–546.

Kerr, O. S., 1992: Two-dimensional instabilities of steady double-diffusive interleaving. *J. Fluid Mech.*, 242, 99–116.

Kerr, O. S., 1995: The effect of rotation on double-diffusive convection in a heated vertical slot. *J. Fluid Mech.*, 301, 345–370.

Kerr, O. S., 2000: Three-dimensional instabilities of steady double-diffusive interleaving. *J. Fluid Mech.*, 418, 297–312.

Kerr, O. S., and J. Y. Holyer, 1986: The effect of rotation on double-diffusive interleaving. *J. Fluid Mech.*, 162, 23–33.

Kerr, R. C., and J. C. Turner, 1982: Layered convection and crystal layers in multicomponent systems. *Nature*, 298, 731–733.

Kerswell, R. R., 1998: Unification of variational principles for turbulent shear flows: the background method of Doering-Constantin and the mean-fluctuation formulation of Howard-Busse. *Physica D*, 121, 175–192.

Kevorkian, J., and J. D. Cole, 1996: *Multiple Scale and Singular Perturbation Methods*. Springer.

Kimura, S., and W. Smyth, 2011: Secondary instability of salt sheets. *J. Mar. Res.*, 69, 57–77.

Kimura, S., W. Smyth, and E. Kunze, 2011: Turbulence in a sheared, salt-fingering-favorable environment: anisotropy and effective diffusivities. *J. Phys. Oceanogr.*, 41, 1144–1159.

Kippenhahn, R., G. Ruschenplatt, and H.-C. Thomas, 1980: The time scale of thermohaline mixing in stars. *Astron. Astrophys.*, 91, 175–180.

Knobloch, E., M. R. E. Proctor, and N. O. Weiss, 1992: Heteroclinic bifurcations in a simple model of double-diffusive convection. *J. Fluid Mech.*, 239, 273–292.

Kolmogorov, A., 1941: The local structure of turbulence in incompressible viscous fluid for very large Reynolds number. *Dokl. Akad. Nauk SSSR*, 30, 301–305.

Kranenborg, E. J., and H. A. Dijkstra, 1998: On the evolution of double-diffusive intrusions into a stably stratified liquid: a study of the layer merging process. *Int. J. Heat Mass Transfer*, 41, 2743–2756.

Krishnamurti, R., 2003: Double-diffusive transport in laboratory thermohaline staircases. *J. Fluid Mech.*, 483, 287–314.

Krishnamurti, R., 2006: Double-diffusive interleaving on horizontal gradients. *J. Fluid Mech.*, 558, 113–131.

Krishnamurti, R., 2009: Heat, salt and momentum transport in a laboratory thermohaline staircase. *J. Fluid Mech.*, 638, 491–506.

Kundu, P. K., and I. M. Cohen, 2008: *Fluid Mechanics*. 4th edn. Elsevier Academic.

Kunze, E., 1987: Limits on growing, finite length salt fingers: a Richardson number constraint. *J. Mar. Res.*, 45, 533–556.

Kunze, E., 1990: The evolution of salt fingers in inertial wave shear. *J. Mar. Res.*, 48, 471–504.

Kunze, E., 2003: A review of salt fingering theory. *Prog. Oceanogr.*, 56, 399–417.

Kuzmina, N., and V. Zhurbas, 2000: Effects of double diffusion and turbulence on interleaving at baroclinic oceanic fronts. *J. Phys. Oceanogr.*, 30, 3025–3038.

Kuzmina, N., B. Rudels, V. Zhurbas, and T. Stipa, 2011: On the structure and dynamical features of intrusive layering in the Eurasian Basin in the Arctic Ocean. *J. Geophys. Res.*, 116, C00D11, doi:10.1029/2010JC006920.

Lambert, R. B., and J. W. Demenkow, 1972: On the vertical transport due to fingers in double diffusive convection. *J. Fluid Mech.*, 54, 627–640.

Lambert, Jr., R. B., and W. Sturges, 1977: A thermohaline staircase and vertical mixing in the thermocline. *Deep-Sea Res.*, 24, 211–222.

Lan, C. W., and C. Y. Tu, 2000: Morphological instability due to double diffusive convection in directional solidification: the pit formation. *J. Cryst. Growth*, 220, 619–630.

Langer, J. S., 1980: Instabilities and pattern formation in crystal growth. *Rev. Mod. Phys.*, 52, 1–28.

Langer, N., 1991: Evolution of massive stars in the Large Magellanic Cloud: models with semiconvection. *Astron. Astrophys.*, 252, 669–688.

Langer, N., D. Sugimoto, and K. J. Fricke, 1983: Semiconvective diffusion and energy transport. *Astron. Astrophys.*, 126, 207–208.

Large, W. G., J. C. McWilliams, and S. C. Doney, 1994: Oceanic vertical mixing: a review and a model with a nonlocal boundary layer parameterization. *Rev. Geophys.*, 32, 363–403.

Laughlin, G., and F. C. Adams, 1997: Possible stellar metallicity enhancements from the accretion of planets. *Astrophys. J.*, 491, L51–L54.

Lay, T., E. J. Garnero, and Q. Williams, 2004: Partial melting in a thermo-chemical boundary layer at the base of the mantle. *Phys. Earth Planet. Inter.*, 146, 441–467.

Lazier, J. R. N., 1973: Temporal changes in some fresh water temperature structures. *J. Phys. Oceanogr.*, 3, 226–229.

Lazier, J., and H. Sandstrom, 1978: Migrating thermal structure in a freshwater thermocline. *J. Phys. Oceanogr.*, 8, 1070–1079.

Leconte, J., and G. Chabrier, 2012: A new vision on giant planet interiors: the impact of double diffusive convection. *Astron. Astrophys.*, 540, A20.

Ledwell, J. R., A. J. Watson, and C. S. Law, 1993: Evidence for slow mixing across the pycnocline from an open ocean tracer-release experiment. *Nature*, 364, 701–703.

Lee, J. H., and K. J. Richards, 2004: The three-dimensional structure of the interleaving layers in the western equatorial Pacific Ocean. *Geophys. Res. Lett.*, 31, L07301.

Lee, J. H., and G. Veronis, 1991: On the difference between tracer and geostrophic velocities obtained from C-SALT data. *Deep-Sea Res.*, 38, 555–568.

Leibovich, S., 1983: The form and dynamics of Langmuir circulation. *Annu. Rev. Fluid Mech.*, 15, 391–427.

Leitch, A. M., 1990: Free convection in laboratory models. *Earth-Sci. Rev.*, 29, 369–383.

Levine, M. D., 1990: Internal waves under the Arctic pack ice during the Arctic Internal Wave Experiment: the coherence structure. *J. Geophys. Res.*, 95, 7347–7357.

Lewis, M. R., W. G. Harrison, N. S. Oakey, D. Hebert, and T. Platt, 1986: Vertical nitrate fluxes in the oligotrophic ocean. *Science*, 234, 870–873.

Liang, Y., F. M. Richter, and E. B. Watson, 1994: Convection in multicomponent silicate melts driven by coupled diffusion. *Nature*, 369, 390–392.

Liang, Y., 1995: Axisymmetric double-diffusive convection in a cylindrical container: linear stability analysis with applications to molten CaO-Al2O3-SiO2. In A. Brandt and H. Fernando, eds., *Double-Diffusive Convection*. Geophysical Monograph, 94, American Geophysical Union, pp. 115–124.

Linden, P. F., 1971: Salt fingers in the presence of grid-generated turbulence. *J. Fluid Mech.*, 49, 611–624.

Linden, P. F., 1974: Salt fingers in a steady shear flow. *Geophys. Fluid Dyn.*, 6, 1–27.

Linden, P. F., 1978: The formation of banded salt finger structure. *J. Geophys. Res.*, 83, 2902–2912.

Linden, P. F., and T. G. L. Shirtcliffe, 1978: The diffusive interface in double-diffusive convection. *J. Fluid Mech.*, 87, 417–432.

Love, A. J., C. T. Simmons, and D. A. Nield, 2007: Double-diffusive convection in groundwater wells. *Water Resour. Res.*, 43, W08428, doi:10.1029/2007WR006001.

Lueck, R. G., 1987: Microstructure measurements in a thermohaline staircase. *Deep-Sea Res.*, 34, 1677–1688.

Mack, S. A., and H. C. Schoeberlein, 1993: Discriminating salt fingering from turbulence induced microstructure: analysis of towed temperature–conductivity chain data. *J. Phys. Oceanogr.*, 23, 2073–2106.

Magnell, B., 1976: Salt fingers observed in the Mediterranean outflow region (34° N, 11° W) using a towed sensor. *J. Phys. Oceanogr.*, 6, 511–523.

Malki-Epshtein, L., O. M. Phillips, and H. E. Huppert, 2004: The growth and structure of double-diffusive cells adjacent to a cooled sidewall in a salt-stratified environment. *J. Fluid Mech.*, 518, 347–362.

Malkus, W. V. R., 1954: The heat transport and spectrum of thermal turbulence. *Proc. R. Soc. London. A. Math. Phys. Sci.*, 225, 196–212.

Malkus, W. V. R., and G. Veronis, 1958: Finite amplitude cellular convection. *J. Fluid Mech.*, 4, 225–261.

Manfroi, A. J., and W. R. Young, 1999: Slow evolution of zonal jets on the beta plane. *J. Atmos. Sci.*, 56, 784–800.

Marmorino, G. O., 1987: Observations of small-scale mixing processes in the seasonal thermocline. Part I: Salt fingering. *J. Phys. Oceanogr.*, 17, 1339–1347.

Marmorino, G. O., 1989: Substructure of oceanic salt finger interfaces. *J. Geophys. Res.*, 94, 4891–4904.

Marmorino, G. O., and D. R. Caldwell, 1976: Heat and salt transport through a diffusive thermohaline interface. *Deep-Sea Res.*, 23, 59–67.

Marmorino, G. O., and D. Greenewalt, 1988: Inferring the nature of microstructure signals. *J. Geophys. Res.*, 93, 1219–1225.

Marshall, J., and T. Radko, 2003: Residual-mean solutions for the Antarctic Circumpolar Current and its associated overturning circulation. *J. Phys. Oceanogr.*, 33, 2341–2354.

Marshall, J., and T. Radko, 2006: A model of the upper branch of the meridional overturning of the southern ocean. *Prog. Oceanogr.*, 70, 331–345.

Martin, D., and I. H. Campbell, 1988: Laboratory modeling of convection in magma chambers: crystallization against sloping floors. *J. Geophys. Res.*, 93, 7974–7988.

Martin, B., and X. L. Wu, 1998: Double-diffusive convection in freely suspended soap films. *Phys. Rev. Lett.*, 80, 1892–1895.

Maruyama, S., K. Tsubaki, K. Taira, and S. Sakai, 2004: Artificial upwelling of deep seawater using the perpetual salt fountain for cultivation of ocean desert. *J. Oceanogr.*, 60, 563–568.

Maruyama, S., T. Yabuki, T. Sato, *et al.*, 2011: Evidences of increasing primary production in the ocean by Stommel's perpetual salt fountain. *Deep-Sea Res. I*, 58, 567–574.

Maxworthy, T., 1999: The dynamics of sedimenting surface gravity currents. *J. Fluid Mech.*, 392, 27–44.

May, B. D., and D. E. Kelley, 1997: Effect of baroclinity on double-diffusive interleaving. *J. Phys. Oceanogr.*, 27, 1997–2008.

May, B. D., and D. E. Kelley, 2002: Contrasting the interleaving in two baroclinic ocean fronts. *Dyn. Atmos. Oceans*, 36, 23–42.

McDougall, T. J., 1981: Double-diffusive convection with a nonlinear equation of state. II. Laboratory experiments and their interpretation. *Prog. Oceanogr.*, 10, 91–121.

McDougall, T. J., 1983: Greenland Sea Bottom Water formation: a balance between advection and double-diffusion. *Deep-Sea Res.*, 30, 1109–1117.

McDougall, T. J., 1985a: Double-diffusive interleaving. Part I: Linear stability analysis. *J. Phys. Oceanogr.*, 15, 1532–1541.

McDougall, T. J., 1985b: Double-diffusive interleaving. Part II: Finite amplitude steady state interleaving. *J. Phys. Oceanogr.*, 15, 1542–1556.

McDougall, T. J., 1987: Thermobaricity, cabbeling, and water-mass conversion. *J. Geophys. Res.*, 92, 5448–5464.

McDougall, T. J., 1991: Interfacial advection in the thermohaline staircase east of Barbados. *Deep-Sea Res.*, 38, 357–370.

McDougall, T. J., and B. R. Ruddick, 1992: The use of ocean to quantify both turbulent mixing and salt-fingering. *Deep-Sea Res.*, 39, 1931–1952.

McDougall, T. J., and J. R. Taylor, 1984: Flux measurements across a finger interface at low values of the stability ratio. *J. Mar. Res.*, 42, 1–14.

McDougall, T. J., and J. A. Whitehead, 1984: Estimates of the relative roles of diapycnal, isopycnal and double-diffusive mixing in Antarctic Bottom Water in the North Atlantic. *J. Geophys. Res.*, 89, 10479–10483.

McGillicuddy, Jr., D. J., A. R. Robinson, D. A. Siegel, *et al.*, 1998: Influence of mesoscale eddies on new production in the Sargasso Sea. *Nature*, 394, 263–266.

McIntyre, M. E., 1970: Diffusive destabilization of the baroclinic circular vortex. *Geophys. Fluid Dyn.*, 1, 19–57.

McWilliams, J. C., 1998: Oceanic general circulation models. In E. P. Chassignet and J. Verron, eds., *Ocean Modeling and Parameterization*. Kluwer Academic, pp. 1–44.

Mei, C. C., and M. Vernescu, 2010: *Homogenization Methods for Multiscale Mechanics*. World Scientific Publishing.

Merryfield W. J., 1995: Hydrodynamics of semiconvection. *Astrophys. J.*, 444, 318–337.

Merryfield, W. J., 2000: Origin of thermohaline staircases. *J. Phys. Oceanogr.*, 30, 1046–1068.

Merryfield, W. J., 2002: Intrusions in double-diffusively stable arctic waters: evidence for differential mixing? *J. Phys. Oceanogr.*, 32, 1452–1451.

Merryfield W. J., 2005: Ocean mixing in 10 steps. *Science*, 308, 641–642.

Merryfield W. J., and M. Grinder, 2002: Salt fingering fluxes from numerical simulations (unpublished manuscript).

Merryfield, W. J., G. Holloway, and A. E. Gargett, 1999: A global ocean model with double-diffusive mixing. *J. Phys. Oceanogr.*, 29, 1124–1142.

Miles, J. W., 1961: On the stability of heterogeneous shear flows. *J. Fluid. Mech.*, 10, 496–508.

Mirouh, G. M., P. Garaud, S. Stellmach, A. L. Traxler, and T. S. Wood, 2012: A new model for mixing by double-diffusive convection (semi-convection). I. The conditions for layer formation. *Astrophys. J.*, 750, 61, doi:10.1088/0004–637X/750/1/61.

Molcard, R., and A. J. Williams, 1975: Deep stepped structure in the Tyrrhenian Sea. *Mem. Soc. R. Sci. Liege*, 6, 191–210.

Molemaker, M. J., and H. A. Dijkstra, 1997: The formation and evolution of a diffusive interface. *J. Fluid Mech.*, 331, 199–229.

Monin A. S., and Ozmidov, R. V. 1985: *Turbulence in the Ocean*. D. Reidel.

Montague, N. L., and L. H. Kellogg, 2000: Numerical models of a dense layer at the base of the mantle and implications for the geodynamics of D″. *J. Geophys. Res.*, 105, 11101–11114.

Montague N. L., L. H. Kellogg, and M. Manga, 1998: High Rayleigh number thermochemical models of a dense boundary layer in D″. *Geophys. Res. Lett.*, 25, 2345–2348.

Moore, D. R., and N. O. Weiss, 1990: Dynamics of double convection. *Philos. Trans. Phys. Sci. Eng.*, 332, 121–134.

Morell, J. M., J. E. Corredor, and W. J. Merryfield, 2006: Thermohaline staircase in a Caribbean eddy and mechanisms for staircase formation. *Deep-Sea Res. II*, 53, 128–139.

Mueller, R. D., W. D. Smyth, and B. Ruddick, 2007: Shear and convective turbulence in a model of thermohaline intrusions. *J. Phys. Oceanogr.*, 37, 2534–2549.

Muench, R. D., H. J. S. Fernando, and G. R. Stegen, 1990: Temperature and salinity staircases in the northwestern Weddell Sea. *J. Phys. Oceanogr.*, 20, 295–304.

Mueth, D. M., J. C. Crocker, S. E. Esipov, and D. G. Grier, 1996: Origin of stratification in creaming emulsions. *Phys. Rev. Lett.*, 77, 578–581.

Munk, W. H., 1966: Abyssal recipes. *Deep-Sea Res.*, 13, 707–730.

Murray, J. W., Z. Top, and E. Ozsoy, 1991: Hydrographic properties and ventilation of the Black Sea. *Deep-Sea Res.*, 38, S663–S689.

Nagasaka, M., H. Nagashima, and J. Yoshida, 1995: Double diffusively induced intrusions into a density gradient. In A. Brandt and H. Fernando, eds., *Double-Diffusive Convection*. Geophysical Monograph, 94, American Geophysical Union, pp. 81–87.

Narusawa, U., and Y. Suzukawa, 1981: Experimental study of double-diffusive cellular convection due to a uniform lateral heat flux. *J. Fluid Mech.*, 113, 387–405.

Nash, J. D., and J. N. Moum, 1999: Estimating salinity variance dissipation rate from conductivity microstructure measurements. *J. Atmos. Ocean. Technol.*, 16, 263–274.

Naslund, H. R., and A. R. McBirney, 1996: Mechanisms of formation of igneous layering. *Dev. Petrol.*, 15, 1–43.

Neal, V. T., and S. Neshyba, 1973: Microstructure anomalies in the Arctic Ocean. *J. Geophys Res.*, 78, 2695–2701.

Neal, V. T., S. Neshyba, and W. Denner, 1969: Thermal stratification in the Arctic Ocean. *Science*, 166, 373–374.
Neshyba, S., V. Neal, and W. W. Denner, 1971: Temperature and conductivity measurements under ice island T-3. *J. Geophys. Res.*, 76, 8107–8120.
Newell, T. A., 1984: Characteristics of a double-diffusive interface at high density stability ratios. *J. Fluid Mech.*, 149, 385–401.
Newman, F. C., 1976: Temperature steps in Lake Kivu: a bottom heated saline lake. *J. Phys. Oceanogr.*, 6, 157–163.
Nguyen Thi, H., B. Billia, and H. Jangothhian, 1989: Influence of thermosolutal convection on the solidication front during upwards solidification. *J. Fluid Mech.*, 204, 581–597.
Nield, D. A., 1967: The thermohaline Rayleigh–Jeffries problem. *J. Fluid Mech.*, 29, 545–558.
Nield, D. A., 1968: Onset of thermohaline convection in a porous medium. *Water Resour. Res.*, 4, 553– 560.
Niino, H., 1986: A linear stability theory of double-diffusive horizontal intrusions in a temperature-salinity front. *J. Fluid Mech.*, 171, 71–100.
Noguchi, T., and H. Niino, 2010a: Multi-layered diffusive convection. Part 1. Spontaneous layer formation. *J. Fluid Mech.*, 651, 443–464.
Noguchi, T., and H. Niino, 2010b: Multi-layered diffusive convection. Part 2. Dynamics of layer evolution. *J. Fluid Mech.*, 651, 465–481.
Oakey, N. S., 1988: Estimates of mixing inferred from temperature and velocity microsctructure. In J. Nihoul and B. Jamart, eds., *Small-Scale Turbulence and Mixing in the Ocean*. Elsevier, pp. 239–248.
Oldenburg, C. M., and F. J. Spera, 1991: Numerical modeling of solidification and convection in a viscous pure binary eutectic system. *Int. J. Heat Mass Transfer*, 34, 2107–2121.
Oldenburg, C. M., F. J. Spera, D. A. Yuen, and G. Sewell, 1989: Dynamic mixing in magma bodies: thory, simulations, and implications. *J. Geophys. Res.*, 94, 9215–9236, doi:10.1029/JB094iB07p09215.
Osborn, T. R., 1980: Estimates of the local rate of vertical diffusion from dissipation measurements. *J. Phys. Oceanogr.*, 10, 83–89.
Osborn, T. R., 1991: Observations of the “Salt Fountain”. *Atmos.-Ocean*, 29, 340–356.
Osborn, T. R., and C. S. Cox, 1972: Oceanic fine structure. *Geophys. Fluid Dyn.*, 3, 321–345.
Oschlies, A., H. Dietze, and P. Kahler, 2003: Salt-finger driven enhancement of upper ocean nutrient supply. *Geophys. Res. Lett.*, 30, 2204, doi:10.1029/2003GL018552.
Ouni, M., A. Guizani, H. Lu, and A. Belghith, 2003: Simulation of the control of a salt gradient solar pond in the south of Tunisia. *Solar Energy*, 75, 95–101.
Ozgokmen, T. M., O. E. Esenkov, and D. B. Olson, 1998: A numerical study of layer formation due to fingers in double-diffusive convection in a vertically-bounded domain. *J. Mar. Res.*, 56, 463–487.
Ozsoy, E., and S. Besiktepe, 1995: Sources of double diffusive convection and impacts on mixing in the Black Sea. In A. Brandt and H. J. S. Fernando, eds., *Double-Diffusive Convection*. Geophysical Monograph, 94, American Geophysical Union, pp. 261–274.
Ozsoy, E., Z. Top, G. White, and J. W. Murray, 1991: Double diffusive intrusions, mixing and deep convective processes in the Black Sea. In E. Zdar and J. Murray, eds., *Black Sea Oceanography*. NATO ASI series C, 351, Kluwer Academic, pp. 17–42.
Ozsoy, E., U. Unluata, and Z. Tor, 1993: The evolution of Mediterranean water in the Black Sea: interior mixing and material transport by double diffusive instructions. *Prog. Oceanogr.*, 31, 275–320.

Padman, L., and T. M. Dillon, 1987: Vertical heat fluxes through the Beaufort Sea thermohaline staircase. *J. Geophys. Res.*, 92, 10799–10806.

Padman, L., and T. M. Dillon, 1988: On the horizontal extent of the Canada Basin thermohaline steps. *J. Phys. Oceanogr.*, 18, 1458–1462.

Padman, L., and T. M. Dillon, 1989: Thermal microstructure and internal waves in the Canada Basin diffusive staircase. *Deep-Sea Res.*, 36, 531–542.

Paparella, F., 2000: Fingering convection: interplay of small and large scales. *Ann. NY Acad. Sci.*, 898, 144–158.

Paramo, P., and W. S. Holbrook, 2005: Temperature contrasts in the water column inferred from amplitude-versus-offset analysis of acoustic reflections. *Geophys. Res. Lett.*, 32, L24611, doi:10.1029/2005GL024533.

Parsons, J. D., J. W. M. Bush, and J. P. M. Syvitski, 2001: Hyperpycnal plume formation from riverine outflows with small sediment concentrations. *Sedimentology*, 48, 465–478.

Pedlosky, J., 1979: *Geophysical Fluid Dynamics*. Springer-Verlag.

Perkin, R. G. and E. L. Lewis, 1984: Mixing in the West Spitsbergen current. *J. Phys. Oceanogr.*, 14, 1315–1325.

Pezzi, L. P., and K. J. Richards, 2003: Effects of lateral mixing on the mean state and eddy activity of an equatorial ocean. *J. Geophys. Res.*, 108, 3371, doi:10.1029/2003JC001834.

Phillips, O. M., 1972: Turbulence in a strongly stratified fluid: is it unstable? *Deep-Sea Res.*, 19, 79–81.

Pojman, J. A., and I. R. Epstein, 1990: Convective effects on chemical waves. 1. Mechanisms and stability criteria. *J. Phys. Chem.*, 94, 4966–4972.

Pojman, J. A., I. P. Nagy, and I. R. Epstein, 1991: Convective effects on chemical waves. 3. Multicomponent convection in the iron(II)–nitric acid system. *J. Phys. Chem.*, 95, 1306–1311.

Pojman, J. A., R. Craven, A. Khan, and W. West, 1992: Convective instabilities in traveling fronts of addition polymerization. *J. Phys. Chem.*, 96, 7466–7472.

Polzin, K., 1996: Statistics of the Richardson number: mixing models and fine structure. *J. Phys. Oceanogr.*, 26, 1409–1425.

Polzin, K. L., E. Kunze, J. M. Toole, and R. W. Schmitt, 2003: The partition of fine-scale energy into internal waves and geostrophic motions, *J. Phys. Oceanogr.*, 33, 234–248.

Posmentier, E. S., 1977: The generation of salinity finestructure by vertical diffusion. *J. Phys. Oceanogr.*, 7, 298–300.

Priestley, C. H. B., 1954: Convection from a large horizontal surface. *Aust. J. Phys.*, 7, 176.

Prikasky, I., 2007: Direct numerical simulations of the oscillatory diffusive convection and assessment of its climatologic impact. Thesis. Naval Postgraduate School, Monterey, California.

Proctor, M. R. E., 1981: Steady subcritical thermohaline convection. *J. Fluid Mech.*, 105, 507–521.

Proctor, M. R. E., and Holyer, J. Y., 1986: Planform selection in salt fingers. *J. Fluid Mech.*, 168, 241–253.

Proctor, M. R. E., and N. O. Weiss, 1982: Magnetoconvection. *Rep. Prog. Phys.*, 45, 1317–1379.

Radko, T., 2003: A mechanism for layer formation in a double-diffusive fluid. *J. Fluid Mech.*, 497, 365–380.

Radko, T., 2005: What determines the thickness of layers in a thermohaline staircase? *J. Fluid Mech.* 523, 79–98.

Radko, T., 2007: Mechanics of merging event for a series of layers in a stratified turbulent fluid. *J. Fluid Mech.*, 577, 251–273.

Radko, T., 2008: The double-diffusive modon. *J. Fluid Mech.*, 609, 59–85.

Radko, T., 2010: Equilibration of weakly nonlinear salt fingers *J. Fluid Mech.*, 645, 121–143.

Radko, T., 2011: Mechanics of thermohaline interleaving: beyond the empirical flux laws. *J. Fluid Mech.*, 675, 117–140.

Radko, T., and I. Kamenkovich, 2011: Semi-adiabatic model of the deep stratification and meridional overturning. *J. Phys. Oceanogr.*, 41, 757–780.

Radko, T., and D. P. Smith, 2012: Equilibrium transport in double-diffusive convection. *J. Fluid Mech.*, 692, 5–27.

Radko, T., and M. E. Stern, 1999: Salt fingers in three dimensions. *J. Mar. Res.*, 57, 471–502.

Radko, T., and M. E. Stern, 2000: Finite amplitude salt fingers in a vertically bounded layer. *J. Fluid Mech.*, 425, 133–160.

Radko, T., and M. E. Stern, 2011: Finescale instabilities of the double-diffusive shear flow. *J. Phys. Oceanogr.*, 41, 571–585.

Rayleigh, Lord, 1883: Investigation of the character of the equilibrium of an incompressible heavy fluid of variable density. *Proc. London Math. Soc.*, 14, 170–177.

Rica, T., E. Popity-Toth, D. Horvath, and A. Toth, 2010: Double-diffusive cellular fingering in the horizontally propagating fronts of the chlorite–tetrathionate reaction. *Physica D*, 239, 831–837.

Richards, K. J., and H. T. Banks, 2002: Characteristics of interleaving in the western equatorial Pacific. *J. Geophys. Res.*, 107, 1–12.

Richardson, L. F., 1920: The supply of energy from and to atmospheric eddies. *Proc. R. Soc. London A*, 97, 354–373.

Robertson, R., L. Padman, and M. D. Levine, 1995: Fine structure, microstructure, and vertical mixing processes in the upper ocean in the western Weddell Sea. *J. Geophys. Res.*, 100, 18517–18535.

Robinson, A., and H. Stommel, 1959: The oceanic thermocline and the associated thermohaline circulation. *Tellus*, 11, 295–308.

Rosenblum, E., P. Garaud, A. Traxler, and S. Stellmach, 2011: Turbulent mixing and layer formation in double-diffusive convection: three-dimensional numerical simulations and theory. *Astrophys. J.*, 731, doi:10.1088/0004–637X/731/1/66.

Ruddick, B., 1983: A practical indicator of the stability of the water column to double-diffusive activity. *Deep-Sea Res.*, 30, 1105–1107.

Ruddick, B., 1992: Intrusive mixing in a Mediterranean salt lens: intrusion slopes and dynamical mechanisms. *J. Phys. Oceanogr.*, 22, 1274–1285.

Ruddick, B., 2003: Sounding out ocean fine structure. *Science*, 301, 772–773.

Ruddick, B., and O. Kerr, 2003: Oceanic thermohaline intrusions: theory. *Prog. Oceanogr.*, 56, 483–497.

Ruddick, B., and K. Richards, 2003: Oceanic thermohaline intrusions: observations. *Prog. Oceanogr.*, 56, 499–527.

Ruddick, B., and J. S. Turner, 1979: The vertical length scale of double-diffusive intrusions. *Deep-Sea Res.*, 26A, 903–913.

Ruddick, B., and L. Zhang, 1989: The mythical thermohaline oscillator? *J. Mar. Res.*, 47, 717–746.

Ruddick, B. R., O. M. Phillips, and J. S. Turner, 1999: A laboratory and quantitative model of finite-amplitude thermohaline intrusions. *Dyn. Atmos. Oceans*, 30, 71–99.

Ruddick, B., H. Song, C. Dong, and L. Pinheiro, 2009: Water column seismic images as maps of temperature gradient. *Oceanography*, 22, 192–206.
Ruddick, B. R., N. S. Oakey, and D. Hebert, 2010: Measuring lateral heat flux across a thermohaline front: a model and observational test. *J. Mar. Res.*, 68, 523–539.
Rudels, B., G. Bjork, R. D. Muench, and U. Schauer, 1999: Double-diffusive layering in the Eurasian Basin of the Arctic Ocean. *J. Mar. Syst.*, 21, 3–27.
Rudels, B., N. Kuzmina, U. Schauer, T. Stipa, and V. Zhurbas, 2009: Double-diffusive convection and interleaving in the Arctic Ocean: distribution and importance. *Geophysica*, 45, 199–213.
Rudnick, D. L., and R. Ferrari, 1999: Compensation of horizontal temperature and salinity gradients in the ocean mixed layer. *Science*, 283, 656–659.
Saito, H., T. Suga, K. Hanawa, and N. Shikama, 2011: The transition region mode water of the North Pacific and its rapid modification. *J. Phys. Oceanogr.*, 41, 1639–1658.
Sajja, U. K., and S. D. Felicelli, 2011: Modeling freckle segregation with mesh adaptation. *Metall. Mater. Trans. B.*, 42, 1118–1129.
Salpeter, E. E., and D. J. Stevenson, 1976: Heat transport in a stratified two-phase fluid. *Phys. Fluids*, 19, 502–509.
Samelson, R. M., and G. K. Vallis, 1997: Large-scale circulation with small diapycnal diffusion: the two-thermocline limit. *J. Mar. Res.*, 55, 223–275.
Sample, A. K., and A. Hellawell, 1984: The mechanism of formation and prevention of channel segregation during alloy solidification. *Metall. Trans. A*, 15, 2163–2173.
Samuel, H., and C. G. Farnetani, 2003: Thermochemical convection and helium concentrations in mantle plumes. *Earth Planet. Sci. Lett.*, 207, 39–56.
Sarazin, J. R., and A. Hellawell, 1988: Channel formation in Pb-Sn, Pb-Sb and Pb-Sn-Sb alloy ingots and comparison with the system NH_4Cl-H_2O. *Metall. Trans. A*, 19, 1861–1871.
Schaefer R. J., and S. R. Coriell, 1984: Convection-induced distortion of a solid–liquid interface. *Metall. Trans. A*, 15, 2109–2115.
Schladow, S. G., E. Thomas, and J. R. Koseff, 1992: The dynamics of intrusions into a thermohaline stratification. *J. Fluid Mech.*, 236, 127–165.
Schmid, M., A. Lorke, C. Dinkel, G. Tanyileke, and A. Wuest, 2004: Double-diffusive convection in Lake Nyos, Cameroon. *Deep-Sea Res. I*, 51, 1097–1111.
Schmid, M., M. Busbridge, and A. Wuest, 2010: Double-diffusive convection in Lake Kivu. *Limnol. Oceanogr.*, 55, 225–238.
Schmitt, R. W., 1979a: The growth rate of supercritical salt fingers. *Deep-Sea Res.*, 26A, 23–44.
Schmitt, R. W., 1979b: Flux measurements on salt fingers at an interface. *J. Mar. Res.*, 37, 419–436.
Schmitt, R. W., 1981: Form of the temperature-salinity relationship in the Central Water: evidence for double-diffusive mixing. *J. Phys. Oceanogr.*, 11, 1015–1026.
Schmitt, R. W., 1983: The characteristics of salt fingers in a variety of fluid systems, including stellar interiors, liquid metals, oceans, and magmas. *Phys. Fluids*, 26, 2373–2377.
Schmitt, R. W., 1990: On the density ratio balance in the Central Water. *J. Phys. Oceanogr.*, 20, 900–906.
Schmitt, R. W., 1994a. Triangular and asymmetric salt fingers. *J. Phys. Oceanogr.*, 24, 855–860.
Schmitt, R. W., 1994b: Double diffusion in oceanography. *Annu. Rev. Fluid Mech.* 26, 255–285.

Schmitt, R. W., 1995a: The salt finger experiments of Jevons (1857) and Rayleigh (1880). *J. Phys. Oceanogr.*, 25, 8–17.

Schmit, R. W., 1995b: Why didn't Rayleigh discover salt fingers? In A. Brandt and H. Fernando, eds., *Double-Diffusive Convection*. Geophysical Monograph, 94, American Geophysical Union, pp. 3–10.

Schmitt, R. W., 1999: Spice and the Demon. *Science*, 283, 498–499.

Schmitt, R. W., 2012: Finger puzzles. *J. Fluid Mech.*, 692, 1–4.

Schmitt, R. W., and D. L. Evans, 1978: An estimate of the vertical mixing due to salt fingers based on observations in the North Atlantic central water. *J. Geophys. Res.*, 83, 2913–2919.

Schmitt, R. W., and D. T. Georgi, 1982: Finestructure and microstructure in the North Atlantic Current. *J. Mar. Res.*, 40 (Suppl.), 659–705.

Schmitt, R. W., R. G. Lueck, and T. M. Joyce, 1986: Fine- and microstructure at the edge of a warm-core ring. *Deep-Sea Res.*, 33, 1665–1689.

Schmitt, R. W., H. Perkins, J. D. Boyd, and M. C. Stalcup, 1987: C-SALT: an investigation of the thermohaline staircase in the western tropical North Atlantic. *Deep-Sea Res.*, 34, 1697–1704.

Schmitt, R. W., J. R. Ledwell, E. T. Montgomery, K. L. Polzin, and J. M. Toole, 2005: Enhanced diapycnal mixing by salt fingers in the thermocline of the tropical Atlantic. *Science*, 308, 685–688.

Schneider, M. C., and C. Beckermann, 1995: Formation of macrosegregation by multicomponent thermosolutal convection during the solidification of steel. *Metall. Mater. Trans. A*, 26, 2373–2388.

Schoofs, S., F. J. Spera, and U. Hansen, 1999: Chaotic thermohaline convection in low-porosity hydrothermal systems. *Earth Planet. Sci. Lett.*, 174, 213–229.

Schoofs, S., R. A. Trompert, and U. Hansen, 2000: The formation and evolution of layered structures in porous media: effects of porosity and mechanical dispersion. *Phys. Earth Planet. Inter.*, 118, 205–225.

Shaw, W., T. Stanton, and J. Morison, 2013: Dynamic and double-diffusive instabilities in a weak pycnocline: Part I, Observations of heat flux and diffusivity in the vicinity of Maud Rise, Weddell Sea. *J. Phys. Oceanogr.*, submitted.

Shcherbina, A. Y., M. C. Gregg, M. H. Alford, and R. R. Harcourt, 2009: Characterizing thermohaline intrusions in the North Pacific Subtropical Frontal Zone. *J. Phys. Oceanogr.*, 39, 2735–2756.

Sheen, K., N. White, and R. Hobbs, 2009: Estimating mixing rates from seismic images of oceanic structures. *Geophys. Res. Lett.*, 36, L00D04.

Shen, C. Y., 1989: The evolution of the double-diffusive instability: salt fingers. *Phys. Fluids A*, 1, 829–844.

Shen, C. Y., 1993: Heat-salt finger fluxes across a density interface. *Phys. Fluids A*, 5, 2633–2643.

Shen, C. Y., 1995: Equilibrium salt-fingering convection. *Phys. Fluids*, 7, 706–717.

Shen, C. Y., and R. W. Schmitt, 1995: The wavenumber spectrum of salt fingers. In A. Brandt and H. Fernando, eds., *Double-Diffusive Convection*. Geophysical Monograph, 94, American Geophysical Union, pp. 305–312.

Shen, C. Y., and G. Veronis, 1997: Numerical simulations of two-dimensional salt fingers. *J. Geophys. Res.*, 102, 23131–23143.

Shih, L. H., J. R. Koseff, G. N. Ivey, and J. H. Ferziger, 2005: Parametrization of turbulent fluxes and scales using homogeneous sheared stably stratified turbulence simulations. *J. Fluid Mech.*, 525, 193–214.

Shirtcliffe, T. G. L., 1973: Transport and profile measurements of the diffusive interface in double diffusive convection with similar diffusivities. *J. Fluid Mech.*, 57, 27–43.
Shirtcliffe, T. G. L., and J. S. Turner, 1970: Observations of the cell structure of salt fingers. *J. Fluid Mech.*, 41, 707–719.
Simeonov, J., and M. Stern, 2004: Double-diffusive intrusions on a finite-width thermohaline front. *J. Phys. Oceanogr.*, 34, 1724–1740.
Simeonov, J., and M. E. Stern, 2007: Equilibration of two-dimensional double-diffusive intrusions. *J. Phys. Oceanogr.*, 37, 625–643.
Simeonov, J., and M. E. Stern, 2008: Double-diffusive intrusions in a stable salinity gradient "heated from below". *J. Phys. Oceanogr.*, 38, 2271–2282.
Smith, K. S., and R. Ferrari, 2009: The production and dissipation of compensated thermohaline variance by mesoscale stirring. *J. Phys. Oceanogr.*, 39, 2477–2501.
Smith, K. S., and J. M. Klinck, 2002: Water properties on the west Antarctic Peninsula continental shelf: a model study of effects of surface fluxes and sea ice. *Deep-Sea Res. II*, 49, 4863–4886.
Smyth, W. D., 2008: Instabilities of a baroclinic, double-diffusive frontal zone. *J. Phys. Oceanogr.*, 38, 840–861.
Smyth, W. D., and S. Kimura, 2007: Instability and momentum transport in a double-diffusive, stratified shear layer. *J. Phys. Oceanogr.*, 11, 1551–1556.
Smyth, W. D., and S. Kimura, 2011: Mixing in a moderately sheared salt-fingering layer. *J. Phys. Oceanogr.*, 41, 1364–1384.
Smyth, W. D., and B. Ruddick, 2010: Effects of ambient turbulence on interleaving at a baroclinic front. *J. Phys. Oceanogr.*, 40, 685–712.
Song, H., L. M. Pinheiro, B. Ruddick, and F. C. Teixeira, 2011: Meddy, spiral arms, and mixing mechanisms viewed by seismic imaging in the Tagus Abyssal Plain (SW Iberia). *J. Mar. Res.*, 69, 827–842.
Spear, D. J., and R. E. Thomson, 2012: Thermohaline staircases in a British Colombia Fjord. *Atmos.-Ocean*, 50, 127–133.
Spiegel, E. A., 1969: Semiconvection. *Comments Astrophys. Space Phys.*, 1, 57–61.
Spiegel, E. A., 1972: Convection in stars. II. Special effects. *Annu. Rev. Astron. Astrophys.*, 10, 261–304.
Spinelli, J. E., I. L. Ferreira, and A. Garcia, 2004: Influence of melt convection on the columnar to equiaxed transition and microstructure of downward unsteady-state directionally solidified Sn–Pb alloys. *J. Alloys Compounds*, 384, 217–226.
Spinelli, J. E., M. D. Peres, and A. Garcia, 2005: Thermosolutal convective effects on dendritic array spacings in downward transient directional solidification of Al–Si alloys. *J. Alloys Compounds*, 403, 228–238.
Spruit, H. C., 1992: The rate of mixing in semiconvective zones. *Astron. Astrophys.*, 251, 131–138.
Sreenivas, K. R., O. P. Singh, and J. Srinivasan, 2009: On the relationship between finger width, velocity, and fluxes in thermohaline convection. *Phys. Fluids*, 21, 026601.
St. Laurent, L., and R. W. Schmitt, 1999: The contribution of salt fingers to vertical mixing in the North Atlantic tracer release experiment. *J. Phys. Oceanogr.*, 29, 1404–1424.
Stamp, A. P., G. O. Hughes, R. I. Nokes, and R. W. Griffiths, 1998: The coupling of waves and convection. *J. Fluid Mech.*, 372, 231–271.
Stancliffe, R., E. Glebbeek, R. Izzard, and O. Pols, 2007: Carbon-enhanced metal-poor stars and thermohaline mixing. *Astron. Astrophys.*, 464, L57–L60.
Stellmach, S., A. Traxler, P. Garaud, N. Brummell, and T. Radko, 2011: Dynamics of fingering convection II: The formation of thermohaline staircases. *J. Fluid Mech.*, 677, 554–571.

Stern, M. E., 1960: The "salt-fountain" and thermohaline convection. *Tellus*, 12,172–175.
Stern, M. E., 1967: Lateral mixing of water masses. *Deep-Sea Res.*, 14, 747–753.
Stern, M. E., 1968: *T-S* gradients on the micro-scale. *Deep-Sea Res.*, 15, 245–250.
Stern, M. E., 1969: Collective instability of salt fingers. *J. Fluid Mech.*, 35, 209–218.
Stern, M. E., 1975: *Ocean Circulation Physics*. Academic Press.
Stern, M. E., 1982: Inequalities and variational principles in double-diffusive turbulence. *J. Fluid Mech.*, 114, 105–121.
Stern, M. E., 2003: Initiation of a doubly diffusive convection in a stable halocline. *J. Mar. Res.*, 61, 211–233.
Stern, M. E., and Radko, T., 1998: The self propagating quasi-monopolar vortex. *J. Phys. Oceanogr.*, 28, 22–39.
Stern, M. E., and J. Simeonov, 2002: Internal wave overturns produced by salt fingers. *J. Phys. Oceanogr.*, 32, 3638–3656.
Stern, M. E., and J. Simeonov, 2004: Amplitude equilibration of sugar–salt fingers. *J. Fluid Mech.*, 508, 265–286.
Stern, M. E., and J. Simeonov, 2005: The secondary instability of salt fingers. *J. Fluid Mech.*, 533, 361–380.
Stern, M. E., and J. S. Turner, 1969: Salt fingers and convecting layers. *Deep-Sea Res.*, 16, 497–511.
Stern, M. E., T. Radko, and J. Simeonov, 2001: 3D salt fingers in an unbounded thermocline with application to the Central Ocean. *J. Mar. Res.*, 59, 355–390.
Stevenson, D. J., 1985: Cosmochemistry and structure of the giant planets and their satellites. *Icarus*, 62, 4–15.
Stevenson, D. J., and E. E. Salpeter, 1977: The dynamics and helium distribution in hydrogen-helium fluid planets. *Astrophys. J. (Suppl.)*, 35, 239–261.
Stommel, H., 1986: A thermohaline oscillator. *Ocean Modelling*, 72, 5–6.
Stommel, H., 1993: A conjectural regulating mechanism for determining the thermocline structure of the oceanic mixed layer. *J. Phys. Oceanogr.*, 23, 142–150.
Stommel, H., and K. N. Fedorov, 1967: Small scale structure in temperature and salinity near Timor and Mindanao. *Tellus*, 19, 306–325.
Stommel, H., A. B. Arons, and D. Blanchard, 1956: An oceanographic curiosity: the perpetual salt fountain. *Deep-Sea Res.*, 3, 152–153.
Straus, J. M., 1972: Finite amplitude doubly diffusive convection. *J. Fluid Mech.*, 56, 353–374.
Swift, S. A., A. S. Bower, and R. W. Schmitt, 2012: Vertical, horizontal, and temporal changes in temperature in the Atlantis II and Discovery hot brine pools, Red Sea. *Deep-Sea Res. I*, 64, 118–128.
Tabor, H., 1980: Non-convecting solar ponds. *Philos. Trans. R. Soc. London A*, 295, 423–433.
Tabor, H., 1981: Solar ponds. *Solar Energy*, 27, 181–194.
Tait, R. I., and M. R. Howe, 1968: Some observations of thermo-haline stratification in the deep ocean. *Deep-Sea Res.*, 15, 275–280.
Tait, R. I., and M. R. Howe, 1971: Thermohaline staircase. *Nature*, 231, 178–179.
Tait, S., and C. Jaupart, 1992: Compositional convection in a reactive crystalline mush and melt differentiation. *J. Geophys. Res.*, 97, 6735–6756.
Takao, S., and U. Narusawa, 1980: An experimental study of heat and mass transfer across a diffusive interface. *Int. J. Heat Mass Transfer*, 23, 1283–1285.
Talley, L. N., and J. Y. Yun, 2001: The role of cabbeling and double diffusion in setting the density of the North Pacific Intermediate Water salinity minimum. *J. Phys. Oceanogr.*, 31, 1538–1549.

Tanny, J., and A. B. Tsinober, 1988: The dynamics and structure of double-diffusive layers in sidewall-heating experiments. *J. Fluid Mech.*, 196, 135–156.

Tanny, J., Z. Hard, and A. Tsinober, 1995: Thermal diffusion phenomena in thick fluid layers. In A. Brandt and H. Fernando, eds., *Double-Diffusive Convection*. Geophysical Monograph, 94, American Geophysical Union, pp. 31–39.

Taunton, J. W., E. N. Lightfoot, and T. Green, 1972: Thermohaline instability and salt fingers in a porous medium. *Phys. Fluids*, 15, 748–753.

Taylor, J., 1988: The fluxes across a diffusive interface at low values of the density ratio. *Deep-Sea Res.*, 35, 555–567.

Taylor, J., 1991: Laboratory experiments on the formation of salt fingers after the decay of turbulence. *J. Geophys. Res.*, 96, 12497–12510.

Taylor, J., 1993: Anisotropy of salt fingers. *J. Phys. Oceanogr.*, 23, 554–565.

Taylor, J., and P. Bucens, 1989: Laboratory experiments on the structure of salt fingers. *Deep-Sea Res.*, 36, 1675–1704.

Taylor, J. R., and G. Veronis, 1996: Experiments on doubly-diffusive sugar-salt fingers at high stability ratio. *J. Fluid Mech.*, 321, 315–333.

Terrones, G., and A. J. Pearlstein, 1989: The onset of convection in a multicomponent fluid layer. *Phys. Fluids*, 1, 845–853.

Thangam, S., A. Zebib, and C. F. Chen, 1982: Double-diffusive convection in an inclined fluid layer. *J. Fluid Mech.*, 116, 363–378.

Theado, S., and S. Vauclair, 2012: Metal-rich accretion and thermohaline instabilities in exoplanet-host stars: consequences on the light elements abundances. *Astrophys. J.*, 744, I23.

Thompson, A. F., and G. Veronis, 2005: Diffusively-driven overturning of a stable density gradient. *J. Mar. Res.*, 63, 291–313.

Thorpe, S. A., 2005: *The Turbulent Ocean*. Cambridge University Press.

Thorpe, S. A., P. K. Hutt, and R. Soulsby, 1969: The effect of horizontal gradients on thermohaline convection. *J. Fluid Mech.*, 38, 375–400.

Timmermans, M.-L., J. Toole, R. Krishfield, and P. Winsor, 2008: Ice-Tethered Profiler observations of the double-diffusive staircase in the Canada Basin thermohaline. *J. Geophys. Res.*, 113, C00A02.

Toggweiler, J. R., and B. Samuels, 1998: On the ocean's large scale circulation near the limit of no vertical mixing. *J. Phys. Oceanogr.*, 28, 1832–1852.

Toole, J. M., 1981: Anomalous characteristics of equatorial thermocline finestructure. *J. Phys. Oceanogr.*, 11, 871–876.

Toole, J., and D. Georgi, 1981: On the dynamics of double diffusively driven intrusions. *Prog. Oceanogr.*, 10, 123–145.

Toole, J., K. Polzin, and R. Schmitt, 1994: Estimates of diapycnal mixing in the abyssal ocean. *Science*, 264, 1120–1123.

Toole, J. M., R. A. Krihfield, M.-L. Timmermans, and A. Proshutinsky, 2011: The ice-tethered profiler: Argo of the Arctic. *Oceanogr.*, 24, 126–135.

Toramaru, A., and M. Matsumoto, 2012: Numerical experiment of cyclic layering in a solidified binary eutectic melt. *J. Geophys. Res.*, 117, B02209, doi:10.1029/2011JB008204.

Toth, T., D. Horvath, and A. Toth, 2007: Thermal effects in the density fingering of the chlorite–tetrathionate reaction. *Chem. Phys. Lett.*, 442, 289–292.

Toyama, K., and T. Suga, 2012: Roles of mode waters in the formation and maintenance of central water in the North Pacific. *J. Oceanogr.*, 68, 79–92.

Traxler, A., S. Stellmach, P. Garaud, T. Radko, and N. Brummel, 2011a: Dynamics of fingering convection I: Small-scale fluxes and large-scale instabilities. *J. Fluid Mech.*, 677, 530–553.

Traxler, A., P. Garaud, and S. Stellmach, 2011b: Numerically determined transport laws for fingering ("thermohaline") convection in astrophysics. *Astrophys. J. Lett.*, 728, L29.

Trevelyan, P. M. J., C. Almarcha, and A. De Wit, 2011: Buoyancy-driven instabilities of miscible two-layer stratifications in porous media and Hele-Shaw cells. *J. Fluid Mech.*, 670, 38–65.

Tsinober, A. B., Y. Yahalom, and D. J. Shlien, 1983: A point source of heat in a stable salinity gradient. *J. Fluid Mech.*, 135, 199–217.

Tsubaki, K., S. Maruyama, A. Komiya, and H. Mitsugashira, 2007: Continuous measurement of an artificial upwelling of deep sea water induced by the perpetual salt fountain. *Deep-Sea Res. I*, 54, 75–84.

Tsuchiya, M., 1986: Thermostads and circulation in the upper layer of the Atlantic Ocean. *Prog. Oceanogr.*, 16, 235–267.

Turner, J. S., 1965: The coupled turbulent transports of salt and heat across a sharp density interface. *Int. J. Heat Mass Transfer*, 8, 759–767.

Turner, J. S., 1967: Salt fingers across a density interface. *Deep-Sea Res.*, 14, 599–611.

Turner, J. S., 1969: A physical interpretation of hot brine layers in the Red Sea. In E. T. Degens and D. A. Ross eds., *Hot Brines and Recent Heavy Metal Deposits in the Red Sea*. Springer.

Turner, J. S., 1978: Double-diffusive intrusions into a density gradient. *J. Geophys. Res.*, 83, 2887–2901.

Turner, J. S., 1979: *Buoyancy Effects in Fluids*. Cambridge University Press.

Turner, J. S., 1985: Multicomponent convection. *Annu. Rev. Fluid Mech.*, 17, 11–44.

Turner, J. S., 2010: The melting of ice in the Arctic Ocean: the influence of double-diffusive transport of heat from below. *J. Phys. Oceanogr.*, 40, 249–256.

Turner, J. S., and I. H. Campbell, 1986: Convection and mixing in magma chambers. *Earth-Sci. Rev.*, 23, 255–352.

Turner, J. S., and C. F. Chen, 1974: Two-dimensional effects in double-diffusive convection. *J. Fluid Mech.*, 63, 577–592.

Turner, J. S., and H. Stommel, 1964: A new case of convection in the presence of combined vertical salinity and temperature gradients. *Proc. Natl. Acad. Sci.*, 52, 49–53.

Turner, J. S., and G. Veronis, 2000: Laboratory studies of double-diffusive sources in closed regions. *J. Fluid Mech.*, 405, 269–304.

Turner, J. S., and G. Veronis, 2004: The influence of double-diffusive processes on the melting of ice in the Arctic Ocean: laboratory analogue experiments and their interpretation. *J. Mar. Syst.*, 45, 21–37.

Turner, J. S., T. G. L. Shirtcliffe, and P. G. Brewer, 1970: Elemental variations of transport coefficients across density interfaces in multiple-diffusive systems. *Nature*, 228, 1083–1084.

Ulrich, R. K., 1972: Thermohaline convection in stellar interiors. *Astrophys. J.*, 172, 165–178.

Vallis, G. K., 2000: Large-scale circulation and production of stratification: effects of wind, geometry, and diffusion. *J. Phys. Oceanogr.*, 30, 933–954.

Vauclair, S., 2004: Metallic fingers and metallicity excess in exoplanets' host stars: the accretion hypothesis revisited. *Astrophys. J.*, 605, 874–879.

Veronis, G., 1965: On finite amplitude instability in thermohaline convection. *J. Mar. Res.*, 23, 1–17.

Veronis, G., 1967: Analogous behavior of homogeneous, rotating fluids and stratified, non-rotating fluids. *Tellus*, 19, 326–336.

Veronis, G., 1968: Effect of stabilizing gradient of solute on thermal convection. *J. Fluid Mech.*, 34, 315–336.

Veronis, G., 2007: Updated estimate of double diffusive fluxes in the C-SALT region. *Deep-Sea Res. I*, 54, 831–833.

Voller, V. R., C. R. Swaminathan, and B. G. Thomas, 1990: Fixed grid techniques for phase change problems: a review. *Int. J. Numer. Methods Eng.*, 30, 875–898.

Walin, G., 1964: Note on stability of water stratified by both salt and heat. *Tellus*, 18, 389–393.

Walsh, D., and E. Carmack, 2002: A note on evanescent behavior of Arctic thermahaline intrusions. *J. Mar. Res.*, 60, 281–310.

Walsh, D., and B. Ruddick, 1995: Double-diffusive interleaving: the influence of nonconstant diffusivities. *J. Phys. Oceanogr.*, 25, 348–358.

Walsh, D., and B. Ruddick, 1998: Nonlinear equilibration of thermohaline intrusions. *J. Phys. Oceanogr.*, 28, 1043–1070.

Walsh, D., and B. R. Ruddick, 2000: Double-diffusive interleaving in the presence of turbulence: the effect of a nonconstant flux ratio. *J. Phys. Oceanogr.*, 30, 2231–2245.

Walton, I. C., 1982: Double-diffusive convection with large variable gradients. *J. Fluid Mech.*, 125, 123–135.

Webb, D. J., and N. Suginohara, 2001: Vertical mixing in the ocean. *Nature*, 409, 37.

Welander, P., 1982: A simple heat-salt oscillator, *Dyn. Atmos. Oceans*, 6, 233–242.

Welander, P., 1989: A new type of double-diffusive instability? *Tellus*, 41A, 66–72.

Wells, M. G., and R. W. Griffiths, 2002: Localized stirring in a field of salt-fingers. *Dyn. Atmos. Oceans*, 35, 327–350.

Wells, M. G., and R. W. Griffiths, 2003: Interaction of salt finger convection with intermittent turbulence. *J. Geophys. Res.*, 108, 3080, doi:10.1029/2002JC001427.

Wells, M. G., R. W. Griffiths, and J. S. Turner, 2001: Generation of density fine structure by salt fingers in a spatially periodic shear. *J. Geophys. Res.*, 106, 7027–7036.

Whalen, A. D., 1971: *Detection of Signals in Noise*. Academic Press, pp. 124–153.

Whitfield, D. W. A., G. Holloway, and J. Y. Holyer, 1989: Spectral transform simulations of finite amplitude double-diffusive instabilities in two dimensions. *J. Mar. Res.*, 47, 241–265.

Williams, III, A. J., 1974: Salt fingers observed in the Mediterranean outflow. *Science*, 185, 941–943.

Williams, III, A. J., 1981: The role of double diffusion in a Gulf Stream frontal intrusion. *J. Geophys. Res.*, 86, 1917–1928.

Wilson, J. R., and R. W. Mayle, 1993: Report on the progress of supernova research by the Livermore group. *Phys. Rep.*, 227, 97–111.

Wilson, J. R., R. Mayle, S. E. Woosley, and T. Weaver, 1986: Stellar core collapse and supernova. *Ann. NY Acad. Sci.*, 470, 267–293.

Worster, M. G., 2004: Time-dependent fluxes across double-diffusive interfaces. *J. Fluid Mech.*, 505, 287–307.

Worthem, S., E. Mollo-Christensen, and F. Ostapoff, 1983: Effects of rotation and shear on doubly diffusive instability. *J. Fluid Mech.*, 133, 297–319.

Wunsch, C., and R. Ferrari, 2004: Vertical mixing, energy, and the general circulation of the oceans. *Annu. Rev. Fluid Mech.*, 36, 281–314.

You, Y., 2002: A global ocean climatological atlas of the Turner angle: implications for double-diffusion and water mass structure. *Deep-Sea Res.*, 49, 2075–2093.

Zalts, A., C. El Hasi, D. Rubio, A. Urena, and D'Onofrio, 2008: Pattern formation driven by an acid-base neutralization reaction in aqueous media in a gravitation field. *Phys. Rev. E*, 77, 015304(R).

Zangrando, F., and H. J. S. Fernando, 1991: A predictive model for migration of double-diffusive interfaces. *Solar Energy*, 113, 59–65.

Zhang, J., R. W. Schmitt, and R. X. Huang, 1998: Sensitivity of GFDL Modular Ocean Model to the parameterization of double-diffusive processes. *J. Phys. Oceanogr.*, 28, 589–605.

Zhao, F.-Y., D. Liu, and G.-F. Tang, 2008: Natural convection in an enclosure with localized heating and salting from below. *Int. J. Heat Mass Transfer*, 51, 2889–2904.

Zhurbas, V. M., and I. S. Oh, 2001: Can turbulence suppress double-diffusively driven interleaving completely? *J. Phys. Oceanogr.*, 31, 2251–2254.

Zhurbas, V. M., and R. V. Ozmidov, 1983: Formation of stepped fine structure in the ocean by thermohaline intrusions. *Izv. Atmos. Ocean. Phys.*, 19, 977–982.

Zodiatis, G., and G. P. Gasparini, 1996: Thermohaline staircase formations in the Tyrrhenian Sea. *Deep-Sea Res.*, 43, 655–678.

Index